T0250376

CAD
for CONTROL SYSTEMS

CAD
for CONTROL
SYSTEMS

edited by

DEREK A. LINKENS
University of Sheffield
Sheffield, England

Marcel Dekker, Inc.   **New York • Basel • Hong Kong**

Library of Congress Cataloging-in-Publication Data

CAD for control systems / edited by Derek A. Linkens.
    p. cm.
  Includes bibliographical references and index.
  ISBN 0-8247-9060-X
  1. Automatic control. 2. Computer-aided design. I. Linkens, D. A.
TJ213.C22 1993
629.8--dc20
                                        93-1186
                                           CIP

The publisher offers discounts on this book when ordered in bulk quantities. For more information, write to Special Sales/Professional Marketing at the address below.

This book is printed on acid-free paper.

Copyright © 1993 by Marcel Dekker, Inc. All Rights Reserved.

Neither this book nor any part may be reproduced or transmitted in any form or by any means, electronic or mechanical, including photocopying, microfilming, and recording, or by any information storage and retrieval system, without permission in writing from the publisher.

Marcel Dekker, Inc.
270 Madison Avenue, New York, New York 10016

Current printing (last digit):
10 9 8 7 6 5 4 3 2 1

PRINTED IN THE UNITED STATES OF AMERICA

Preface

From the earliest days of the use of computers in engineering, the design of control systems has made heavy demands on such facilities because of its algorithmic-intensive nature. Not only have extensive number-crunching resources been required, particularly for design via optimization, but also graphical displays have been necessary to support the analytical features inherent in many systems methodologies. Thus, interactive computing has been an essential ingredient for encouraging creativity in the design of real-time feedback systems for many years. In recent times, a number of major symposia devoted to Computer-Aided Control Systems Design (CACSD) have been organized by IFAC. The proceedings of these symposia are frequently referred to in this book, and most of the chapter authors have presented their work at such meetings. However, although such proceedings are a valuable source of information, they do not attempt to put together a set of contributions that illustrate the present state of the art and future directions for the field of CACSD. By gathering together an international selection of experts in the area of CACSD, this volume attempts to present a wide spectrum of statement and opinion which is vital for the continuing success and development of a systems engineering methodology, based on dynamic analysis and design principles.

In attempting to provide up-to-date coverage of CAD principles and tools we have arranged the book loosely into four parts: "Modeling and Identification," "Simulation Languages," "Design Techniques and Tools," and "CAD Environments and HCI."

The first section covers a wide range of interests in modeling. Chapter 1 deals with the needs for modeling and simulation in future systems and includes a survey of past and present provisions. It also considers the role of object-oriented modeling, which is covered in Chapter 2. The integration of AI concepts into modeling and simulation is considered in Chapter 3, which demonstrates the need for a multiparadigm architecture with coupling strengths of varying degrees. Chapter 4 deals with the relatively new topic of qualitative, as distinct from quantitative, modeling. It is shown that this also has links with the area of intelligent systems because of its emphasis on reasoning with imprecise operators. A further related emphasis is given in Chapter 5 to the area of symbolic modeling and reasoning. The remaining two chapters in this section are concerned with model identification. Chapter 6 deals with the highly developed algorithmic area of linear system parameter identification. This computing-intensive topic has received much research attention in recent years and provides a burgeoning and bewildering set of algorithms tools for the uninitiated. Thus, this field also provides the necessary motivation for intelligent advice and selection. The final chapter deals with qualitative aspects of nonlinear identification and modeling with a strong emphasis on nonlinear structures. Here, the use of interactive computer graphics is paramount.

The next part is devoted to simulation languages and facilities. Chapter 8 describes the development and utilization of a block-oriented simulation language PSI. Its evolution from an academic tool to a full commercial product is described, showing how industry-led features have been incorporated steadily into the system. In contrast, Chapter 9 is concerned with an equation-oriented language, SPEEDUP, whose primary target has been the process control industry, particularly that of petrochemicals. Chapter 10 describes an unusual but very valuable venture which has brought together a set of linear and nonlinear benchmark models obtained from a wide range of industrial settings. These models are available as equations and also as validated code in the de facto industry-standard language, ACSL. In addition, typical model responses are available for use by people wishing to utilize the models in their own computer-based environment. The models are available in the public domain, and in certain cases directly by e-mail.

The third part deals with design techniques and tools. Chapter 11 describes the internationally accepted MATLAB software, together with some of its many front-end derivatives. No other control software has been used as extensively and creatively as MATLAB, a fact which is attributable to its flexibility and extendability. Other attempts have been made to capitalize on the MATLAB concept to provide wider design and analysis features, and one such approach is discussed in Chapter 12. For some time, it has been realized that complex design packages for multivariable systems are often difficult to understand and utilize for novice or infrequent users. The MAID program described in Chapter 13 acts as an intelligent front end to one such package and provides both design advice and concept descriptions and glossaries. The computing-intensive realm of design via optimization is covered in Chapter 14 and again illustrates the complexity presented by such mathematically taxing approaches. The field of nonlinear systems analysis and design is considered in Chapter 15, with emphasis on CAD manifestations of describing-function derivatives. Such aspects are rich in phenomenological behavioral characteristics as is also evident in Chapter 7.

The final part is devoted to environments for CACSD and the associated human–computer interaction (HCI) problems. Matrix$_x$ is claimed to be an integrated environment for modeling, identification and design and has been used quite extensively in industry. It is described in Chapter 16. A recent major attempt to provide a comprehensive environment with improved database and intelligent support features is the MEAD project, which is covered in Chapter 17. A similar UK-based project was the ECSTACY software harness infrastructure which was intended for the incorporation of academic and industrial component tools. It is described in Chapter 18. Another integrated toolbox facility called CADACS is covered in Chapter 19, whereas the use of ADA for real-time toolbox construction is expounded in Chapter 20. Of fundamental importance to an integrated environment for CACSD is the nature of the core database for holding structural and parameter details of the underlying models required for simulation and design. This subject is addressed in Chapter 21, which emphasizes a declarative approach to such databasing, as also is indicated for human dialoguing programs. Two approaches to HCI aspects of CACSD are detailed in the final chapters. Both model and data input and output features are described with an emphasis on the particularly demanding requirement of analysis and project management for dynamic systems.

Derek A. Linkens

Contents

Contributors

Mats Andersson   *Lund Institute of Technology, Lund, Sweden*

Karl Johan Åström   *Lund Institute of Technology, Lund, Sweden*

D. P. Atherton   *The University of Sussex, Brighton, England*

H. A. Barker   *University of Wales at Swansea, Swansea, Wales*

S. A. Billings   *University of Sheffield, Sheffield, England*

J. M. Boyle   *Oxford Computer Consultants, Oxford, England*

François E. Cellier   *University of Arizona, Tucson, Arizona*

M. Chen   *University of Wales at Swansea, Swansea, Wales*

T. P. Crummey   *University of Wales, Bangor, Wales*

P. J. Fleming   *University of Sheffield, Sheffield, England*

C. M. Fonseca   *University of Sheffield, Sheffield, England*

Dean K. Frederick   *Rensselaer Polytechnic Institute, Troy, New York*

Katsuhisa Furuta   *Tokyo Institute of Technology, Tokyo, Japan*

P. J. Gawthrop   *University of Glasgow, Glasgow, Scotland*

P. W. Grant   *University of Wales at Swansea, Swansea, Wales*

J. O. Gray   *University of Salford, Salford, England*

Dan Groshans   *Integrated Systems, Inc., Santa Clara, California*

Naren K. Gupta   *Integrated Systems, Inc., Santa Clara, California*

Percival H. Hammond   *University of Wales at Swansea, Swansea, Wales*

B. R. Haynes   *University of Leeds, Leeds, England*

Steve P. Houtchens   *Integrated Systems, Inc., Santa Clara, California*

C. P. Jobling   *University of Wales at Swansea, Swansea, Wales*

Masanobu Koga   *Tokyo Institute of Technology, Tokyo, Japan*

Roy Leitch   *Heriot-Watt University, Edinburgh, Scotland*

P. H. M. Li   *University of Salford, Salford, England*

Derek A. Linkens   *University of Sheffield, Sheffield, England*

Lennart Ljung   *Linköping University, Linköping, Sweden*

J. M. Maciejowski   *Cambridge University, Cambridge, England*

Sven Erik Mattsson   *Lund Institute of Technology, Lund, Sweden*

David Q. Mayne   *University of California, Davis, California*

Hannah Michalska*   *Imperial College of Science, Technology and Medicine, London, England*

---

*Current affiliation: McGill University, Montreal, Canada

N. Munro   *University of Manchester Institute of Science and Technology, Manchester, England*

Mohammad Rahbar   *AspenTech UK Ltd., Cambridge, England*

Magnus Rimvall   *General Electric Corporate R&D, Schenectady, New York*

Hunt A. Sutherland   *General Electric Corporate R&D, Schenectady, New York*

C. Y. Tan   *Cambridge University, Cambridge, England*

James H. Taylor   *Odyssey Research Associates, Ithaca, New York*

P. Townsend   *University of Wales at Swansea, Swansea, Wales*

H. Unbehauen   *Ruhr-University Bochum, Bochum, Germany*

P. P. J. van den Bosch   *Delft University of Technology, Delft, The Netherlands*

1

Integrated Continuous-System Modeling and Simulation Environments

François E. Cellier

University of Arizona
Tucson, Arizona

1.1 INTRODUCTION

The field of simulation software was last reviewed by me in 1983 [7]. A lot has happened since 1983. At that time, most continuous-system simulations were still performed on either CDC or IBM mainframes. Many engineers still wrote their simulation software in Fortran because the simulation languages of that era were not available at the site (mainframe software tended to be quite expensive), or not implemented on the particular hardware platform, or too slow for the intended purpose, or too restricted in their modeling capabilities.

Today's Engineering Workstations place more number-crunching power and a larger memory allocation on the average engineer's personal desk than the mainframes of one decade ago had to offer to an entire enterprise. We have seen a trend toward standardization of operating system software across different hardware vendors with, since the design of the RISC architectures, a strong trend toward accepting UNIX as the "universal" operating system language, and C as the "universal" programming language. We have seen standardization of graphics software with X-Windows becoming the de facto standard of low-level graphics, and Open Look and Motif the (unfortunately still two) de facto standards of higher-level graphic functions. We have seen a standardization of the ASCII representation of graphics in the form of the Postscript language, which, for the first time, allows engineers and scientists to make their papers (including figures) available electronically to their colleagues around the globe

1

by placing them in so-called "anonymous FTP" accounts. We have finally seen the general acceptance of the object-oriented (OO) programming paradigm as a means of managing large pieces of code in a modular fashion with C++ emerging as the most widely used OO programming language.

One decade ago, the operating system kernels offered on mainframes were extremely rudimentary. This was a deliberate choice since computer manufacturers wanted to make a large percentage of the scarce computer cycles and memory cells available to the end user, keeping the overhead of the operating software (both in terms of CPU cycles and occupied memory) as small as they could get away with. Today's trend is just the opposite. The operating system software is made as comfortable to use as possible, irrespective of how much resources the operating system consumes. The time of the engineer is a considerably more precious and scarcer commodity than either CPU cycles or memory chips. After all, the Engineering Workstation is idling most of the time, waiting for its slow single interactive user to issue the next command.

For this reason, the implementation of flexible integrated software environments was unthinkable at the time when my last survey was written. To be more precise, the first integrated simulation environment, TESS [39], was in its early design phase [38] around the time when my last review was written. However, the first version of TESS, released in 1985, offered a rather crude environment (operating) language, rudimentary graphics only, a painfully slow and not very robust database, and was generally a far cry from what can be achieved today. TESS deserves credit though for being visionary in predicting what simulationists would ask for in terms of simulation support software in the years to come.

In light of the rapid development of computer technology over the past decade, I was delighted when I was asked to undertake a new effort of surveying the state-of-the-art of continuous-system simulation software. However, whereas my previous review focused on features and capabilities of individual simulation languages, the current review places its emphasis on integrated modeling and simulation software environments, stating what has been achieved so far, and daring to predict what the near future might bring in addition.

1.2 SIMULATION SOFTWARE

Many of the simulation languages that were reviewed in 1983 are still in use. If anything, they have become more popular than ever. ACSL [31] is still the most widely used continuous-system simulation language on the market, and for good reasons. It provides flexible model specification capabilities, excellent integration algorithms, and both the ACSL preprocessor and the ACSL run-time system are satisfyingly robust.

One of the major reasons why I did not and could not use ACSL in my research projects at the time of my last review was ACSL's lack of capabilities to handle discontinuities properly [6]. However, shortly after my last review, the *schedule* statement was introduced into ACSL, which now allows one to handle discontinuous models adequately. This feature is still not implemented in an optimal fashion because many of the built-in discontinuous functions (such as the *step* function) have not been recoded to make use of the new facility, but this does not prevent me from using ACSL; it only prevents me from using those built-in functions.

At the time of my last review, ACSL had been fairly new on the market and its preprocessor still contained an unhealthy number of bugs. However, Mitchell & Gauthier offer excellent software support. When I report a problem to them, I usually obtain a fix within 24 to 72 hours. In the mean time, ACSL has matured tremendously. Its preprocessor is now mostly bug-free. Over the past 2 years, I discovered only one new true bug in the ACSL compiler, which was related to a table overflow with handling an unearthly large model (ACSL provided for 10,000 generic variable names, whereas my program needed more). As usual, I received a bug fix within less than a day.

One decade ago, the interface between ACSL's run-time software and its integration algorithms, particularly the Gear algorithm, still had a few problems. However, in the meantime these have been fixed, and I have not discovered any new integration problems with ACSL in a long time.

The availability of ultrafast Engineering Workstations (45 Mips or more) makes it now feasible to apply ACSL even to very large and numerically difficult problems such as the solution of two-dimensional parabolic partial differential equations discretized using the method-of-lines approach.

The initial version of ACSL (running on VAX/VMS systems only) was still fairly expensive (around $10,000) though not as outlandishly expensive as some of its competitors. However, healthy competition and plummeting hardware prices have driven the price of the ACSL software down to a level where it is comfortably affordable to anyone who needs it. On a PC-based system, an educational version of ACSL now sells for a few hundred dollars.

The only remaining major weakness of ACSL is its inability to handle algebraic loops adequately. Although ACSL provides for an implicit loop solver, this tool is totally inadequate and inappropriate.

One of the major achievements of CSSL-type languages [2], such as ACSL, is their equation sorter. Users can group equations together in a fashion that is convenient from a modeling point of view, rather than having to worry about properties of the underlying numerical solution algorithms that may call for a drastically different statement sequence. The simulation preprocessor sorts the model equations into an executable sequence. This facility is no big deal as long as the user plays around with models consisting of 20 equations, but it becomes most essential when the size of the model grows to several hundreds or even thousands of equations, as this is now commonly the case.

Unfortunately, algebraic loops often cut across many different subsystems, forcing the user to group equations together that are involved in an algebraic loop even if they logically belong to different subsystems. Algebraic loops are nasty from a numerical point of view, but the user should not have to worry about them. Use of ACSL's implicit loop solver forces the user to again think about the properties of the underlying numerical algorithm, which is exactly what we tried to prevent by introducing an equation sorter.

Also, algebraic loops can be quite formidable in size, especially when modeling chemical systems. A student of mine once formulated a dynamic model of a 50-tray distillation column. He ended up with an algebraic loop involving exactly 2573 equations! Agreed, my student was still fairly inexperienced when he wrote this model. A more experienced modeler would probably have been able to produce a more manageable model. However, the example is still quite realistic. ACSL's implicit loop solver is a rather inefficient tool for handling large algebraic loops. In such a situation, it may be more appropriate to employ an implicit numerical integration scheme, a so-called DAE-solver [5] that is able to handle problems of the type

$$\mathbf{f}(\mathbf{x}, \ \dot{\mathbf{x}}, \ \mathbf{u}, \ t) = 0.0 \tag{1.1}$$

in place of the traditionally used explicit numerical integration schemes (the so-called ODE-solvers) that handle problems of the type

$$\dot{\mathbf{x}} = \mathbf{f}(\mathbf{x}, \ \mathbf{u}, \ t) \tag{1.2}$$

I recommended strongly that Mitchell & Gauthier add one or several DAE-solvers to their run-time package and modify the code generator of the ACSL preprocessor to automatically invoke the DAE-solver when algebraic loops are detected in the model (maybe while issuing a warning message to the user indicating the problem).

Notice, however, that the described weakness is not unique to ACSL, but is one that all currently available CSSL-type languages have in common.

There are a few special-purpose simulation software systems that are worth mentioning. There exist some systems geared toward simulating chemical reaction dynamics, such as DIVA [25] and SpeedUp [32 and Chapter 9]. Contrary to the traditional CSSL-type systems, these systems employ DAE-solvers instead of ODE-solvers in their run-time software. DIVA has been successfully applied to real-time simulations of bulky chemical processes such as distillation columns.

There also exist special-purpose tools for the simulation of analog electronic circuitry. Most prevalent among those are the various dialects of Spice. As in the case of the chemical simulation systems, circuit simulators employ implicit integration techniques to get around the algebraic loop problem. Circuit simulators have been around for quite some time and could profit tremendously from a reimplementation using modern DAE-solvers in place of the fairly primitive Newton iteration schemes employed in currently available versions.

Furthermore, there exist special-purpose simulators for mechanical manipulators (robots). Also these systems are plagued by the same disease. Each constraint (coupling between neighboring limbs) introduces nasty algebraic loops among outputs of integrators. I shall not discuss these highly specialized simulators here in any further detail.

There exist many real-time training simulators for commercial airplanes, nuclear power plants, and some other complex industrial or military processes. These software systems are highly specialized, and a discussion of the solution techniques employed in their design does not contribute much to a general survey such as this.

There exist a few simulation languages geared toward truly combined continuous and discrete simulation, such as COSMOS [23] and SYSMOD [41]. Combined continuous/discrete simulation [6] requires more than just event handling. Such systems require at least waiting queues and enhanced capabilities for dealing with random numbers and distribution functions, but to be used comfortably, they also require mechanisms for process descriptions. Most of these systems, e.g., SIMAN [33] and SLAM [36], grew out of the discrete-event simulation world. Such systems offer only rudimentary facilities for continuous-system simulation. SYSMOD [41] evolved from the continuous simulation world and offers only a limited set of facilities for discrete-event simulation. COSMOS [23] is the only simulation system currently on the market that offers a fairly well-balanced palette of both continuous and discrete simulation capabilities. However, a more detailed discussion of these software systems is beyond the scope of this survey.

DESIRE [24] offers special facilities for modeling and efficiently simulating artificial neural networks and fuzzy control systems. For these types of applications, DESIRE is clearly the language of choice.

Finally, there exist special-purpose software systems for the qualitative description of continuous-time processes. The most prevalent among those systems is QSIM [26]. There also exist tools for mixed quantitative and qualitative simulation of continuous-time processes [12]. However, also these tools are too specialized to be discussed in more detail in a general survey such as this.

Although there exist still plenty of good reasons why special solutions may be needed for the simulation of special processes, ACSL has become the major workhorse for simulating effectively and efficiently large classes of continuous systems. ACSL has its largest customer base among control engineers.

The statement made in the previous paragraph is somewhat subjective. It is obviously influenced by my own exposure to and experience with the ACSL language. There exist several other simulation languages, such as DESIRE [24] and Simnon [17], with customer bases that are at least of the same order of magnitude. However, I have very good reasons for recommending ACSL, reasons that go beyond matters of personal preference and style. In the past, I have used many simulation languages and usually gave them up after some time because they were not flexible enough. Whenever I wanted to model a type of system for which the software was not originally intended, I had to invent tricks over tricks to convince the software to do what I wanted it to do. ACSL is the first simulation language that I found that does not constrain me. When I recently decided to implement mixed quantitative and qualitative models in ACSL [12], I was able to achieve this goal quickly (with less than two weeks of work) and without a need to invent dirty tricks. I do not know of any other simulation language of which I could say the same.

1.3 MODELING SOFTWARE

Most simulation models coded in a CSSL-type language were still fairly short one decade ago with larger models usually requiring ad hoc solutions (mostly large and poorly maintainable Fortran programs). The situation has changed drastically by now. ACSL programs containing 10,000 lines of code are no longer a rarity. Unfortunately, ACSL's model description capabilities, although far superior to Fortran, are still not adequate for dealing with such large-scale applications. Such applications call for the object-oriented (OO) programming paradigm. Simple subsystems should be describable as atomic objects. Objects can be interconnected to form ever larger molecular objects.

CSSL-type macros do not provide for an adequate mechanism to encapsulate objects. This can be demonstrated by means of the simple circuit problem shown in Figure 1.1. A block diagram of this simple electrical circuit is shown in Figure 1.2. This circuit can be encoded in the following ACSL program:

```
Program Circuit

    Constant R1 = 100.0, R2 = 20.0, C = 0.1E - 6, L = 1.5E - 3
    Constant tmx = 0.01

    u0 = f(t)

    iC = uR1/R1
    uR2 = R2 * iL
    uC = INTEG(iC/C, 0.0)
    iL = INTEG(uL/L, 0.0)

    uR1 = u0 − uC
    uL = u0 − uR2
    i0 = iC + iL

    term(t.ge.tmx)

End
```

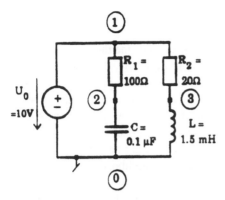

Figure 1.1 Simple electrical circuit.

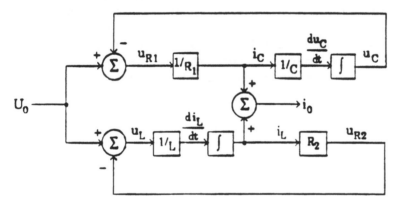

Figure 1.2 Block diagram of electrical circuit.

The first equation describes the input to the circuit, i.e., the voltage source; the next four equations describe the circuit elements themselves, i.e., the objects, whereas the last three equations describe the circuit topology, i.e., the interconnections between the objects (Kirchhoff's laws).

Notice that this circuit contains two objects of type resistor, and, yet, the equations describing these two objects look different. In the case of resistor R_1, the voltage drop across the resistor seems to "cause" a current flow, whereas in the case of resistor R_2, the current flow through the resistor seems to "cause" a voltage drop.

Clearly, the observed "causalities" are purely computational and have nothing to do with the physics of the problem. This example demonstrates that, although an equation sorter is a step in the right direction, it is obviously insufficient for truly modular object-oriented continuous-system modeling [13].

For this purpose, we require more advanced formula manipulation capabilities [11].

It is rather inconvenient that the ACSL user must determine the (numerically) correct causalities of dissipative elements, or more generally, the causalities of all energy transducers (transformers exhibit exactly the same problem as resistors). It would be much nicer if objects, such as a resistor, could be described once and for all in terms of their *physical* properties and their interactions with the environment. In the case of the resistor, such an approach would call for a description of the resistor itself (Ohm's law) and a separate description of how this equation interacts with other equations of the neighboring components (Kirchhoff's laws).

However, object-oriented continuous-system modeling [13] is much more than just a matter of convenience. State-space models suggest that each state variable changes with time according to some law that is expressed in the corresponding state equation. But why does this happen? The voltage across a capacitor does not change with time unless it has a good reason for doing so. Physics is strictly a matter of trade. The only tradable goods are mass, energy, and momentum. Consequently, it would be much safer if the modeling environment were to enable the user to formulate mass balances and energy balances rather than state equations. If a state equation is formulated incorrectly, a CSSL-type simulation language will happily accept the incorrect equation and trade it for beautiful looking multicolored graphs that may even seem plausible [9].

The modeling language Dymola [8,11,15] incorporates these concepts. In Dymola, a resistor can be described as follows:

```
model type resistor
    cut WireA(Va/i), WireB(Vb/ − i)
    main path P < WireA − WireB >
    local u
    parameter R = 1.0
        u = Va − Vb
        u = R * i

end
```

Ohm's law is described in the usual way. It involves the parameter R, which has a default value of 1.0, the local variable u, and the terminal variable i. The *cut* and *path* declarations are used to describe the interface to the outside world. Additional equations are formulated to specify the relations between the local variables and the terminal variables.

Of course, the chosen approach also calls for a general mechanism to describe the couplings between different interconnected objects. In Dymola, the above circuit can, for example, be represented as follows:

model circuit

> **submodel** (vsource) U0
> **submodel** (resistor) R1 (R = 100.0), R2 (R = 20.0)
> **submodel** (capacitor) C(C = 0.1E-6)
> **submodel** (inductor) L(L = 1.5E-3)
> **submodel** Common
> **node** n0, n1, n2, n3
> **input** u
> **output** y1, y2
>
> **connect** Common **at**   n0,
>   U0   **from** n0 **to** n1,
>   R1   **from** n1 **to** n2,
>   C   **from** n2 **to** n0,
>   R3   **from** n1 **to** n3,
>   L   **from** n3 **to** n0
>
> U0.V = u
> y1 = C.u
> y2 = L.i

end

The *submodel* declaration instantiates objects from classes. For example, two objects of type *resistor* are instantiated, one named $R1$ with a parameter value of $R = 100.0\ \Omega$ and the other named $R2$ with a parameter value of $R = 20.0\ \Omega$. The *connect* statement is used to describe the interconnections between objects. Notice that the connecting equations (Kirchhoff's laws) are not explicitly formulated at all. They are automatically generated at compile time from the topological description of the interconnections.

Upon entering the model, Dymola immediately instantiates all submodels (objects) from the model types (classes). It then extracts the formulated equations from these objects and expands them with the coupling equations that are being generated from the description of the interconnections between objects. For the above example, the result of this operation is the following.

U0   V = Vb − Va
R1   u = Va − Vb
             u = R * i
R2   u = Va − Vb
             u = R * i
C   u = Va − Vb
             C * der(u) = i
L   u = Va − Vb
             L * der(i) = u

Common   V = 0
circuit   U0.V = u
               y1 = C.u
               y2 = L.i
               R1.Vb = C.Va
               C.i = R1.i
               R.1Va = R2.Va
               U0.Vb = R1.Va
               R2.i + R1.i = U0.i
               R2.Vb = L.Va
               L.i = R2.i
               C.Vb = L.Vb
               U0.Va = C.Vb
               Common.V = U0.Va

The first 10 of these equations are extracted from the submodels. The next 3 equations are extracted from the circuit model. The last 10 equations represent Kirchhoff's laws. These equations are automatically generated from the connect statements that describe the interconnections between the objects.

The *partition* command in Dymola solves the causality assignment problem [11]. It also eliminates trivial equations of the type

$$a = b \qquad\qquad (1.3)$$

The result of this operation is as follows:

Common   [L.Vb] = 0
U0   circuit.u = [R2.Va] − L.Vb
C   u = [Va] − L.Vb
R1   [u] = R2.Va − C.Va
               u = R * [i]
C   C * [der(u)] = R1.i
R2   [u] = R * L.i
               u = Va − [L.Va]
L   [u] = Va − Vb
               L * [der(i)] = u
circuit   L.i + R1.i = [U0.i]
               [y1] = C.u
               [y2] = L.i

In each equation, the variable to be solved for is marked by square brackets. Notice the different causalities for the two resistors.

At this point, further formula manipulation can be used to solve the equations in order to generate a state-space model. Dymola has rules about the inverse of certain functions and handles the case of several linear occurrences of the unknown variable. Solving the following equation for x

$$\exp\left(a + \sin\left(\frac{[x]}{b} + c[x] - d\right)(\exp(e) + 1)\right)^2 - f = 2g \tag{1.4}$$

gives the result

$$x = \frac{\arcsin((\ln(sqrt(2g + f)) - a)/(\exp(e) + 1)) + d}{1/b + c} \tag{1.5}$$

For the above circuit example, the result of the command

> output solved equations

is as follows:

```
Common    L.Vb = 0
UO        R2.Va = circuit.u + L.Vb
C         Va = u + L.Vb
R1        u = R2.Va - C.Va
          i = u/R
C         der(u) = R1.i/C
R2        u = R * L.i
          L.Va = Va - u
L         u = Va - Vb
          der(i) = u/L
circuit   UO.i = L.i + R1.i
          y1 = C.u
          y2 = L.i
```

Finally, the state-space model can be automatically encoded as a text file in any one of a series of simulation languages. For example, the commands

> language acsl
> output program

automatically generate the following ACSL program:

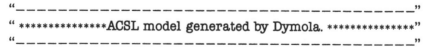

```
"_____"
" **************ACSL model generated by Dymola. **************"
"_____"

PROGRAM circuit

    INITIAL

        CONSTANT ...
            R1XR = 100.0, C = 0.1E-6, L = 1.5E-3, ...
            R2XR = 20.0
```

CONSTANT tmax = 0.01

CONSTANT ...
u = 10.0

END $ "of INITIAL"

DYNAMIC

DERIVATIVE

"————————————————————————Submodel: Common"
LXVb = 0
"————————————————————————Submodel: U0"
R2XVa = u + LXVb
"————————————————————————Submodel: C"
CXVa = CXu + LXVb
"————————————————————————Submodel: R1"
R1Xu = R2XVa − CXVa
R1Xi = R1Xu/R1XR
"————————————————————————Submodel: C"
CXu = INTEG(R1Xi/C, 0.0)
"————————————————————————Submodel: R2"
R2Xu = R2XR * LXi
LXVa = R2XVa − R2Xu
"————————————————————————Submodel: L"
LXu = LXVa − LXVb
LXi = INTEG(LXu/L, 0.0)
"————————————————————————Submodel: circuit"
U0Xi = LXi + R1Xi
y1 = CXu
y2 = LXi

END $ "of DERIVATIVE"

termt (t.ge.tmx)

END $ "of DYNAMIC"

END $ "of PROGRAM"

Notice that Dymola is not a simulation language in its own right. Dymola can be viewed as a sophisticated *macro processor* because it can be used as a frontend to a simulation language and thereby (among other things) assumes the role of its macro processor. Dymola can also be viewed as a *model generator* because it can generate models for a variety of different simulation languages. The currently supported languages are ACSL [31], DESIRE [24], and Simnon [17]. However, the most adequate interpretation is to view Dymola as a *modeling language*. Dymola has been designed to facilitate the object-oriented formulation of models of complex continuous systems.

Dymola offers the following features:

1. Modular formulation of atomic continuous-system models (objects)
2. Hierarchical composition and interconnection of atomic and molecular objects into objects of ever-increasing complexity with automatic generation of all coupling equations
3. Hierarchical data structures for connections (wires can be grouped into cables, and cables can be grouped into trunks)
4. Support of both across and through variables in connections (the values of all across variables connected to a node are the same, whereas the values of all through variables connected to a node add up to zero) with automatic generation of all coupling equations
5. Object instantiation (multiple objects can be instantiated from a single class)
6. Class inheritance (subclasses can inherit declarations of variables and equations from their parent classes)
7. Equation sorting and solving (equations automatically sorted into an executable sequence, and each equation automatically solved for the correct variable)
8. Index reduction (algebraic loops among state variables automatically reduced to algebraic loops among auxiliary variables by means of symbolic differentiation)
9. Linear algebraic loop solving (algebraic loops isolated and, if linear in the involved variables, automatically solved by means of formula manipulation)
10. Nonlinear function inversion (analytic functions automatically inverted during equation solving as needed)

Although Dymola was the first modeling language on the market, it is no longer alone. Omola [29] has been recently added to the language zoo. Omola's functionality is basically equivalent to that of Dymola. Omola offers features similar to those available in Dymola, except for linear algebraic loop solution, a feature that is not currently offered in Omola. Contrary to Dymola which translates models into ODE form (reduction to index 0), Omola translates models into DAE form (reduction to index 1). Dymola supports the generation of simulation programs in a variety of simulation languages, such as ACSL, whereas Omola comes with its own underlying simulation system, OmSim. OmSim contains several DAE-solvers. Dymola is now a commercially available product, whereas Omola and OmSim are still experimental systems.

1.4 GRAPHICAL MODEL EDITOR

Traditionally, models were always entered as text files. However, it is legitimate to ask whether this is the most convenient way to encode models. Atomic models are described by sets of equations, and there is nothing wrong with encoding those as text files. After all, atomic models are usually quite small anyway. But would it not be more convenient if each object could be associated with an icon on the screen (to be designed interactively by use of a so-called *icon editor*) and interconnections between objects could be described by means of graphical connections between icons?

Graphical model editors were slow to come. TESS [39] offered a so-called network editor (for discrete-event models) in 1985, but the networks were flat. The network editor did not come with an icon editor to encapsulate subnetworks as molecular objects. The first commercially available graphical model editors for continuous systems were EASE+ [20], a generic model editor that comes with a programmable target interface, i.e., can be used to generate models for a variety of simulation languages, System-Build [22], a graphical simulation language added to the MATRIX$_x$ [21] software, and a model editor incorporated in Boeing's EASY-5 simulation software [4]. Meanwhile, new graphical model editors are thrown onto the simulation market monthly. Model-C is a competitor of System-Build added to the CTRL-C [40] software; SimuLink was recently added to Matlab [28]; and there also exists meanwhile a block diagram editor for ACSL [31] called ProtoBlock. Simnon [17] offers a block diagram editor called ISEE-Simnon.

Why this sudden avalanche of new products? In the past, the development of graphics software was hampered by inadequate hardware and operating system support and a heavy hardware dependency. Products, such as EASE+, had to be reimplemented from scratch for each new hardware platform to which they were ported. Consequently, these systems were very expensive. Moreover, pixel graphics is bus-intensive. It is not meaningful to supply a fancy graphical model editor for a terminal that is connected to a main frame computer by a 1200-baud modem line. Only the availability of ultrafast and cheap Engineering Workstations made graphical model editors attractive. The very recent standardization of graphics software with X for the low-level functions and Open Look and Motif for the higher-level functions make the development of new graphics systems much simpler, faster, and thereby cheaper. Moreover, X and its widget toolboxes are hardware-independent, i.e., porting the software once developed to a new platform has become a relatively easy task.

Evidently there is a market for graphical model editors. A company that does not offer such a product is no longer competitive on the market. Consequently, new products are rushed out onto the market as fast as the simulation software producers can throw them together.

Although most of these new products look slick and professional, quality has suffered a bit under the hurry. The major problem is that all of these systems are simple block diagram editors, i.e., they are not truly modular. They allow one to draw on the screen a circuit as shown in Figure 1.2, but not one as shown in Figure 1.1. (For electronic circuits, there do exist so-called schematic capture programs, such as WorkView [42], that can be used to draw schematics on the screen and that then generate Spice programs, but these are special-purpose products for electronic circuits only.) Most of the block diagram editors meanwhile offer an icon editor, i.e., they are at least hierarchical in nature. However, if two of these blocks share 10 variables, 10 connecting lines must be drawn between their icons. There is no support for hierarchical data structures. Also, connections are strictly unidirectional, i.e., the user must specify which of the blocks is responsible for computing each of the shared variables.

Many of the graphical model editors on the market as of today are self-contained. They come with a very simple rudimentary simulation system integrated into the software. The block diagram is simply interpreted. In some cases, the model equations are numerically solved by a simple forward-Euler algorithm; other products offer at least a fourth-order Runge–Kutta algorithm. Years of development that went into today's commercial simulation software were simply thrown away to be able to offer a self-contained product that can be sold cheaply and does not require collaboration with a competitor company. The products are proudly presented at conference exhibitions, and all vendors demonstrate happily that they can solve the Van der Pol oscillator problem on them.

Some designers were a little wiser and decided not to reinvent the wheel. Instead of making their system self-contained, they offer a frontend to an existing simulation language, such as ACSL, thereby inheriting years of engineering that went into the design of robust simulation software. Molecular objects are translated into simulation macros. This approach is better, but still not good enough. The problem is that all the shortcomings of the macro solution are inherited. As explained earlier, macros are not truly modular.

What should have been done was to design the graphical model editor as a frontend to a modeling language, such as Dymola or Omola, rather than as a frontend to a simulation language. With this approach, composition knowledge could be expressed in terms of so-called *stylized block diagrams* [8]. Figure 1.3 shows a stylized block diagram of our simple electrical circuit. Each interconnection between blocks may represent multiple variables. In the above example, each interconnection represents exactly two variables, one of the across-type (the potential) and one of the through-type (the current). Connections between blocks are nondirectional, i.e., the user is relieved of the burden of having to solve the causality assignment problem manually. The code generated from the stylized block diagram is the previously shown Dymola circuit model. Of course, blocks in a stylized block diagram do not have to be square boxes. It is

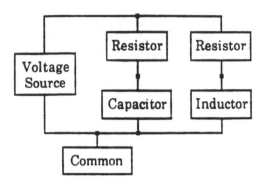

Figure 1.3 Stylized block diagram of electrical circuit.

perfectly compatible with the concept of a stylized block diagram to offer an icon editor for the custom design of blocks. With this additional facility, the circuit diagram of Figure 1.1 can be interpreted as a stylized block diagram.

Decomposition knowledge and taxonomic knowledge are encoded in a separate window by use of a so-called *system entity structure* [43]. Figure 1.4 depicts a cable reel system for the deployment of deep-sea fiber-optic communication cables.

A system entity structure for the cable reel system is shown in Figure 1.5. A decomposition that is indicated by a single vertical bar denotes a decomposition into parts. Each such decomposition is associated with a stylized block

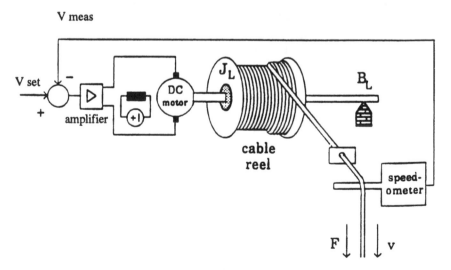

Figure 1.4 Functional diagram of a cable reel system.

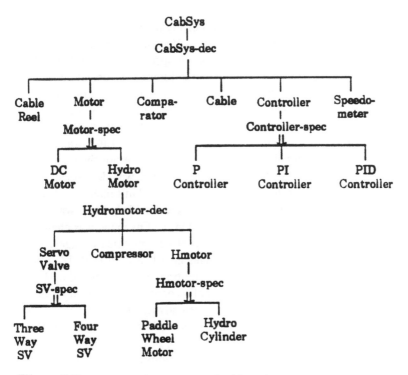

Figure 1.5 System entity structure of cable reel system.

diagram. Double-clicking on the CabSys-dec object opens up a new window that shows the stylized block diagram of Figure 1.6. Stylized block diagrams are hierarchical. For example, the motor block of Figure 1.6 contains another stylized block diagram. Double-clicking on the Hydromotor-dec object opens up another

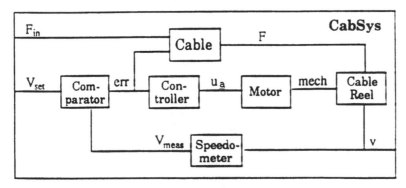

Figure 1.6 Stylized block diagram of cable reel system.

window that shows the stylized block diagram of Figure 1.7. A decomposition that is indicated by a double vertical bar denotes a decomposition into variants, a so-called specialization. For example, the motor is specialized into either a DC motor or a hydraulic motor. Double-clicking on the Motor-spec object opens up a new window with a rule-base editor to describe under what conditions the DC motor should be chosen and when the hydraulic motor should be selected.

Unfortunately, the above described system is currently vaporware. There exists a graphical frontend for Dymola, called LICS [16], but LICS was coded 10 years ago, at a time when neither the computer hardware nor the operating software were ripe to support such a development. LICS was developed on a VAX/11-780. It was controlled by a mouse with a home-built interface! While LICS was running, all other log-ins to the VAX were disabled. LICS was later ported over to a Silicon Graphics IRIS Workstation, and its name was changed to HIBLIZ [18]. However, the software was ported with minimal changes, and also the new version is heavily hardware-dependent. LICS (and HIBLIZ) offer stylized block diagrams (without icon editor) and translate into Dymola. Hierarchical decomposition is supported by means of a zoom/pan feature rather than the system entity structure approach that has been advocated in this chapter. A complete reimplementation of this software is needed. Using X and Motif, this should be a much simpler task than the development of the original system.

Of course, icon editors can be used not only to custom-design boxes, but also to custom-design connections between boxes. In this way, bond graph [10] editors also could be designed as special versions of stylized block diagrams, in the same way as the previously shown circuit diagram can be interpreted as a special type of a stylized block diagram. There are currently a few bond graph editors on the market; however, those that I have had an opportunity to experience and work with are all built in a fairly amateurish fashion employing ad hoc programming techniques, i.e., they were designed in total ignorance of basic computer science principles. A professionally built bond graph editor could be a very valuable element in an integrated modeling and simulation environment.

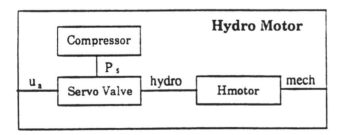

Figure 1.7   Stylized block diagram of the hydromotor.

1.5 SIMULATION ANIMATION

During simulation, users would like to see what is going on. Traditional CSSL-type simulation software does not offer any run-time display capabilities. The simulation is run first, and thereafter, the results can be viewed. This approach makes a lot of sense in a main frame/terminal environment with limited resources both in terms of CPU cycles and memory allocation. Today, this approach is obviously outdated. Some simulation systems, such as DESIRE [24], offer, at least, a run-time display. A few variables can be selected that are displayed graphically while the simulation run is executed. However, more can be done and has been achieved. In the context of graphical model editors, some systems offer a "probe" feature that allows the user to double-click on any connection in the block diagram during the simulation run with the effect that a new window pops up in which the selected variable is displayed. Some systems (such as SimuLink) offer an "oscilloscope icon" that can be attached to any block diagram connection for preselected run-time displays. A "floating oscilloscope" can be used to select an additional signal to be displayed on the spot.

However, even this facility is kind of primitive. It would be nice to be able to interactively design a cockpit of an aircraft with meters, needles, and gauges, and associate these with signals of the simulation software. Such a system has been built on top of KEE (running on Symbolics computers). However, the types of graphical elements needed to populate the animation depend strongly on the application area. Chemical process control requires quite different types of animations involving pictures of distillation columns, valves, and pipes. Simnon [17] offers a real-time simulator for process control operator training, called Simosa. Quite obviously, each application area has its special needs, and, yet, recent advances in computer graphics make it now feasible to develop quite general graphics systems that are flexible enough to custom-design screens for diverse types of applications. However, these advances are so recent that simulation animators that are currently on the market do not exploit them yet. Consequently, all commercially available simulation animators are invariably outdated. A new generation of simulation animators will quite certainly replace them in the near future.

1.6 POSTANALYSIS SOFTWARE

Besides the on-line simulation animator, there is still a place for off-line analysis of simulation results. After the simulation has been performed, users like to perform statistical analysis on simulation data, apply Fast Fourier Transforms (FFTs) to simulation trajectories, redisplay some of the curves for a closer look, etc. Graphical postanalysis of simulation results is a feature that traditional

CSSL-type simulation software has offered for many years. However, the post-analysis support offered was usually limited to the display of trajectory behavior. This is clearly insufficient.

Already a few years ago, interfaces were created between ACSL on the one hand and CTRL-C [40] and Matlab [28] on the other. This is how I run *all* my ACSL simulations. I never make use of the so-called "ACSL run-time commands" (which are not run-time commands in any true sense, but rather pre-analysis and postanalysis commands—ACSL does not provide for any interactivity during the execution of a simulation run). Instead, I kick off the ACSL simulation from within either the CTRL-C or Matlab environment and import the simulation trajectories back into the environment in the form of vectors and matrices. These simulation data can then be flexibly manipulated in many ways. It is possible to look at a subset of data only (zoom), to apply a logarithmic transformation on the data without rerunning the simulation, to superpose curves in an arbitrary fashion, to apply an FFT transform to the data, to perform statistical analysis on them, etc.

It turns out that flexible postanalysis software must make provisions for a full-fledged high-level programming language (such as Matlab or CTRL-C) because it is impossible to foresee all meaningful postanalysis features and cast them once and for all into a fixed set of precoded postanalysis operations to be selected from a menu.

The postanalysis package should be intimately tied in with the *environment language* (to be discussed later in this chapter). CTRL-C [40], Matlab [28], and MATRIX$_x$ [21] are excellent examples of powerful and flexible, yet user-friendly environment languages that can host the postanalysis software.

1.7 DOMAIN MODEL LIBRARIES

Although the previously discussed modeling software provides the capabilities for developing object-oriented modular models of arbitrary continuous-time systems, the end user does not care to develop models for the basic modeling components of his application area, such as transistors, compressors, robot arms, turbines, etc., on his own.

In addition to the modeling software itself, which is domain-independent, an integrated modeling and simulation environment should provide for domain-dependent model libraries.

A modeling/simulation environment for electrical circuit simulation should offer basic models of simple passive components (resistors, capacitors, inductors, transformers), of active components (voltage and current sources), but also of more complex components such as transistors (both BJTs and FETs) and diodes. Transistor models can be fairly sophisticated and quite complex [8]. These models make programs, such as Spice, powerful and valuable.

A modeling/simulation environment for electric power plants should contain basic models of pumps, turbines, heat exchangers, pipes, etc. MMS [19] is a modeling/simulation environment specialized for this purpose. MMS is currently available in two versions. One version contains a library of ACSL macros; the other contains EASY-5 modules. Thus, MMS was designed and implemented as a frontend to a simulation language. However, this decision places many unnecessary constraints on the MMS user. For example, MMS modules are characterized as either resistive modules or storage-type modules. Resistive modules can only be connected to storage-type modules, and vice versa. This rule encapsulates the fact that the user is indirectly responsible for solving the causality assignment problem. MMS would be much more flexible if it were implemented as a frontend to a modeling language rather than as a frontend to a simulation language. One of my students is currently reimplementing MMS in Dymola [45].

A modeling/simulation environment for thermal heating systems should contain basic models of rooms, walls, windows, rockbeds, sunspaces, trompe walls, etc. The end user should be able to make models of buildings by putting these basic models together and should not have to worry about the equations that describe these basic models themselves. Commercial products, such as CALPAS 3 [3] and DOE2 [1], are successful, not because they "know" the thermodynamic equations that describe conductive, convective, and radiative heat flow, but because they protect the end user from having to apply these equations directly. The end user is being offered a set of fairly sophisticated modules from which models of entire buildings can be thrown together within a few hours. Another student of mine is currently implementing a new system of this type on the basis of Dymola [45].

A modeling/simulation environment for chemical process modeling should contain basic models of different types of separation columns (distillation columns, stripping columns, rendering columns), of compressors and condensors, of vapor/liquid separators and oil/water separators, etc. ASCEND [35] and DESIGN-KIT [37] are modeling/simulation environments that have been designed for this purpose. These systems offer a fairly nice touch-and-feel, they are fully object-oriented, and they are carefully designed. ASCEND has been mostly used for steady-state analysis, but the language design is not limited to this type of application. DESIGN-KIT has a well-designed database interface. Both systems are tailored toward chemical process engineering, i.e., they do not make a clear separation between the domain-independent modeling language and the domain-dependent model library. Thus, they are a little less general than Dymola or Omola. I have currently two students working on a Dymola-based implementation of this type of model library. One of my students is working on a bond graph model of a distillation column [44]; the other student is working on a bond graph model of an oxygen production plant for planet Mars.

1.8 DOMAIN DATABASES

However, even availability of a domain model library is not sufficient. Models
cannot be simulated with equations alone; they also require data.

A BJT model in an electronic circuit simulator contains more than 50 parameters [8]. Collections of well-matched sets of parameter values for various types
of commercially available transistors may be equally if not more valuable than
the transistor model itself. There are companies who sell such databases as a
separate product [30]. The end user would like to describe a transistor used in
his circuit simply by providing its part number. Spice supports such a feature.

An electric power plant simulator relies heavily on data characterizing various types of valves. It also depends on steam tables, and many other types of
data. These data items should be physically separated from the models that capture the structural relations among variables, i.e., the model equations.

A chemical process plant simulator should have access to tables of enthalpies,
evaporation temperatures, etc. Thick books have been written that are full of
such tables and other data [34]. A process plant simulator should be able to access a computerized version of Perry's handbook stored in an SQL database
[14], and the modeling language should provide for mechanisms to access such
a database.

In this respect, all of today's modeling languages are deficient. Data are usually hard-coded into the models that use them. Neither Dymola nor Omola offer
special mechanisms for accessing domain databases. However, the problem can
be overcome by shifting the responsibility down to the underlying simulation
software. Many SQL databases are Fortran callable. In ACSL, calls to Fortran
subroutines can be encapsulated in macros that can then be referenced from
within the Dymola program.

TESS offers a built-in database, but TESS is geared toward discrete-event
systems, and moreover, the database is not particularly well-suited to hold permanent data. It is important to distinguish between sharable read-only databases
(e.g., to store Perry's handbook) offering slow data storage but fast data retrieval, and nonsharable read-and-write databases for storing simulation trajectories, note book files, etc., offering fast data storage but slow data retrieval.
The database provided as part of TESS is of the latter kind.

1.9 MODEL IDENTIFICATION AND
PARAMETER ESTIMATION

Even the most complete model database cannot provide numerical values for all
parameters. For example, in a mechanical system, numerical values of masses,
inertias, and spring constants are fairly easy to come by, but numerical values
for friction constants are almost impossible to obtain. Thus, a decent modeling

environment should offer facilities for estimating a subset of the model parameters from measurement data.

Physical parameters can be estimated by means of nonlinear programming packages. Such facilities are already available for ACSL, implemented in the form of two separate optional software tools called OPTDES and SimuSolv. OPTDES is most frequently used for mechanical and control systems, whereas SimuSolv is mostly used for chemical process engineering and pharmacokinetics. Matlab also offers two optimization toolboxes: a nonlinear programming toolbox (that can be used together with SimuLink) and a system identification toolbox for the computation of maximum likelihood estimators of linear systems.

Data to be identified often include statistical parameters of distribution density functions. There exist special programs on the market for just that purpose, e.g., UniFit II [27]. It would be very useful if UniFit II could also be invoked as a Matlab toolbox.

1.10 THE ENVIRONMENT LANGUAGE

At this point, it should be discussed how the various programs that were described in the previous sections of this chapter fit together. We need some sort of "operating system" that connects all these programs. However, traditional operating systems are concerned with file handling. For our purposes, this is too inconvenient. What we need is a system that can manipulate data structures and, most importantly, matrices and vectors. Such systems have been developed and are on the market for a few years. Good candidates for the environment language are CTRL-C [40], Matlab [28], and MATRIX$_x$ [21]. These three systems are very similar. They are easy to use. Their touch-and-feel is that of a comfortable pocket calculator that operates on double precision complex matrices rather than scalars. A high-level programming language to manipulate these matrices was also added. All of these systems offer excellent graphics capabilities for viewing data interactively. Some of these systems offer special-purpose toolboxes for such tasks as statistical analysis and parameter estimation. All three systems offer capabilities for simulating nonlinear models; all of them also offer graphics frontends (block diagram languages).

TESS [39] offers a much more primitive environment language. Its syntax is that of an adventure game: a verb followed by a noun, such as: "build network," or "graph facility," or "report rule." TESS understands four different verbs and about eight different nouns. Most combinations of a verb and a noun form a legal sentence. Each legal sentence invokes a particular program (language) with its own syntax and semantics. For example: "build network" invokes the graphical model editor, and "build icon" invokes the icon editor.

Although this environment language provides for a loose framework linking the various programs that are part of the TESS software suite together, this does not solve the real problems. The different modules (programs) communicate with each other, and the user should have the possibility to exert some control over this interaction. A true environment language, such as Matlab, will provide for the necessary flexibility; the environment language offered in TESS does not.

1.11   THE SOFTWARE ARCHITECTURE

There exist several ways the various software tools in the integrated modeling and simulation environment can interact with each other.

TESS [39] employs a *database architecture*. It is shown in Figure 1.8. In the center of this architecture is the SDL relational database [38]. Built around this database are the various modules that belong to the TESS software suite, such as the network editor, the icon editor, the SLAM simulation language, the postsimulation animator, and the postsimulation statistical analysis program. The different modules share a common core of Fortran service routines. All communications between the different modules go through the database. At the outskirt of the software, the user is protected from having to call each of these programs separately by the TESS environment language.

This architecture is very easy to realize. By demanding that all communications between modules go through the database, each module can be limited to offer exactly three interfaces: (i) a standardized interface to the relational database (ii) another standardized interface to the environment language, and (iii) a nonstandardized interface to the user. Notice that the TESS language does not

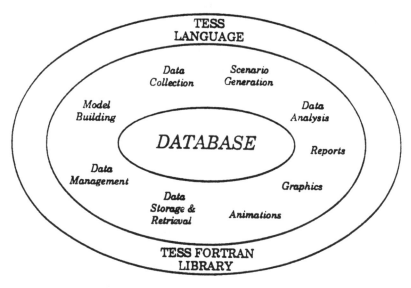

Figure 1.8   TESS architecture.

protect the user from understanding each of the modules individually—it only protects him from understanding the (VMS) operating system.

Although TESS (and its competitor shell, CINEMA) are fairly primitive from the perspective of modern window management systems and CASE architectures, these two discrete-event modeling and simulation environments are interesting from a historical perspective. They clearly represent pioneering efforts into the design of a new generation of modeling and simulation software environments.

Another approach to be considered is the more layered architecture shown in Figure 1.9. In this architecture, the user interacts with a (Matlab-like) environment language that is intimately linked to the read-and-write database (in Matlab, this database is simply the stack). The end user calls the graphical model editor through the environment language. The icon editor and the model library builder can be called in the same fashion, and also the animation editor can be called through the environment language. Once a model has been created, it is compiled down into the textual modeling language (e.g., Omola or Dymola). Library module calls are resolved through the link between the icon library and the model library. The textual OO-model is then compiled further into a simulation program. At this point, the object-orientation is lost. The resulting simulation program is a monolithic unstructured program that is optimized for execution

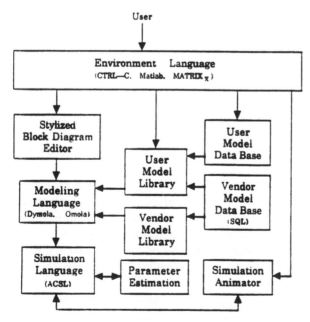

Figure 1.9 Layered environment architecture.

speed, not for human readability. The access to the model database can be re-solved either at compile time (by extracting the necessary data items once and for all from the read-only SQL database and incorporating them directly into the generated simulation program), or at run-time (by creating a link between the simulation program and the SQL database). The former approach is more cum-bersome, but may lead to more computationally efficient run-time code.

The interaction between the simulation program and the animation program should probably be direct, not going through the database. Again, this decision is based on efficiency considerations. However, it is not clear at this point whether it is better to have the simulation program call the animation program like a subroutine, whether the animation program should be in charge and call the simulation program whenever it needs new data, or whether both programs should be implemented as separate tasks with a standardized means of intertask communication (rendezvous). The latter approach may provide for the most modular solution, and in addition, it allows one to run the simulation task and the animation task on two separate processors. These two tasks have to be ex-ecuted simultaneously, and both are computation-intensive.

A system as depicted in Figure 1.9 does not currently exist for any applica-tion domain. However, modern window management and CASE software have made the implementation of such a system feasible and affordable. I am con-vinced that several such systems for different application domains will appear on the simulation software market within the next few years.

Although the field of computer simulation has matured quite a bit over the past decade, modeling and simulation software design can still make for an ex-citing and rewarding research area.

1.12 CONCLUSIONS

In this chapter, a new architecture for an integrated continuous-system modeling and simulation environment has been presented. An overview of currently avail-able software tools that might be used as components of such an integrated soft-ware environment was also given.

This is the third survey of continuous-system simulation software that I wrote. My first survey (1975) focused on concepts and components of continuous-system simulation languages; the second survey (1983) compared features of an extensive list of contemporary continuous-system simulation systems. The third survey (1992) focuses on concepts and components of integrated continuous-system modeling and simulation environments. Finally, I hope to be able to write a fourth survey around the turn of the century in which I intend to compare features of a palette of contemporary modeling and simulation environments.

My software surveys are usually quite critical. I do not mind stepping on var-ious vendors' toes—after all, they are only virtual toes. In the past, my critiques were usually followed quickly by new releases of several of the critiqued products, releases that removed many of the shortcomings that were discussed in my reviews. Whether this is purely accidental, I do not dare to say. However, if it is not accidental, I believe that my software critiques have made a significant contribution to the advancement of simulation science.

This survey has, quite noticeably, a somewhat personal touch. It could not possibly be totally impartial or objective. The survey draws heavily from my own personal experiences with the various software tools. This is the reason why I decided to write this chapter in the active voice rather than in the more traditional passive voice. An account of personal experiences reported in passive voice reads strangely, and use of the active voice softens my statements down since the reader is constantly reminded of the subjective nature of the account. However, to those readers (except for software vendors) who might feel offended by my personal writing style, I present my sincere and humble apologies. I only try to serve my scientific community in the way I can do it best.

REFERENCES

1. *Micro-DOE2 User Manual*, Acrosoft International, Inc., Denver, CO, 1987.
2. Augustin, D. C., Fineberg, M. S., Johnson, B. B., Linebarger, R. N., Sansom, F. J., and Strauss, J. C., The SCi continuous system simulation language (CSSL), *Simulation*, *9*, 281–303 (1967).
3. *CALPAS 3 User Manual*, Berkeley Solar Group, Berkeley, CA, 1982.
4. *EASY5/W—User's Manual*, Boeing Computer Services, Engineering Technology Applications (ETA) Division, Seattle, WA, 1988.
5. Brenan, K. E., Campbell, S. L., and Petzold, L. R., *Numerical Solution of Initial-Value Problems in Differential Algebraic Equations*, North-Holland, Amsterdam, 1989.
6. Cellier, F. E., *Combined Continuous/Discrete System Simulation by Use of Digital Computers: Techniques and Tools*, Ph.D. Dissertation, Diss ETH No. 6483, Swiss Federal Institute of Technology, Zürich, 1979.
7. Cellier, F. E., Simulation software: Today and tomorrow, *Proceedings IMACS Symposium on Simulation in Engineering Sciences*, (J. Burger and Y. Jarny, eds.), Nantes, France, May 9–11, 1983, pp. 426–442.
8. Cellier, F. E., *Continuous System Modeling*, Springer-Verlag, New York, 1991.
9. Cellier, F. E., Bond graphs—The right choice for educating students in modeling continuous-time systems, *Proceedings 1992 Western Simulation MultiConference on Simulation in Engineering Education*, Newport Beach, CA, January 22–24, 1992, pp. 123–127.
10. Cellier, F. E., Hierarchical nonlinear bond graphs: A unified methodology for modeling complex physical systems, *Simulation*, *58(4)*, 230–248 (1992).
11. Cellier, F. E., and Elmqvist, H., The need for automated formula manipulation in object-oriented continuous-system modeling, *Proceedings CACSD '92—IEEE*

Computer-Aided Control System Design Conference, Napa, CA, March 17–19, 1992, pp. 1–8.

12. Cellier, F. E., Nebot, A., Mugica, F., and de Albornoz, A., Combined qualitative/ quantitative simulation models of continuous-time processes using fuzzy inductive reasoning techniques, *Proceedings SICICA '92, IFAC Symposium on Intelligent Components and Instruments for Control Applications*, Málaga, Spain, May 20–22, 1992, pp. 589–593.

13. Cellier, F. E., Zeigler, B. P., and Cutler, A. H., Object-oriented modeling: Tools and techniques for capturing properties of physical systems in computer code, *Proceedings CADCS '91—IFAC Computer-Aided Design in Control Systems Conference*, Swansea, Wales, July 15–17, 1991, pp. 1–10.

14. Date, C. J., *An Introduction to Data Base Systems*, 5th ed., Addison-Wesley, Reading, MA, 1991.

15. Elmqvist, H., *A Structured Model Language for Large Continuous Systems*, Ph.D. Dissertation, Report CODEN: LUTFD2/(TFRT-1015), Dept. of Automatic Control, Lund Institute of Technology, Lund, Sweden.

16. Elmqvist, H., *LICS—Language for Implementation of Control Systems*, Report CODEN: LUTFD2/(TFRT-3179), Dept. of Automatic Control, Lund Institute of Technology, Lund, Sweden.

17. Elmqvist, H., Åström, K. J., Schönthal, T., and Wittenmark, B., *Simnon—User's Guide for MS-DOS Computers*, SSPA Systems, Gothenburg, Sweden, 1990.

18. Elmqvist, H., and Mattson, S. E., Simulator for dynamical systems using graphics and equations for modelling, *IEEE Contr. Syst. Mag.*, 9(1), 53–58, 1989.

19. *Modular Modeling System (MMS): A Code for the Dynamic Simulation of Fossil and Nuclear Power Plants*, Report: CS/NP-3016-CCM, Electric Power Research Institute, Palo Alto, CA, 1983.

20. *EASE+—User's Manual*, Expert-EASE Systems, Inc., Belmont, CA, 1988.

21. *MATRIX$_x$ User's Guide, MATRIX$_x$ Reference Guide, MATRIX$_x$ Training Guide, Command Summary and On-Line Help*, Integrated Systems, Inc., Santa Clara, CA, 1984.

22. *SYSTEM-BUILD User's Guide*, Integrated Systems, Inc., Santa Clara, CA, 1985.

23. Kettenis, D. L., COSMOS: A simulation language for continuous, discrete and combined models, *Simulation*, 58(1), 32–41, (1992).

24. Korn, G. A., *Interactive Dynamic-System Simulation*, McGraw-Hill, New York, 1989.

25. Kröner, A., Holl, P., Marquardt, W., and Gilles, E. D., DIVA—An open architecture for dynamic simulation, *Comput. Chem. Eng.*, 14(11), 1289–1295 (1990).

26. Kuipers, B., and Farquhar, A., *QSIM: A Tool for Qualitative Simulation*, Internal Report: Artificial Intelligence Laboratory, The University of Texas, Austin, TX, 1987.

27. Law, A. M., and Vincent, S. G., *UniFit II User's Manual*, Simulation Modeling and Analysis Company, Tucson, AZ, 1990.

28. Mathworks, Inc., *The Student Edition of MATLAB for MS-DOS or Macintosh Computers*, Prentice-Hall, Englewood Cliffs, NJ, 1992.

29. Mattson, S. E., and Andersson, M., The ideas behind OMOLA, *Proceedings CACSD '92—IEEE Computer-Aided Control System Design Conference*, Napa, CA, March 17–19, 1992, pp. 23–29.

30. *ACCULIB User's Manual*, Mentor Corp., Palo Alto, CA, 1987.
31. Mitchell, E. E. L., and Gauthier, J. S., *ACSL: Advanced Continuous Simulation Language—User Guide and Reference Manual*, Mitchell & Gauthier Assoc., Concord, MA, 1986.
32. Pantelides, C. C., Speed Up—Recent advances in process simulation, *Comput. Chem. Eng.*, *12*(7), 745–755 (1988).
33. Pegden, C. D., *Introduction to SIMAN*, Systems Modelling Corp., State College, PA, 1982.
34. Perry, R. H., Green, D. W., and Maloney, J. O. (eds.), *Perry's Chemical Engineers' Handbook*, 6th ed., McGraw-Hill, New York, 1984.
35. Piela, P. C., Epperly, T. G., Westerberg, K. M., and Westerberg, A. W., ASCEND: An object-oriented computer environment for modeling and analysis—Part 1: The modeling language, *Comput. Chem. Eng.*, *15*(1), 53–72 (1991).
36. Pritsker, A. A. B., *Introduction to Simulation and SLAM-II*, 3rd ed., Halsted Press, New York, 1985.
37. Stephanopoulos, G., Johnston, J., Kriticos, T., Lakshmanan, R., Mavrovouniotis, M., and Siletti, C., DESIGN-KIT: An object-oriented environment for process engineering, *Comput. Chem. Eng.*, *11*(6), 655–674 (1987).
38. Standridge, C. R., and Pritsker, A. A. B., Using data base capabilities in simulation, in *Progress in Modelling and Simulation* (F. E. Cellier, ed.), Academic Press, London, 1982, pp. 347–365.
39. Standridge, C. R., and Pritsker, A. A. B., *TESS—The Extended Simulation Support System*, Halsted Press, New York, 1987.
40. *CTRL-C, A Language for the Computer-Aided Design of Multivariable Control Systems, User's Guide*, Systems Control Technology, Inc., Palo Alto, CA, 1985.
41. *SYSMOD User Manual*, Release 1.0, D05448/14/UM, Systems Designers, plc, Ferneberga House, Farnborough, Hampshire, UK, 1986.
42. *WORKVIEW Reference Guide*, Release 3.0, Viewlogic Systems, Inc., Marlboro, MA, 1988.
43. Zeigler, B. P., *Multifaceted Modelling and Discrete Event Simulation*, Academic Press, London, 1984.
44. Brooks, B. A., and Cellier, F. E., Modeling of a distillation column using bond graphs, *Proceedings 1993 Western Simulation MultiConference on Bond Graph Modeling*, San Diego, CA, January 18–20, 1993, pp. 315–320.
45. Weiner, M., and Cellier, F. E., Modeling and simulation of a solar energy system by use of bond graphs, *Proceedings 1993 Western Simulation MultiConference on Bond Graph Modeling*, San Diego, CA, January 18–20, 1993, pp. 301–306.

2

Object-Oriented Modeling and Simulation

Sven Erik Mattsson, Mats Andersson, and Karl Johan Åström

Lund Institute of Technology
Lund, Sweden

2.1 INTRODUCTION

Modeling is an important aspect of many branches of engineering. A rich variety of models is used for many different purposes. Model development requires skills in many different fields. Different procedures and practices have been developed in different branches of engineering. Dynamic models are of particular interest for control engineers. Since a control system is often composed of parts covering a wide variety of technologies, it is necessary for a control engineer to have a unified approach to modeling all kinds of systems.

Today there is a significant industrial interest in using mathematical models and simulation. There are several reasons for this. Industrial processes are becoming more and more complex. For example, requirements on saving energy and raw material as well as avoiding environmental pollution imply that the systems must contain recirculation loops to win back energy and material. Such loops introduce interactions between various parts of the process which must be taken into account in design and operation. Desire to avoid buffers and to produce just in time introduce also tighter couplings. Increasing demands on reliability, safety and improved control make it necessary to integrate supervisory functions. Mixture of continuous process models and discrete supervision systems also increase complexity.

Consequently, there is a need to deal with complex processes. Prior trial-and-error methods are less effective and there are no feasible analytical tools available for many interesting and challenging problems. On the other hand, the significant decrease in the cost of computation makes mathematical models and simulation more useful as general-purpose tools. Simulation is useful to investigate specific problems, but also to test new approaches. Computer and software technology has now progressed so much that it is possible to make a drastic departure from earlier approaches in modeling and simulation.

This chapter presents results of experiments to develop modeling tools with drastically improved functionality. The key idea is to exploit object-oriented techniques and equation-based models. In this way, it is possible to simplify the modeling procedure and to facilitate reuse of models. A background is given in Section 2.2. Then, in Sections 2.3–2.4, we discuss and illustrate the basic ideas by means of a concrete example. In Sections 2.5–2.8, we focus on dynamic simulation and discuss how object-oriented and equation-based continuous-time models can be used to calculate time behavior.

2.2 TOOLS FOR MODELING AND SIMULATION

A typical modeling task consists of the following steps:

1. Decompose the system into subsystems
2. Describe the interaction between the subsystems
3. Develop behavioral descriptions for the subsystems
4. Combine the different submodels to obtain a complete model
5. Validate the model

The reason for the first step is to reduce the complexity and to obtain tractable subproblems. An attempt was made to find subdivisions that are natural and so that several subsystems can be used in many places. Also, it is important to make subdivisions that give simple interfaces between the submodels. The subdivision is often done hierarchically. In some cases, it is natural to make a subdivision that corresponds to physical components. This is the case, for example, when modeling electrical circuits; the components can be resistors, capacitors, inductors, and transistors. In other cases, it is more suitable to make subdivisions according to more abstract principles. The granularity of the descriptions and the amount of detail required are key issues. In a control problem, we have good guidance by considering the final use of the model. This will typically indicate the frequency ranges that are of interest and the orders of magnitude of the different terms involved. It is also very valuable to guide the granularity of the modeling by results from system identification experiments.

The behavioral descriptions for physical components are obtained from equations that express balance of mass, momentum, and enthalpy. These equations are augmented by constitutive equations like Hooke's, Boyle's, and Arrhenius's laws. The balance equations are typically expressed as ODEs or PDEs. A key problem is the amount of detail required. A typical problem is the approximation of PDE by ODEs. In some cases, it is also useful to give approximations of the behavior, e.g., in terms of aggregate models or qualitative models. After Steps 1–3 have been completed, the pieces are assembled into a complete model. This step can be tedious as well as error-prone. When a complete model is obtained, it remains to validate the model. Sometimes the different submodels can also be validated individually. A model can be used in many different ways. Typical tasks are calculation of equilibria, linearization, generation of simulation code, model reduction, and control design.

2.2.1. Today's Tools

Modeling is unfortunately difficult and time-consuming. Specialized modeling tools have been developed in different branches of engineering; there is SPICE [33] with several available commercial dialects for analog modeling of electrical circuits. VHDL is the IEEE 1076 standard for digital circuits; see, e.g., the textbook by Perry [37]. The Power System Simulator PSS/E from Power Technologies, Inc. is a widely used program for simulation of power grids. The Plant Modelling System Program (PMSP) and the Plant Design Analyser developed by the Central Electric Generating Board in the United Kingdom are other examples of tools for modeling and simulation of power systems. The program MEDYNA [17] is an example of a modeling tool for multibody dynamics. For chemical processes, there is ASPEN [16] with the commercial version ASPEN PLUS from Aspen Technology, Inc. for static modeling. For dynamical modeling, there is SPEEDUP [36], which is now a product of Aspen Technology, Inc. For building simulation and simulation of heating, ventilation, and air conditioning, there are many; for example, DOE-2 [12] and HVACSIM$^+$ [11]. A drawback of available special-purpose programs is that they are closed and rigid. It is not possible for a user to add a missing component. It is also extremely difficult to use different special-purpose tools to model different parts of a complex system and then to merge the parts to make a model of the total system.

For a control engineer, it is of interest to have a unified approach to modeling all kinds of systems. There are very few tools available for general modeling, but there are tools for simulation of some classes of continuous-time or discrete-event systems. Surveys of general-purpose simulation software are given by Kreutzer [25] and Kheir [24]. The journal *Simulation* annually publishes long lists of available simulation software.

In control engineering, modeling has traditionally been performed in connection with simulation. The work has always had a strong component of pencil-and-paper analysis. The purpose of the analysis is to transform the model to a form that is suitable for the simulation program. The value of standardization was recognized early. Standardized ways of representing analog computations were introduced a long time ago. This naturally led to the development of a standard for simulation of continuous-time systems CSSL [43]. An implementation of this standard in the form of ACSL from Mitchell & Gauthier Associates is a very widely used tool. The simulator Simnon [15] developed at our department is another successful commercial system. An advantage of this simulator is that it admits a good structuring of systems.

Use of CSSL tools for simulation is discussed thoroughly by Cellier [8]. These tools assume that a system can be decomposed into block diagram structures with causal interaction. This means that the models are expressed as an interconnection of submodels on explicit state-space form:

$$\frac{dx}{dt} = f(x, u)$$

$$y = g(x, u)$$

where u is input and y is output. It is seldom that a natural decomposition into subsystems lead to such a model. It is often a significant effort in terms of analysis and analytical transformations to obtain a problem in this form. It requires a lot of engineering skills and manpower and it is error-prone. Furthermore, it is also difficult to document models in such a way that models can be reused since there are few structuring mechanisms. The modeling procedure, therefore, is costly and difficult to repeat. Since simulation is used in different ways for feasibility studies, design support, and decision support, the lack of good modeling tools means that different models for the same process have to be developed for each use.

Ordinary differential equations have traditionally been the underlying mathematical framework in continuous-time simulation. However, when developing a model for a physical system, one uses fundamental laws such as mass balances, energy balances, and phenomenological equations. These are either algebraic equations or ordinary differential equations. Thus, differential–algebraic equations are the appropriate framework. Furthermore, in some applications such as power system simulation, it is not feasible to eliminate analytically a large number of algebraic equations introduced by the grid. We need to be able to simulate differential–algebraic systems.

An early modeling language that allowed noncausal submodels and differential equations on a general implicit form was Dymola [14]. In the area of chemical process simulation, there are early tools, e.g., SPEEDUP, supporting the use of differential–algebraic equations. Use of real equations has a long tradition in static modeling of chemical processes.

Discrete-event simulation is a much more heterogeneous field than continuous-time simulation. There is no comprehensive theory for discrete-event dynamical systems and various approaches are used to describe behavior. A formal approach to discrete-event simulation is found in the work of Ziegler [47]. Real-time programming is closely related to describing behavior of discrete-event systems. It would be highly desirable to have a modeling framework that makes it possible to deal with mixed continuous- and discrete-time systems. The system described in this chapter is also an effort in this direction.

2.2.2 The Omola Modeling Language

We have developed a new universal, equation-based, and object-oriented modeling language called Omola [1, 30]. The language is designed to support model development and to facilitate reuse. Omola supports several concepts for model structuring. Models can be decomposed hierarchically with well-defined interfaces that describe interactions. All model components are represented as classes. Inheritance and specialization support easy modification. Omola supports behavioral descriptions in terms of differential–algebraic equations, difference equations, and discrete events. Discrete-event modeling and simulation are discussed in [2]. OmSim is an environment of tools supporting modeling and simulation of Omola models. The prototype implementation of OmSim is written in C++ and based on the X-Window System.

Omola may be considered as a textual format for a model database or model library. Special-purpose modeling tools can use Omola as a common underlying model representation. This will result in more flexible tools since the models are readable and can be modified by the user. It will also be possible to use various special-purpose tools together for defining different parts of a complex system.

2.2.3 Object-Oriented Technology

Object-oriented methods are natural tools for modeling. The development of Simula [5] was motivated by simulation. Use of object-oriented methods for modeling of control systems was discussed in Åström and Kreutzer [3].

A recent review of current research activities within continuous-time modeling and simulation is given in the work of Marquardt [29]. He discusses three object-oriented modeling languages: ASCEND [38], MODEL.LA [42], and Omola. He states that though rather different, they build on the same basic ideas from object-oriented programming and from structured knowledge representation. ASCEND is a language and environment for rapid development of equation-based models. It contains many facilities for static simulation. Much effort has been put into designing a user-friendly interface. However, there are no facilities for defining models graphically. MODEL.LA is a modeling language tailored to chemical applications. The modeling tool has knowledge about chemical models and gives guidance and issues warnings. Simulation is not supported. It would be possible to use Omola to define the model classes of

MODEL.LA. Object-oriented modeling and simulation are also discussed by Cellier, Zeigler, and Cutler [9].

Much effort is spent on developing integrated environments supporting modeling and design. One approach is to build the environment on top of a common database. The database is used to communicate between tools and to keep track of models, data, and the vast amount of intermediate results created during the design process. One such project is the GE-MEAD project [40]. Another example is called DB-Prolog by Maciejowski and Tan [27]. Object-oriented technology is also emerging in database management systems. It seems that an object-oriented database scheme, rather than the traditional relational model, is better suited for representing the special kind of data structures common in design support systems [23].

2.3 COMPOSITION AND DECOMPOSITION

The notion of decomposition is very useful when dealing with large systems. By decomposing a large system into smaller components that can be analyzed and understood separately, it is possible to get an understanding of the total system. Decomposition can, of course, be applied at many levels. The inverse concept, composition, is naturally applied to construct larger systems from simple components. In the same way, composition is used in modeling. A large model can be defined as a composition of available submodels.

There are two other concepts that are closely related to composition and decomposition. These are *abstraction* and *modularization*. The terms are used in software engineering [6] and they are also important in modeling. We will use them with the following specific meanings:

Decomposition is the act of dividing a model definition into a set of interacting submodels.
Abstraction is the ability to focus on important aspects while unimportant details are disregarded. For example, a submodel should be defined so that its interface is separated from the internal description. It should be possible to apply a component only by looking at its interface.
Modularization is decomposition aimed at reuse. The components are made general enough to be applied in different contexts.

Since we are interested in defining and modifying large models in a user-friendly way and since we want to reuse models as much as possible, it is important to use a modeling language that supports model decomposition, abstraction, and modularization.

Omola contains a number of structuring concepts to support decomposition and abstraction. In the work of Nilsson [34] and of Mattsson [31], it is discussed how Omola can be used to model chemical processes and power systems. In the following, a simple mechanical system will be used to illustrate the decomposition concepts.

2.3.1 Mechanical Modeling

As an example to illustrate model decomposition, a mechanical system moving in a two-dimensional space is used. A one-dimensional example is too trivial to illustrate the important concepts, whereas a three-dimensional example would give lengthier component descriptions without adding any fundamental problems. We will start by decomposing the system into simple elements. A small library of reusable components will then be created. Finally, a model of the system will be assembled from library components. The approach can be viewed as a top-down system decomposition followed by a bottom-up model definition.

The system is shown in Figure 2.1. It consists of a cart moving along a horizontal track. A pendulum and a spring are attached to the cart. The other end of the spring is attached to a fix position. It is assumed that the cart and the pendulum are moving without friction. The gravitational forces are directed downward in the picture.

The general approach to modeling outlined in Section 2.2 will be followed. Thus, the model will be divided into components until we have reached a level where the components are simple and general enough to be taken from a library of mechanical objects. Such a library may for example include models of springs, joints, links, etc. The decomposition leads to the graphical representation in Figure 2.2 and to the Omola code in Listing 1. In the cart-and-pendulum model, we have used the submodel classes FixPosition, LinearSpring,

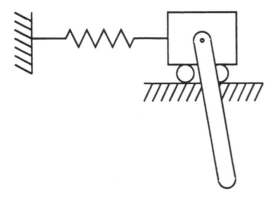

Figure 2.1 The cart-and-pendulum system.

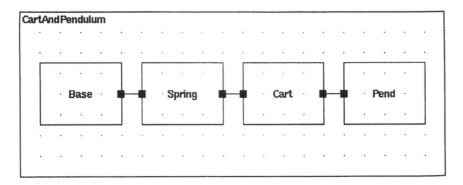

Figure 2.2 Decomposition of the cart-and-pendulum system.

CartX, and Pendulum that have to be defined. In the connections, we assumed that these submodels have terminals named CC1, and C2.

CartAndPendulum ISA Model WITH

submodels:

> Base ISA FixPosition;
> Spring ISA LinearSpring;
> Cart   ISA CartX;
> Pend   ISA Pendulum;

connections:

> Base.C   AT Spring.C1;
> Spring.C2 AT Cart.C1;
> Cart.C2   AT Pend.C;

END;

Listing 1. Omola model for the cart-and-pendulum system.

2.3.2 Terminals

Terminals are used in Omola to describe interaction between submodels. In the mechanical model, the base has one terminal, C, where the spring is attached. The spring has two terminals, C1 and C2 corresponding to each end of the spring. The cart also has two terminals, C1 where the spring is attached and C2 where the pendulum is attached. The pendulum has one terminal, C.

To describe model interaction, it is necessary to give relations between all variables that are involved in the interaction between the submodels. To obtain general submodels that can be combined in a flexible way, it is important to choose terminals carefully. Six quantities are needed to represent interaction in a two-dimensional mechanical system: two position coordinates, one angle, two forces, and one torque. It is chosen to represent positions and forces in a global Cartesian coordinate system. It follows from the principle of force balance that all forces and torques at an interconnection sum up to 0. Such variables are called *through* variables. Positions and angles at terminals of different submodels are normally equal when the systems are connected. These variables are, therefore, called *across* variables [14].

It will now be shown how terminals, appropriate for our example, can be constructed in Omola. The Omola terminal class SimpleTerminal has a number of predefined attributes. One attribute, called direction, defines whether the terminal represents an across or a through variable. The attribute defaults to across but can be changed to either in or out to define the flow direction of a through variable. Notice that an across terminal has no notion of flow direction. Also notice that the attribute values in or out has nothing to do with the causality of the terminal, i.e., whether it is an input or an output variable.

Listing 2 shows the definition of two simple terminal types, ForceTerminal and AngleTerminal, needed in mechanics models. The direction of the force terminal is set to in, to define that the terminal represents a through variable with the direction of positive flow *into* the model. This is interpreted as a force acting from the environment to the component owning the force terminal. The angle terminal represents an across terminal so the direction attribute of AngleTerminal is set to across. Simple terminals also have other attributes. In this example, we have specified the physical quantities and units of measure represented by the terminals. These attributes are used by the OmSim environment to check consistency of terminal connections and to introduce proper scale factors in the connection equations. OmSim has knowledge about quantities and their corresponding SI units.

```
ForceTerminal ISA SimpleTerminal WITH

    direction := 'in;
    quantity  := "force";
    unit      := "N";

END;

AngleTerminal ISA SimpleTerminal WITH

    direction := 'across;
    quantity  := "angle";
    unit      := "rad";

END;
```

Listing 2. Omola definition of ForceTerminal and AngleTerminal.

Two additional types of simple terminals are needed to define a mechanical system terminal. They are TorqueTerminal and PositionTerminal. Their definitions are similar to those in Listing 2.

We are now ready to define the structured terminals used to connect mechanical components. In the following, this particular type of structured terminal will be called *cut* since it represents a cut in a mechanical system. A cut of a component model represents an attachment point which can be connected to other components in the system. The cut is built up from components of the previously defined terminal types. Listing 3 shows the definition of the terminal type Cut.

Cut ISA RecordTerminal WITH

    x, y   ISA PositionTerminal;
    Fx, Fy ISA ForceTerminal;
    a   ISA AngleTerminal;
    T   ISA TorqueTerminal;

END;

Listing 3. Omola definition of Cut.

A connection between two cuts of different mechanical subsystems will result in six equations: three zero-sum equations and three simple equalities. Using a subscript for referring to the two cuts we will get

$$x_1 = x_2$$

$$y_1 = y_2$$

$$a_1 = a_2$$

$$Fx_1 + Fx_2 = 0$$

$$Fy_1 + Fy_2 = 0$$

$$T_1 + T_2 = 0$$

Notice that no causality is implied by these equations. Depending on the total structure of submodels, the equations may be used for computations in either direction.

2.3.3 Component Models

The components of CartAndPendulum will now be defined. FixPosition and LinearSpring are simple enough to be defined as primitive models, i.e., they will not be further decomposed into submodels. They are also general enough to be useful in a library of primitive components. CartX and Pendulum will, however, be further decomposed.

Since CartX is moving along a track without rotation it can be modeled as a particle moving in the *x*-direction. The cart model needs cuts for attaching the spring and the pendulum. The torques at these two cuts are zero. We can now identify three primitive library models that are useful in our example. They are called Particle, SliderX, and Joint. A SliderX constraints the motion of any cut to a line parallel to the *x*-axis. Joint has two cuts and decouples the angle in between with zero torque.

For Pendulum, we will use a library model called Bar which models the homogeneous bar. This model will be examined in detail below.

A graphical representation of the cart model is shown in Figure 2.3 and the Omola code for CartX and Pendulum is shown in Listing 4.

CartX ISA Model WITH

terminals:

C1, C2 ISA Cut;

submodels:

P ISA Particle;
S ISA SliderX;
J ISA Joint;

connections:

C1 AT P.C;
P.C AT S.C;
P.C AT J.C1;
J.C2 AT C2;

END;

Pendulum ISA Bar;

Listing 4. Omola models for the cart and the pendulum.

2.3.4 Primitive Models

We have previously recognized the need for a number of primitive components. Two of these will be defined in the following.

A model of a particle is fairly simple. It has one cut and a parameter, *m*, for the mass. Newton's second law of motion gives two equations, one for each co-ordinate direction:

$$m\ddot{x} = F_x + mg_x$$

$$m\ddot{y} = F_y + mg_y$$

The forces F_x and F_y are elements of the cut. The constants, g_x and g_y represent the gravity (if the x-axis is horizontal, then $g_x = 0$ and $g_y = -g$). We also need to add an equation defining the torque to be zero. The Omola model is shown in Listing 5. Time derivatives are specified by the dot operator. For example, dot(x2) refers to the second time derivative of x. The gravity field is represented by the global variables gx and gy. The notation with a double colon (::gx, etc.) is Omola syntax for referring to a global constant.

Particle ISA Model WITH

terminals:

  C ISA Cut;

parameters:

  m ISA Parameter WITH default := 1.0; END;

equations:

  m*dot(C.x, 2) = C.Fx + m * : :gx;
  m*dot(C.y, 2) = C.Fy + m * : :gy;
  C.T = 0.0;

END;

Listing 5.   Omola model of a particle.

As an example of a slightly more complicated model, we are now going to define Bar. This is a model of a bar with an equally distributed mass and cuts at each end. The dynamics of general plane motion of a rigid body can be represented by a force equation describing the translation of the mass center, and a momentum balance around the mass center. We get

$$m\dot{v} = \sum F$$

$$I_g\dot{\omega} = \sum M_g$$

where $\sum F$ is the sum of force vectors, m is the mass, v is the velocity vector of the mass center, $\sum M_g$ is the sum of torques, I_g is the moment of inertia around the mass center, and ω is the angular velocity. In the model, we need the equations in scalar form. The forces and torques are contributions from the cuts at the ends of the bar. In the force equation, we also have to consider a gravity force. Using subscript indices for referring to variables at the two cuts, we get the motion equations:

$$m\ddot{x}_0 = F_{x1} + F_{x2} + mg_x$$

$$m\ddot{y}_0 = F_{y2} + F_{y2} + mg_y$$

$$\frac{ml^2}{12}\ddot{\varphi} = T_1 + T_2 + \frac{l}{2}((F_{x1} - F_{x2}) \sin \varphi - (F_{y1} - F_{y2}) \cos \varphi)$$

where g_x and g_y are the gravities, x_0 and y_0 are the positions of the mass center, l is the length, and φ is the angle of the bar. Furthermore, the geometrical constraints of the bar model have to be specified in the model. These constraints define the position of the mass center relative to the cut positions, the positions of the two cuts depending on the angle and the equality of the angles at the ends. We get

$$x_0 = \frac{x_1 + x_2}{2}$$

$$y_0 = \frac{y_1 + y_2}{2}$$

$$a_1 = a_2$$

$$y_2 - y_1 = l \sin \varphi$$

$$x_2 - x_1 = l \cos \varphi$$

From these equations, it is a straightforward procedure to define the Omola model for the bar. The mass m and the length l are defined as parameters of the model. The complete code for the bar model is shown in Listing 6.

```
Bar ISA Model WITH
terminals:
  C1, C2 ISA Cut;
parameters:
  m ISA Parameter WITH default := 1.0; END;
  % total mass

  L ISA Parameter WITH default := 1.0; END;
  % nominal length
variables:
  I Type Real := m * L^2/12; % moment of inertia
  x0, y0 TYPE Real;          % center of gravity
equations:
  % Geometric constraints:
  x0 = (C1.x + C2.x)/2;
  y0 = (C1.y + C2.y)/2;
  C2.y-C1.y = L*sin(C1.a);
  C2.x-C1.x = L*cos(C1.a);
  C1.a = C2.a;
```

% Equations of motion:
  C1.Fx + C2.Fx + : :gx∗m = m∗dot(x0, 2);
  C1.Fy + C2.Fy + : :gy∗m = m∗dot(y0, 2);
  I∗dot(C1.a,2) = C1.T + C2.T +
    L/2∗((C2.Fy - C1.Fy)∗cos(C1.a) − (C2.Fx - C1.Fx)∗ sin(C1.a));

END;

Listing 6.   An Omola model of a homogeneous bar.

2.3.5   Graphical Model Composition

The OmSim environment contains a graphical model editor. It allows the user to define structured models by selecting submodels from a library menu and positioning their graphical representation with the mouse. Connections and terminals can also be specified graphically. The editor generates the Omola code from the graphical representation. Submodels are, by default, represented as a rectangular box with a name, but the user can also define a special icon for a model. Figure 2.2 is an example of a screen dump of the block diagram editor showing submodels in the default way. Icons and graphical layouts are represented in Omola and associated with the actual models.

Figure 2.3 shows a graphical representation of the cart-and-pendulum model where special icons have been defined for the mechanical components in the library. The decomposition here is a single-level decomposition, but the behavior of this model is identical to the one presented above.

The diagram shows an abstract representation of the structure of the mechanical system. It does not show the correct spatial relationships between the components, even though the layout can sometimes be made to resemble the real system.

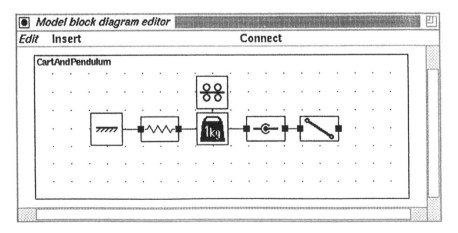

Figure 2.3   Icon diagram of a complete cart-and-pendulum model.

2.3.6 Simulation of an Omola Model

The OmSim environment contains a simulator for Omola models. The procedure of transforming an Omola model into a representation that can be handled by the simulator is discussed in Section 2.5. As a simple illustration of the mechanics example, a simulation of the cart-and-pendulum model is shown in Figure 2.4.

OmSim is based on a set of separate tools having their own windows. Figure 2.3 shows the window of the graphical model editor. Figure 2.4 shows the

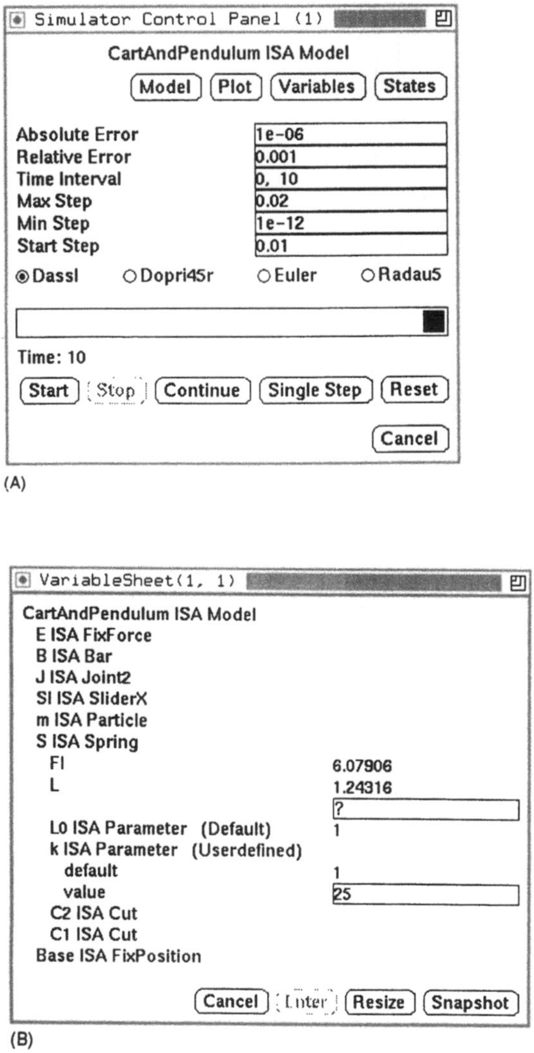

Figure 2.4 OmSim simulating the cart-and-pendulum model.

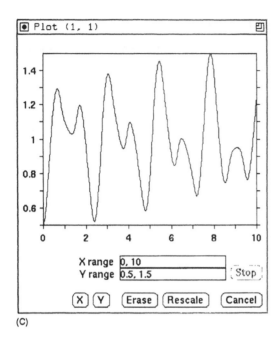

(C)

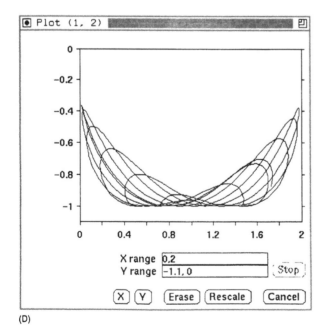

(D)

Figure 2.4 Continued.

control panel window of the simulator, a hierarchical variable browser, and two plot windows. The control panel, shown in the upper left window, gives the possibility to select model to simulate open new plot windows and variable browsers, select integration method, etc. The variable browser, shown in the upper right window, gives access to all variables of the model and the possibility to change parameter and initial values. The left plot window shows the x-coordinate of the cart, whereas the right plot window shows a phase plane of the x- and y-coordinates of the free end of the pendulum.

2.4 MODEL INHERITANCE

Class inheritance is a concept used for defining abstract data structures in some programming languages. It was first introduced in Simula [5] and has become an important concept in object-oriented programming. For an overview of object-oriented terminology and concepts, see, for example, the work of Stefik and Bobrow [41]. In object-oriented programming, a *class* defines an abstract data type, i.e., a collection of data attributes and functions operating on the data. A class can be defined as a *subclass* of another class, called the *super class* or the *base class.* By inheritance, all attributes of the base class are available in the subclass. Additional attributes can be added to the subclass to extend or specialize the definition.

In this section, it will be shown how the concept of subclasses and inheritance can be used to structure models and to build model libraries. In Omola, models are represented by classes and every class is a subclass of some other class (except for one predefined class called Class). A model can, therefore, be defined as a subclass, i.e., a specialization or an extension, of another model.

2.4.1 Inheritance in the Mechanics Library

In Section 2.3.3, we defined a few primitive models that were subclasses of the predefined class Model. Since Model is an empty base class, no inheritance occurred. If we had continued to define a more complete library of mechanical components, we would have discovered that many of the components were quite similar, differing only in a few equations. Inheritance can be used to take advantage of such similarities between different models. By arranging the models as a tree of classes, the similarities between models can be exploited and made explicit. The best way to present an inheritance tree to the user is probably to show it as a graph. Figure 2.5 shows an inheritance tree for a mechanics library. In the following, we will discuss how this inheritance structure is organized.

The root of the inheritance tree of all mechanics models in our library is called MechanicsModel which is an empty specialization of the predefined class Model. The class MechanicsModel is introduced to show the conceptual

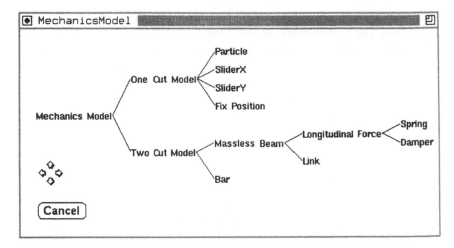

Figure 2.5 Inheritance tree for parts of the mechanics library.

similarity between all models in the library, i.e., the fact that they all represent
mechanical objects. The definition of MechanicsModel is simply:

MechanicsModel ISA Model;

One level up in the inheritance tree we find two classes called OneCutModel
and TwoCutModel. These classes define component models with two types of
interfaces with one or two terminals. The Omola definitions of the two classes
are shown in Listing 7. These interface classes are reused by all models in the
mechanics library, so that the definitions of the interface variables do not have
to be repeated in every model. This helps to maintain coherency in the library.
Since all models share the same interface definitions, they are interchangeable,
at least as far as their terminal interfaces are concerned.

```
OneCutModel ISA MechanicsModel WITH
terminal:
  C ISA Cut;
END;

TwoCutModel ISA MechanicsModel WITH
terminals:
  C1, C2 ISA Cut;
END;

Particle ISA OneCutModel WITH
parameter:
  m ISA Parameter WITH default := 1.0; END; % Mass
```

equations:
    m*dot(C.x, 2) = C.Fx + m * : :gx;
    m*dot(C.y, 2) = C.Fy + m * : :gy;
    C.T = 0.0;

END;

Listing 7. The definitions of OneCutModel, TwoCutModel, and Particle.

The Particle model defined in the previous section can now be redefined as a specialization of OneCutModel. The difference is that OneCutModel is the base model instead of just Model and that the terminal definition is omitted since it is inherited from the base class. The new definition of Particle is shown in Listing 7.

The models with two terminals will now be refined further. To do this, we observe that several objects in a mechanics library use the idealized assumption of zero mass. Massless objects can be used to represent geometric constraints of a system. To represent dynamic behavior, they must be combined with mass objects.

A certain set of constraints applies to all massless objects. These are that the sum of all forces and torques acting on the object is zero. These constraints can be expressed by three equations that can be introduced in a new class, Mass-lessBeam, a subclass of TwoCutModel. In MasslessBeam, we also add a local variable, L, for the distance between the endpoints. The orientation angle is defined as the angle of a straight line between the endpoints of the beam. It is postulated that the orientation angles are equal when two beams are connected. The MasslessBeam class is shown in Listing 8.

MasslessBeam ISA TwoCutModel WITH
variable:
    L Type Real; % length
equations:
    C1.Fx + C2.Fx = 0;
    C1.Fy + C2.Fy = 0;
    C1.T + C2.T − C2.Fx*(C2.y-C1.y) + C2.Fy*(C2.x-C1.x) = 0;
    C2.y-C1.y = L*sin(C1.a);
    C2.x-C1.x = L*cos(C1.a);
    C1.a = C2.a;
END;

LongitudinalForce ISA MasslessBeam WITH
variable:
    F1 Type Real; % longitudinal force
equation:

```
    C2.Fx*cos(C1.a) +  C2.Fy*sin(C1.a) = Fl;
END;
```

Listing 8. The Omola code for MasslessBeam and LongitudinalForce.

A model of a stiff, massless link can now be defined by adding a constant length constraint to MasslessBeam by giving a fix value to the L variable. However, since the user of a link model probably wants to decide the actual length, it is defined as a parameter. This is accomplished simply by redefining the inherited length variable as a parameter with the same name. The Link model is shown in Listing 9.

```
Link ISA MasslessBeam WITH
parameter:
    L ISA Parameter WITH default := 1.0; END; % Length
END;

Spring ISA LongitudinalForce WITH
parameters:
    k ISA Parameter WITH default := 1.0; END; % spring constant
    L0 ISA Parameter WITH default := 1.0; END; % nominal length
equations:
    Fl = k*(L − L0);
END;

Damper ISA LongitudinalForce WITH
parameter:
    d ISA Parameter WITH default := 1.0; END; % damping
equation:
    Fl = d*dot(L);
END;
```

Listing 9. Models for Link, Spring, and Damper.

This example illustrates some possibilities in Omola to redefine inherited attributes. If a class defines an attribute with the same name as an inherited one, the local definition will shadow the inherited definition. Other examples where this can be useful will be given later.

Another way to specialize MasslessBeam is illustrated by the class LongitudinalForce in Listing 8. In this case a new variable, Fl, is introduced for the force in the longitudinal direction. Different kinds of springs and dampers can then be defined by giving different equations for the longitudinal force. The classes Spring and Damper, shown in Listing 9, are typical examples. In the linear spring model the longitudinal force is the product $k(L − L_0)$ where k is a spring constant, L is the actual length, and L_0 is the nominal length. In the linear damper model the longitudinal force is instead given by $d(dL/dt)$, where d is the damping coefficient. Different kinds of nonlinear springs and dampers can be defined in a similar way.

2.4.2 Inheritance for Model Reuse

It will now be shown how inheritance can be utilized for reuse of complete system models. The modification of the cart-and-pendulum model, defined in Section 2.3, is used as an illustration.

First, we will show how a slightly different cart-and-pendulum can be defined by using the original model as a base class. The model CartAndPendulum has four submodels taken from the library. Assume we want to define a model where only the pendulum component is different while the rest of the system remains unchanged. This can be done with a minimum of effort by defining a new model, CAP2, as a subclass of CartAndPendulum. In CAP2, we only have to redefine the pendulum component called Pend. With a new pendulum model called Pend2 the code becomes

```
CAP2 ISA CartAndPendulum WITH
  Pend ISA Pend2;
END;
```

The new model inherits all submodels and connections from CartAndPendulum, but the inherited submodel Pend is replaced by a local definition using a different pendulum model. This works well as long as the new component has terminals that are similar to the original one.

Inheritance can also be used to extend a structured model. For example, CartAndPendulum models an autonomous system. It might be of interest to create a version that can be controlled by an external force. It is then necessary to add terminals corresponding to a control input and a measurement output. We also need a device for adding a controlled force in the x-direction. Listing 10 shows the model of the controlled force device and the extended cart-and-pendulum model called CAP3. In the new models we have assumed that InputTerminal and OutputTerminal are terminal classes available in the library.

```
ForceX ISA OneCutModel WITH
  F ISA InputTerminal;
  C.Fx = F;
  C.Fy = 0;
  C.T = 0;
END;
```

```
CAP3 ISA CartAndPendulum WITH
terminals:
  In ISA InputTerminal;
  Out ISA OutputTerminal;
submodels:
  Fx ISA ForceX;
connections:
  In AT Fx.F;
  Out AT Cart.Cl.x;
  Fx.C AT Cart.Cl;
END;
```

Listing 10. A controllable version of the cart-and-pendulum.

2.4.3 Summary

We have demonstrated how a library of mechanical models can be organized using the concept of subclassing and inheritance. The inheritance hierarchy was explained in a top-down order. Starting with the most general classes, we successively specialized them in several layers until we reached the level of fully specified mechanical components. This is, however, not the way a library is initially developed. Normally, the library developer has only vague ideas about a possible inheritance hierarchy. When a library is developed from scratch, the developer starts by defining a few fully specified model components. Then he or she discovers that some of these models have parts that are similar. These parts are then taken out and put in a common base class. This may go on in many steps and redesigns until a satisfying inheritance tree has emerged.

It is a difficult task to develop a well-structured and reusable model library from scratch. Once it is done, it is a much easier task to extend it in several directions and to use the models as components in other models.

2.4.4 Inheritance Versus Composition

We have studied two concepts for models structuring: composition and subclassing with inheritance. These concepts are complementary. In most cases, it is easy to decide intuitively if a new model should be defined by combining available models by composition or by subclassing. A systematic method for choosing between composition and inheritance is still an open research issue. We will, therefore, summarize the differences between the two concepts.

Model composition is a well-known concept that is supported by several commercial simulation languages. The methodology for model structuring is based on the fact that most commercial simulation languages only allow causal submodels with defined input and output terminals. For example, a mechanics library with terminals like we have studied here cannot be represented in such a system. Our model library depends on a system that can handle noncausal differential and algebraic equations.

Model decomposition is a way to handle complexity by dividing a large system into smaller components. A component model abstracts behavior and makes it accessible through a small set of terminals and parameters. It should be possible to reuse a model component without concern for its internal definition. On the other hand, reusing a model as a base class gives full access to its internal structure. It is flexible but it requires more knowledge about the model.

Two different groups of model developers may be considered. One group are modeling experts that develop new models and libraries from scratch. Another group are application engineers who are mainly using simulation as a tool. They do not want to develop models from scratch but they need libraries that are easy to combine into large systems. Inheritance is a way for the library designer to organize a library such that it becomes easier to find what is needed and to add new elements.

2.5 SIMULATION

A model can be used in many different ways. In this and the following three sections, we will discuss the use of Omola models for dynamic simulation. The goal is to calculate time behavior. For simplicity, we will focus on continuous-time simulation and assume that behavior is described by differential–algebraic equation (DAE) systems. Combined discrete-event and continuous-time simulation is discussed by Andersson [2].

For all realistic continuous-time simulation problems, numerical solvers are needed. Different approaches to numerical solution will be the topic of this section. The Omola description of the simulation problem must be transformed to fit the numerical solvers. This procedure, which includes analysis of consistency and completeness, is called compilation for simulation and it will be discussed in Section 2.6. Unfortunately, today's numerical DAE-solvers are not able to solve all mathematically well-defined problems. For example, they fail to solve high-index problems. Some of the difficulties can be detected and removed by combined symbolic and numeric techniques. Solution of high-index problems will be the topic of Section 2.7. Finally, in Section 2.8 we will discuss ways to facilitate numerical solution by symbolic manipulation.

There are two extreme approaches to solving a simulation problem: simultaneous or modular solution. In the simultaneous approach, all equations are collected and solved as one global problem. The idea of modular solution is to calculate the behavior locally for each component and then include the effects of interaction by iteration or by external modification of the states at each step of integration. Neither the simultaneous approach nor the modular approach is superior with respect to efficiency and flexibility. Thus, both approaches as well as combinations are of interest.

2.5.1 The Modular Approach

The modular approach has been very successful in discrete-event simulation. Object-oriented discrete-event simulation is a typical manifestation. The objects are actors and they communicate by means of message passing during a simulation run. In Sections 2.3–2.4, we have been discussing object-oriented modeling which is quite another thing. Object-orientation has been a means to organize and structure knowledge about system structure and behavior. The Omola models presented in Sections 2.3–2.4 cannot be actors in a simulation since they do not know how to calculate their behavior. This problem can be redressed by providing each translated component with a simulation method. It is, however, more difficult to cope with the interactions. The idea to use message passing is well-suited for discrete-time simulation but it is not a good approach for continuous-time simulation, where interaction means that variables should be equal for all times. Thus, it is necessary to communicate frequently, which means that efficiency is lost.

Interaction can be handled in other ways. Waveform relaxation is one approach (see, e.g., [46]) where each iteration is a complete run and where the subcomponents are simulated individually and independently of each other, taking the inputs to be the time histories of the other subcomponents from the previous iteration. Another approach is to modify the states of the submodels after each integration step with respect to the interactions. Söderlind, Eriksson, and Bring [44] have developed such a modular solver for differential–algebraic equations. In an object-oriented interpretation, we may say that the subcomponents do not communicate with each other but with their owner. The task of simulating a composite model is to supervise the simulation of its components and make proper corrections according to the interactions of its subcomponents.

The advantage with the modular approach is that it gives great flexibility. Different integration methods can be used for different components. The algorithms can be parallelized. It is also easy to switch submodels during simulation. This is useful when simulating different phases of a batch process or when simulating the effects of catastrophic malfunctions. A drawback is that there is no high-quality modular DAE-solver available today. Tight interactions between submodels may cause convergence problems and even lead to instabilities. The modular approach has been used frequently in static simulation of chemical process plants. The user is then supposed to use his domain knowledge to cut connections properly and to supply good initial guesses.

2.5.2 The Simultaneous Approach

Simultaneous solution is the prevailing approach. The advantage of the simultaneous simulation approach is that it allows a thorough analysis and optimization for simulation efficiency. For example, it is possible to deal with high-index problems. A drawback is that the analysis may take some time. However, this time is hopefully gained back many times over in the simulation runs. Another drawback is that it is laborious to change model structure during simulation. Basically, a total recompilation is needed. Due to combinatorial explosion, it may be infeasible to analyze every possible model structure in advance.

To be able to use multiprocessor systems for parallel computations, there is interest in ways of decomposing the numerical solution calculations. This decomposition need not to be identical to that implied by the component hierarchy of the model. It is possible to make a decomposition that is more suited for numeric solution.

In the following, we will focus on the simultaneous simulation approach.

2.5.3 Numerical Solvers

We have assumed that the behavior of primitive models are described by a combination of algebraic equations and ordinary differential equations. Since the equations implied by the connections are algebraic, this means that the simulation problem to be solved is a differential–algebraic equation (DAE) system. There are numerical DAE-solvers which treat problems of the type

$$g(t, x, \dot{x}) = 0$$

if they are provided with a routine that calculates the residual

$$\Delta = g(t, x, z)$$

when the arguments t, x, and z are known. It is straightforward to generate a residual routine from an Omola problem description. Unfortunately, today's numerical solvers fail to solve high-index problems so that problems must be taken care off.

A good numerical DAE-solver is DASSL developed by Petzold (see [7]). The code for DASSL is in public domain. It has a reputation of being one of the best and most robust solvers for DAEs. The program DASSL converts a DAE into a nonlinear algebraic equation system by approximating derivatives with backward differences. For example, the simplest discretization is backward Euler,

$$\dot{x}_{n+1} \approx \frac{x_{n+1} - x_n}{h}$$

where h is the step size. This approximation turns the DAE problem into the implicit recursion

$$g\left(t_n, x_n, \frac{x_n - x_{n-1}}{h}\right) = 0$$

DASSL uses backward differences of order up to 5. The version DASRT can detect zero crossings of user-defined indicator functions. This is useful for continuous-time simulation with discrete events.

One drawback with multistep methods like DASSL is that discontinuities force the methods to restart with low-order approximations. If one-step discretizations of the Runge–Kutta type is used, this problem is avoided. Since many real problems are discontinuous or involve sampled parts or other discrete-event elements, there is an interest in using implicit Runge–Kutta methods. One available high-quality numerical solver is RADAU5 [22]. It assumes that the problem is on the form $B\dot{y} = h(t, y)$, where B must be a constant matrix, but it may be singular. The problem $g(t, x, \dot{x}) = 0$ can always be transformed to the form $B\dot{y} = h(t, y)$ by introduction of algebraic variables to represent derivatives which appear nonlinearly or have time-varying coefficients.

Another approach is to try to transform the problem analytically into explicit state-space form $\dot{x} = f(t, x)$ and to use a numerical ODE solver. This is not a general method since the transformation into explicit state-space form requires in solution of nonlinear equations. It is, however, a feasible approach in many applications. Cellier and Elmqvist [10] discuss generation of state-space models for special classes of systems.

There are ongoing work to provide simulator engines with several different numerical solvers having a well-defined uniform low-level problem format. Two examples are ANDECS-DSSIM [19] and DIVA [26]. These engines can also treat discrete events. A special feature of DIVA is that it has solvers that exploit sparse matrix techniques to allow simulation of large problems with thousands of equations.

2.6 MODEL COMPILATION

Before a model represented in a high-level language like Omola can be simulated, it has to be transformed into a representation suitable for the numerical integration routine. This transformation which is called model compilation also involves extensive checking for errors in the high-level description.

It is practical to include extensive error checking in the compilation since it is advantageous to catch model errors as early as possible. At this stage, it is also easier to give good error messages related to the original model formulation. The question of model correctness can be divided into two parts:

1. Is the model syntactically and semantically correct according to the rules of the modeling language?
2. Does the model represent a complete and consistent problem from a mathematical point of view?

The first question of model correctness has a number of aspects. Lexical and syntactical analysis according to a formally specified grammar is a standard problem in computer science. There are also a number of semantical checks that are similar to those in programming language compilers. They involve scope rules for names and type consistency of expressions. In Omola, there are also a number of rules, specific to dynamic modeling, that determines interpretation and consistency of terminals and connections. For example, a simple terminal has attributes for direction, causality, name of quantity, etc. The compilation procedure transforms terminal connections into mathematical equations depending on these attributes. Inconsistencies may be detected due to illegal combinations of attributes.

After the model has passed the first set of checks, it can be instantiated into a set of variables and equations. The rest of the compilation procedure involves:

1. Check for structural defects.
2. Order variables and equations into a sequence of subproblems.
3. Sort out time-invariant parts.
4. Derive the differentiated index 1 problem.
5. Sort the equations in computational order.
6. Check consistency of user specified causalities.
7. Reduce index.
8. Make additional partitioning and other symbolic manipulations to facilitate the numerical computations.
9. Output a result suitable for simulation.

In Steps 1–6, the variables and the equations are permuted to find basic structural properties. Permutations are nice since they do not change the equations themselves. It is easy to give good error messages, since the equations are in their original form. When steps 1–6 are performed a DAE solver can handle the equations. It may be advantageous to also carry out an index reduction. Index reduction and other ways to facilitate the numerical solution are discussed in the two sections following this one.

It is useful recall the operations carried out when using CSSL-languages. Check for structural defects means to check that each variable is assigned a value exactly once, i.e., appears exactly once as a left hand side variable. The assignment statements are then sorted in computational order. In CSSL-tools the procedure stops by issuing error messages, if algebraic loops are detected.

2.6.1 Structural Properties

It must be realized that it is not feasible or even possible to perform complete consistency checks of a problem. A numerical DAE-solver must test singularities along the calculation. However, much can be done to detect and sort out singular problems beforehand. In a systematic approach, it is advisable to first analyze the structure, i.e., focus on which variables that appear in each equation rather than how they appear. In addition to detecting defects, the aim of the structural analysis is to decompose the problem into subproblems, which can be analyzed in turn. This approach is very advantageous for large problems.

To explain the ideas, let us first consider the nonlinear algebraic equation system

$$h(x, p) = 0$$

where x are the unknowns to be calculated and p are parameters. To investigate if this problem is singular, a common approach is to determine if the Jacobian $J = \partial h / \partial x$ is singular. It is easier to analyze a more general problem. If all non-zero elements of J are replaced by independent variables, it is easy to check if this matrix always is singular independent of the numerical values of the nonzero entries. Following Brenan, Campbell, and Petzold [7], p. 21, such a matrix is called structurally singular. A structurally singular matrix is of course singular, but a structurally nonsingular matrix may be singular.

A matrix is structurally nonsingular if and only if it is possible to permute the rows to obtain a matrix with a nonzero diagonal. Another more graph-oriented formulation is that it should be possible to form a set of ordered pairs of variables and equations such that each variable x_j and each equation $h_i = 0$ are only members of one pair and for each pair $(x_j, h_i = 0)$; the variable x_j appears in the expression h_i. Such a set is often called an output set. There are well-known, efficient procedures to find an output set; see, for example, [13]. These algorithms can be implemented in about one page of Pascal or C++ code.

Consequently, to check if $h(x, p) = 0$ is structurally singular, we need not know the values of the elements of the Jacobian. We only need to know which elements are nonzero. An element J_{ij} can be nonzero only if the variable x_j appears in the expression $h_i(x, p)$. It is convenient to introduce the structure Jacobian or incidence (occurrence) matrix of variable dependencies such that for each element i, j, the value is zero if x_j does not appear in the expression h_i, otherwise it is one.

To further investigate the properties of the problem we shall make use of the Block Lower Triangular (BLT) partitioning. The unknowns and the equations are then permuted to make the structure Jacobian block lower triangular. A BLT partitioning reveals the structure of a problem. It decomposes a problem into subproblems which can be solved in sequence, starting with the first block. There are efficient procedures for constructing BLT partitions with minimum-sized diagonal blocks. A BLT partition is done in two steps. In the first step, an output set is constructed. The second step of the BLT partition procedure is to order the pairs of the output set to make the structural matrix BLT. The basic algorithm was given by Tarjan [45] (see also [13]). It is only a bit more complex than the algorithms for constructing output sets.

The method for investigating singularities of algebraic equation systems can be applied to differential–algebraic systems by using the following observation. We should not distinguish between the appearance of x_i and the appearances of its derivatives; the appearances of $\dot{x}_i$, $\ddot{x}_i$, etc., are considered as appearances of x_i. Thus, to check if the first-order DAE problem $g(t, x, \dot{x}) = 0$ is structurally nonsingular, we check if the algebraic problem $g(t, z, z) = 0$ is structurally nonsingular when considering z to be unknown.

If a problem is structurally singular, it is desirable to find the reasons. If there are variables which have not been assigned an equation, it means that equations are missing. However, it is not possible to constrain just any variable further since a variable may be determined by the equations that are given already. To find out which variables should be constrained, each unassigned variable is assigned a fictitious equation in which all unknowns appear. A BLT partitioning is performed. The fictitious equations will all be collected in the last block. This means that variables of the other blocks can (at least structurally) be determined, whereas the variables of the last block are only partially constrained. The equations to be added must at least include one of these variables. Unfortunately, it is often the case that many variables end up in the last block, thus giving little hint to the user. By assuming that the undeterminacy is caused by missing connections or inputs at the highest level, we can assist the user by sorting the variables according to component hierarchy. If the user can specify or guess causalities, a new analysis may give further assistance.

The problem when there are redundant equations is dual. To analyze it, create a fictitious variable for each redundant equation and make them appear in each equation. Then make a BLT partitioning. To remove the overdeterminacy, the user should select among the equations of the first block.

2.6.2 Sorting Out Time-Invariant Parts

Let us now assume that no structural singularities were found and that the problem has been BLT partitioned. Variables that only change their values when parameters change values are easily sorted out. A diagonal block of a BLT problem is implicitly time invariant if the following conditions hold;

1. Time does not appear explicitly in the equations of the block.
2. None of the unknowns in the block appear differentiated in the block.
3. All other variables appearing in the block are time invariant.

Constants can be sorted out in an analogue way.

2.6.3 Decomposition of Time-Varying Parts

The BLT decomposition has split the problem into a sequence of subproblems, and time-invariant parts have been sorted out. It is possible to simulate the time-varying subproblems sequentially by using the results from the previous simulations as inputs. The usefulness of this approach to decomposition should not be overestimated. For many real applications, there will, typically, be one large block due to feedback. In front of the large block, there are typically a number of small blocks calculating inputs. After the large block, there are small blocks for calculation of outputs.

2.6.4   The Differentiated Index 1 Problem

Consider the DAE problem

$$g(t, x, \dot{x}) = 0$$

A natural way to transform this problem into explicit state-space form is to try to solve for $\dot{x}$ algebraically. The Jacobian with respect to $\dot{x}$ is, however, often singular. To transform the problem to state-space form, we must, therefore, differentiate some equations a number of times.

The minimum number of times that all or part of $g(t, x, \dot{x}) = 0$ must be differentiated with respect to t to determine $\dot{x}$ as a continuous function of x, and t defines the *index* [7]. For example, an ODE on state-space form, $\dot{x} = f(t, x)$ is index 0, and the problem defined by $\dot{x} = f(t, x, y)$ and $g(t, x, y) = 0$ is index 1 if the Jacobian $\partial g/\partial y$ is nonsingular. An alternative approach which is more convenient for our purposes is to allow higher-order derivatives as well as purely algebraic variables. The index is at most 1 if it is possible to determined the highest-order derivatives as continuous functions of time and lower derivatives. The index is 0 if there are no pure algebraic variables.

The index is of significant interest when using numerical DAE-solvers since there is no reliable general-purpose software for solving high-index problems. The available numerical DAE-solvers may solve some index 2 problems, but they fail when the index is greater.

The index can be reduced by differentiating the equations a number of times; see, e.g., [7], p. 33. To find the differentiated index 1 problem, we can use the algorithm of Pantelides [35]. This algorithm establishes the minimum number each equation has to be differentiated to make the differentiated problem structurally nonsingular with respect to its highest-order derivatives.

Pantelides's algorithm can be viewed as an extension of the algorithms for constructing output sets. It ends in a finite number of steps, provided the DAE problem is structurally nonsingular. Pantelides's algorithm is easy to implement. To find out which derivatives appear in an equation that is differentiated m times, we need only know the orders of the derivatives in the original, undifferentiated equation and increment the orders with m. Thus, it is not at all necessary to perform any differentiation when implementing the algorithm.

High-index problems are difficult to solve because they will, in principle, involve differentiations. The order of the derivatives required is the same as the index of the problem. The numerical difficulties thus increase with the index. Available DAE-solvers are designed to integrate, i.e., calculate x from $\dot{x}$, but if the index is greater than 1, they also should differentiate, i.e., calculate $\dot{x}_i$ from x_i for some components, which is quite another numerical problem. In numerical integration, it is assumed that we can make the errors arbitrarily small by taking sufficiently small steps. An integration routine typically estimates the errors by taking steps of different sizes and comparing the results. In numerical differentiation, it is well-known that the step cannot be taken infinitely small. There is a finite, optimal value. Thus, the error estimates calculated by the DAE-solvers behave irregularly and today's DAE-solvers cannot handle the situation, but they have to exit with some error message. Methods for removing the difficulties by symbolic manipulation are outlined in the next section.

2.6.5 Sorting in Computational Order

After having applyed Pantelides's algorithm, it is natural to make a new BLT partitioning with respect to highest-order derivatives to decompose the problem into smaller subproblems. This partition is a useful starting point for index reduction, for transformation to explicit state-space form, and for consistency checking of user-defined causalities.

2.6.6 Checking Consistency of User-Defined Causalities

In Omola, equations are used to describe behavior. It is important to note that generally the causality of a terminal (input or output) is not defined by the model designer, but inferred from the use of the model. If the causality is known, it is, however, useful to specify it to allow automatic consistency checking. Examples, where the causality is known, are terminals representing measurement values and the terminals of a controller. Specifications of causalities are, at this stage, translated into assignment expressions. To check causality consistency, we check each block of the partitioned problem. For consistency, all assignments must appear in scalar blocks, where the unknown of the block is the left-hand-side variable of its assigned equation.

2.7 INDEX REDUCTION

To facilitate the numerical solution of the DAE problem, it is of interest to reduce the index to 1. It is often less satisfactory to solve the index 1 problem obtained by differentiating the equations because its set of solutions is larger. The algebraic relations of the DAE are only implicit in the differentiated problem as solution invariants. Unless linear, these invariants are generally not preserved under discretization. As a result, the numerical solution drifts off the algebraic constraints. Let us consider a classical example in the DAE literature.

EXAMPLE 1—THE PENDULUM

The equations of motion of a planar pendulum of length L and mass m are

$$x^2 + y^2 - L^2 = 0 \tag{7.1a}$$

$$m\ddot{x} + \frac{\lambda}{L} = 0 \tag{7.1b}$$

$$m\ddot{y} + \frac{\lambda}{L} + mg = 0 \tag{7.1c}$$

where x is the horizontal position, y is the vertical position, λ is the force in the string, and g is the gravitational constant, assuming that the positive direction of the gravitational force is opposite to the direction of the y-axis. The problem is index 3. It is easy to see that the index is greater than 1 since the length constraint (7.1a) does not contain any highest-order derivative, $\ddot{x}$, $\ddot{y}$, or λ. But they appear in the length-constraint differentiated twice:

$$2x\dot{x} + 2y\dot{y} = 0 \tag{7.1a'}$$
$$2x\ddot{x} + 2\dot{x}^2 + 2y\ddot{y} + 2\dot{y}^2 = 0 \tag{7.1a''}$$

Equations (7.1a''), (7.1b), and (7.1c) constitute an index 1 problem since it is possible to solve continuously for the highest-order derivatives. Note that this problem has a larger solution set than the original problem. If we select the initial values of x, $\dot{x}$, y, and $\dot{y}$ to fulfill Eqs. (7.1a') and (7.1a), we have, from a mathematical point of view, an equivalent problem. However, the numerical solution procedures do not give exact solutions, which means that the length constraint may drift.

To eliminate drift, so-called constraint stabilization techniques have been devised [4, 21]. Another approach is try to obtain a low-index formulation, with a solution set identical to that of the original problem. This can be achieved by augmenting the system as the index reduction proceeds: All original equations and their successive derivatives are retained in the process. The result is an overdetermined, but consistent index 1 DAE. Like invariants, however, consistency is generally lost when the system is discretized. Therefore, special projection techniques are required for the numerical solution; see, e.g., [18].

An index reduction technique which overcomes the latter complication has been developed by us. Let us illustrate the idea on the pendulum problem. For small oscillations around the equilibrium point $x = 0$ and $y = -L$, Eq. (7.1a'') can be used as an implicit definition of $\ddot{y}$. To avoid overdeterminacy, a new algebraic variable, say w, is introduced to represent $\ddot{y}$ wherever it occurs in system (7.1). Similarly, we can use Eq. (7.1a') and replace $\dot{y}$ by v. This yields the augmented but determined system, which is index 1 and mathematically equivalent to Eq. (7.1):

$$x^2 + y^2 - L^2 = 0 \tag{7.2a}$$

$$2x\dot{x} + 2yv = 0 \tag{7.2a'}$$

$$2x\ddot{x} + 2\dot{x}^2 + 2yw + 2v^2 = 0 \tag{7.2a''}$$

$$m\ddot{x} + \frac{\lambda}{L} = 0 \tag{7.2b}$$

$$mw + \frac{\lambda}{L} + mg = 0 \tag{7.2c}$$

Problem (7.2) has five equations and five unknowns: x, y, λ, v, and w, which all are algebraic except for x which appears differentiated twice; the problem has a single degree of freedom.

The original problem has been augmented with (7.1a'') and (7.1a'). For each differentiated equation appended to the original system, one "new" dependent variable was introduced. The introduced variables represent derivatives, and we call them *dummy derivatives*. Dummy derivatives are purely algebraic variables and are not subject to discretization. We know that $w = \ddot{y}$, $v \equiv \dot{y}$, but this is not explicit in the transformed problem.

In the general case, our index reduction procedure consists of two major steps. First Pantelides's algorithm is used to find the differentiated index 1 problem and then dummy derivatives are selected. Selection of dummy derivatives is discussed by Mattsson and Söderlind [32]. Finally, we would like to point out that in many cases it is not all necessary to derive the differentiated equations by symbolic differentiation. The residuals of the differentiated equations are much more efficiently calculated by use of automatic differentiation (see, e.g., [39] or [20]). Automatic differentiation is a way to organize the computations, so a numerical value of the analytic derivative is calculated with essentially the same effort as it takes to calculate the function itself.

2.7.1 Discussion

An index greater than 1 implies that there are algebraic relations between dynamic variables. The formulation of the problem is, in some sense, overparameterized; there are too many dynamic variables. The index is not a problem invariant since the index of a problem may be changed by transformations. Thus, it may be argued that high-index models although they are mathematically well-behaved in some sense are bad models. If we are going to make a model from scratch, we can avoid dynamic overparameterization. However, if our aim is to support modular and incremental modeling in general and to allow a user to make a model by combining library models, then high-index problems are unavoidable. Library models must be general to support a flexible use and a user must be allowed to constrain them.

2.8 WAYS TO FACILITATE THE NUMERICAL SOLUTION

In this section we will outline and discuss some possible uses of symbolic manipulation to facilitate numerical solution. The possibilities to implement and to integrate symbolic manipulation have increased significantly during the last few years, since symbolic manipulation tools like Maple and Mathematica now have defined interfaces for communication with other programs. Thus it is possible to use them as symbolic manipulation engines.

2.8.1 Identification of Outputs

We will start with a simple decomposition. Consider a problem which is decomposed into two parts as $f(t, x, \dot{x}) = 0$ and $g(t, x, \dot{x}, y) = 0$. We can first let the DAE-solver solve the problem $f(t, x, \dot{x}) = 0$ and then solve y from the algebraic equation $g(t, x, \dot{x}, y) = 0$. It is easy to make this decomposition by analyzing the BLT-partitioned problem. The vector y contains variables that are typically introduced by the model developer for display and plotting purposes, which means that the associated equation is a simple assignment.

2.8.2 Tearing and Hiding of Algebraic Variables

If an algebraic variable v belongs to a scalar block and it is possible to solve for it analytically, it is not necessary to introduce v as an unknown to the numerical DAE-solver. It is possible to hide v since it is possible to calculate v at any call of the residual routine from the information available. It is not advisable to eliminate v by substitutions since that could mean that we have to "calculate v" several times in the residual. Auxiliary variables are often introduced to denote an expression that appears several times.

The idea can be extended to blocks of algebraic variables as long as we can solve for the variables. Furthermore, the idea can also be used when a block has both algebraic and dynamic variables. If it is possible to solve for an algebraic variable, it can be hidden. In more general terms, the variables and equations are divided into two sets so that it is easy to solve for the variables in the first set if the variables of the other set are known. This kind of partitioning is called tearing. Here the intention is to hide the variables of the first set and let the DAE-solver treat the variables of the second set. There are many algorithms for tearing. Unlike the situation for output assignment and BLT partitioning, there are no clear winners. We refer to the textbooks by Mah [28] and Duff, Erisman, and Reid [13] for algorithms and further discussion.

2.8.3 Various Partitions

We have discussed the BLT partition with respect to appearing highest-order derivative. However, when solving the problem $g(t, x, \dot{x}) = 0$ using backward

Euler discretization, the solver must treat the problem $g(t_n, x_n, (x_n - x_{n-1})/h) = 0$ and solve for x_n. Consequently, it is the structure of $g(t, z, z) = 0$ with respect to z that is of interest. This applies for other implicit discretizations as well. Unfortunately, today's DAE-solvers with a few exceptions are not designed to exploit special structures. DASSL can exploit banded structures which is a common feature of DAEs derived by discretizing partial differential equations.

Different kinds of partitionings are of interest when using multicomputer systems. Partitioning into various kinds of forms such as band form, block tridiagonal form, doubly-bordered block diagonal form, and bordered block triangular form are discussed by Duff, Erisman, and Reid [13].

2.8.4 Conditioning

The symbolic manipulation should not make the work for the numerical solver worse. Condition numbers and pivoting are important concepts when solving equation systems. By restricting manipulations to be local to each block, much is gained since it is, in general, a bad idea to use an equation to solve for a variable belonging to another diagonal block. When converting a problem into explicit-state space form, Gauss elimination is not a good approach to triangularize the block. It is impossible to make a proper pivoting when only having symbolic expressions. Furthermore, to get a well-conditioned system it may be necessary to change the pivoting along the simulation run. In the worst case it may happen that the triangularization causes division by zero. For small or sparse linear equation systems it is feasible to use Cramer's rule and calculate the inverse by calculating the determinant and minors.

2.8.5 Singularities

It may happen that a subproblem is singular or becomes singular during a simulation of a run. If the subproblem is algebraic, it means that the whole problem really is singular. So we need not worry about divisions by zero if we solve a scalar block for an algebraic variable or if we have solved a linear equation system by diagonalizing the subproblem properly since division then implies that our problem is singular. This approach requires that we give the user an error message that his problem is singular rather than that division by zero was trapped.

If a block with dynamic variables turns out to be singular with respect to highest-order derivatives, this implies that the index is greater than 1 or that the problem is singular. Symbolic manipulation is required to reduce the index to 1 (for some comments, see [32]). An index increase during simulation implies that the system has lost degrees of freedom. If the variables did not fulfill this new constraint before it turned up, then the variables must change their values instantly to fulfill the constraint. If the new constraint references more than one variable, the model itself does not specify how the adaptation should be done. Consequently, we can, in most cases, say that index increases are signs of bad modeling.

2.8.6 Accuracy

User-specified relative or absolute error bounds on the solution ought to apply to all variables. Consider the sorting out of outputs. Even if the DAE-solver produces a solution for x within the error bounds of $f(t, x, \dot{x}) = 0$, it is not certain that we get the same error bounds for y when solving it from $g(t, x, \dot{x}, y) = 0$. It may require a much more accurate value of x. This calls for an integrated solution of the DAE problem and the algebraic problem. Such numerical tools do not exist today. The DAE-solvers are not designed to exploit this kind of problem structure. Similarly, we cannot guarantee any error bounds for hidden variables. If accuracy is a real concern, the complete problem must be submitted to the numerical DAE-solver.

We would like to point out that today's simulation tools which are of CSSL type do not check error bounds on algebraic variables. If you use fixed-step algorithms, there is no error checking at all. If variable step-size algorithms are used, the user can specify error bounds on the states.

2.8.7 Sparsity

Many problems are originally sparse, but if we try to eliminate algebraic variables or try to transform them to explicit state-space form, they may easily become dense. For large problems, this implies that orders of more storage is needed since the calculation code becomes larger and sparse matrix techniques cannot be used. The computation times increase also drastically.

REFERENCES

1. Andersson, M., *Omola—An Object-Oriented Language for Model Representation*, Licentiate Thesis TFRT-3208, Department of Automatic Control, Lund Institute of Technology, Lund, Sweden, 1990.
2. Andersson, M., Discrete event modelling and simulation in Omola, *Proceedings of the 1992 IEEE Symposium on Computer-Aided Control System Design*, CADCS '92, Napa, CA, March 17–19, 1992, pp. 262–268.
3. Åström, K. J., and Kreutzer, W., System representations, *Proceedings of the IEEE Control Systems Society Third Symposium on Computer-Aided Control Systems Design (CACSD)*, Arlington, VA, September 24–26, 1986.
4. Baumgarte, J., Stabilization of constraints and integrals of motion in dynamical systems, *Computat. Meth. Appl. Mechan.*, *1*, 1–16 (1972).
5. Birtwistle, G. M., Dahl, O-J., Myhrhaug, B., and Nygaard, K., *Simula Begin*, Auerbach, Philadelphia, 1973.

6. Booch, G., *Software Engineering with Ada*, The Benjamin/Cummings Publishing Company, Inc, San Francisco, 1983.
7. Brenan, K. E., Campbell, S. L., and Petzold, L. R., *Numerical Solution of Initial-Value Problems in Differential–Algebraic Equations*, North-Holland, Amsterdam, 1989.
8. Cellier, F. E., *Continuous System Modeling*, Springer-Verlag, New York, 1991.
9. Cellier, F. E., Ziegler, B. P., and Cutler, A. H., Object-oriented modeling: Tools and techniques for capturing properties of physical systems in computer code, *Proceedings of the IFAC Symposium Computer Aided Design in Control Systems, Swansea, UK, 15–17 July 1991*, Pergamon Press, Elsford, NY, 1991, pp. 1–10.
10. Cellier, F. E., and Elmqvist, H., The need for automated formula manipulation in object-oriented continuous-system modeling, *Proceedings of the 1992 Symposium on Computer-Aided Control System Design, Napa, California, March 17–19, 1992*, IEEE Control Systems Society, New York, 1992, pp. 1–8.
11. Clark, D. R., *HVACSIM + Building Systems and Equipment Simulation Program Reference Manual*, U.S. Department of Commerce, National Bureau of Standards, Washington, DC, 1985.
12. DOE-2, *DOE-2 Reference Manual*, LBL-8706 Rev. 2, Los Alamos National Laboratory, Los Alamos, New Mexico.
13. Duff, I. S., Erisman, A. M., and Reid, J. K., *Direct Methods for Sparse Matrices*, Clarendon Press, Oxford, 1986.
14. Elmqvist, H., *A Structured Model Language for Large Continuous Systems*, Ph.D. Thesis, TFRT-1015, Department of Automatic Control, Lund Institute of Technology, Lund, Sweden, 1978.
15. Elmqvist, H., Åström, K. J., and Schönthal, T., *Simnon—User's Guide for MS-DOS Computers*, Department of Automatic Control, Lund Institute of Technology, Lund, Sweden, 1986.
16. Evans, L. B., Boston, J. F., Britt, H. I., Gallier, P. W., Gupta, P. K., Joseph, B., Mahalec, V., Ng, E., Seider, W. D., and Yagi, H., ASPEN: An advanced system for process engineering, *Comput. Chem. Eng.*, *3*, 319–327 (1979).
17. Führer, C., Kortuem, W., Wallrapp, O., and Bausch-Gall, I., MEDYNA—A simulation tool for mechanical systems and its interface to simulation languages, *Proceedings of the 11th IMACS World Congress*, Oslo, Norway, August 5–9, 1985.
18. Führer, C., and Leimkuhler, B. J., A new class of generalized inverses for the solution of discretized Euler–Lagrange equations, in *Real-Time Integration Methods for Mechanical System Simulation*, (E. Haug and R. Deyo, eds.), NATO ASI Series Vol. F 69, Springer-Verlag, New York, 1990, pp. 143–154.
19. Gaus, N., and Otter, M., Dynamic simulation in Concurrent Control Engineering, *Proceedings of the IFAC Symposium Computer Aided Design in Control Systems, Swansea, UK, 15–17 July 1991*, Pergamon Press, Elmsford, NY, 1991, pp. 123–126.
20. Griewank, A., The chain rule revisited in scientific computing, *SIAM News*, May 1991.
21. Gear, C. W., Gupta, G. K., and Leimkuhler, B. J., Automatic integration of the Euler– Lagrange equations with constraints, *J. Computat. Appl. Math.*, *12–13*, 77–90 (1985).

22. Hairer, E., Lubich, C., and Roche, M., *The Numerical Solution of Differential–Algebraic Systems by Runge–Kutta Methods*, Lecture Notes in Mathematics No. 1409, Springer-Verlag, Berlin, 1989.

23. Hope, S., Fleming, P. J., and Wolff, J. G., Object-oriented database support for computer-aided control system design, *Proceedings of the IFAC Symposium Computer Aided Design in Control Systems, Swansea, UK, 15–17 July 1991*, Pergamon Press, Elmsford, NY, 1991, pp. 200–204.

24. Kheir, N. A. (ed.), *Systems Modeling and Computer Simulation*, Marcel Dekker, Inc, New York, 1988.

25. Kreutzer, W., *System Simulation—Programming Styles and Languages*, Addison-Wesley, Reading MA, 1986.

26. Kröner, A., Holl, P., Marquardt, W., and Gilles, E. D., DIVA—An open architecture for dynamic simulation, *Comput. Chem. Eng.* *14*(11), 1289–1295 (1990).

27. Maciejowski, J. M., and Tan, C. Y., Control Engineering Environments in DB-Prolog, *Proceedings of the 1992 IEEE Symposium on Computer-Aided Control System Design, CADCS '92*, Napa, CA, 1992, March 17–19, 1992, pp. 55–61.

28. Mah, R. S. H., *Chemical Process Structures and Information Flows*, Butterworths, Boston, 1990.

29. Marquardt, W., Dynamic process simulation—Recent progress and future challenges, *Proc. Chemical Process Control—CPCIV, CACHE*, Elsevier, New York, 1991.

30. Mattsson, S. E., and Andersson, M., A kernel for system representation, In *Automatic Control—11th Triennial World Congress, Tallinn, 1990*, (Ü. Jaaksoo and V. I. Utkin, eds.), Pergamon Press, Elmsford, NY, 1990, Vol. V, pp. 373–378.

31. Mattsson, S. E., Modelling of power systems in Omola for transient stability studies, *Proceedings of the 1992 IEEE Symposium on Computer-Aided Control System Design, CADCS '92*, Napa, CA, March 17–19, 1992, pp. 30–36.

32. Mattsson, S. E., and Söderlind, G., Index reduction in differential–algebraic equations using dummy derivatives, *SIAM J. Scient. Statist. Comput.* 14(3), 1993.

33. Nagel, L. W., *SPICE2: A Computer Program to Simulate Semiconductor Circuits*, Memorandum No. ERL-M520, Electronics Research Laboratory, College of Engineering, University of California, Berkeley, CA, 1975.

34. Nilsson, B., Object-oriented chemical process modelling in Omola, *Proceedings of the 1992 IEEE Symposium on Computer-Aided Control System Design, CADCS '92*, Napa, CA, March 17–19, 1992, pp. 165–172.

35. Pantelides, C. C., The consistent initialization of differential-algebraic systems, *SIAM J. Scient. Statist. Comput.* 9(2), 213–231 (1988).

36. Pantelides, C. C., SpeedUp—Recent advances in process simulation, *Comput. Chem. Eng.* 12(7), 745–755 (1988).

37. Perry, D. L., *VHDL*, McGraw-Hill, New York, 1991.

38. Piela, P. C., Epperly, T. G., Westerberg, K. M., and Westerberg, A. W., ASCEND: An object-oriented computer environment for modeling and analysis, *Comput. Chem. Eng.* 15(1), 53–72 (1991).

39. Rall, L. B., *Automatic Differentiation—Techniques and Applications*, Lecture Notes in Computer Science, No. 120, Springer-Verlag, Berlin, 1981.

40. Rimvall, M., and Taylor, J. H., Data-driven supervisor design for CACE package integration, *Proceedings of the IFAC Symposium Computer Aided Design in Control Systems, Swansea, UK, 15–17 July 1991*, Pergamon Press, Elmsford, NY, 1991, pp. 33–38.
41. Stefik, M., and Bobrow, D. G., Object-oriented programming: Themes and variations, *AI Mag.*, *6*(4), 40–62 (1986).
42. Stephanopoulus, G., Henning, G., and Leone, H., MODEL.LA: A modeling language for process engineering—I. The formal framework, *Comput. Chem. Eng.* *14*(8), 813–846 (1990).
43. Strauss, J. C. (ed.), The SCi continuous system simulation language (CSSL), *Simulation*, Vol. 9, 281–303 (Dec. 1967).
44. Söderlind, G., Eriksson, L. O., and Bring, A., *Numerical Methods for the Simulation of Modular Dynamical Systems*, Institute of Applied Mathematics, ITM, Stockholm, Sweden, 1988.
45. Tarjan, R. E., Depth first search and linear graph algorithms, *SIAM J. Comput. 1*, 146–160 (1972).
46. White, J. K., and Sangiovanni-Vincentelli, A., *Relaxation Techniques for the Simulation of VLSI Circuits*, Kluwer Academic Publishers, Boston, MA, 1987.
47. Zeigler, B. P., *Object-Oriented Simulation with Hierarchical, Modular Models—Intelligent Agents and Endomorphic Systems*, Academic Press, San Diego, CA, 1990.

3

An Artificial Intelligence Approach to Environments for Modeling and Simulation

Derek A. Linkens

University of Sheffield
Sheffield, England

3.1 INTRODUCTION

From surveys of industrial users, it is known that simulation software is the most widely used CACSD methodology used for complex engineering system design [2,28]. This has been elicited from both European and Japanese studies and is largely due to the inherent nonlinearity encountered in practical engineering systems. The design tools based on mathematical analysis and synthesis are only well-suited to the linear situation, and hence, recourse is most commonly made to simulation which caters to both nonlinear and mixed mode representations (e.g., continuous, digital, or discrete models).

The traditional approach to simulating a dynamic system has been to satisfy a given set of technical requirements, using either a simulation language or a general-purpose programming language. Simulation was mainly performed to aid understanding of the system dynamics. This was usually performed by a small number of people who had the expertise to develop models, write programs, and run the simulation. A more modern approach to simulation calls for a broader perspective. This approach is characterized by the need to satisfy the demands of both technical management and the engineering designers, who will be working in an interactive manner using a personal workstation. In turn, this calls for a knowledge-based environment for modeling and simulation. The concept of an "environment" is appealing as it supports the whole range of activities involved in a simulation study.

A knowledge-based environment for modeling and simulation (KEMS) was developed as a result of a study to find an "ideal" simulation language for simulating complex systems. Several languages, such as ACSL, SIMNON, PSI, etc., were evaluated. The study showed that existing simulation languages not only lack the facilities for simulating some aspects of such systems but also require skilled personnel to develop and drive a simulation study. It was also found that the improvements in hardware technology, the emergence of artificial intelligence (AI), and maturity in database systems and widespread use of them in recent years could play an important role in future modeling and simulating software [10]. For further details of the impact of AI in control systems engineering generally, see a review paper by Linkens [12]. The incorporation of AI techniques into systems modeling has been described for discrete-event systems [14], for mixed systems [7], for steady-state chemical processes [5], and for hybrid symbolic/numerical simulation [4,6].

Utilizing the above concepts, a functional specification for an intelligent simulation environment was prepared [17] and is shown schematically in Figure 3.1. Within this framework, a number of expert systems are involved, linked together in a federated manner. The incorporation of expert system methodologies within the modeling and simulation cycle is described by Lehmann [9]. The core of the environment was perceived to be the modeling methodology, for which a frame-based knowledge representation is particularly appropriate and is equivalent to the object-oriented paradigm referred to in other chapters. The desirability of producing reusable modeling code, akin to software engineering principles, has been noted by Ziegler and de Wael [29]. The prototype software commenced, therefore, with the modeling phase.

3.2 FRAME-BASED ENVIRONMENT FOR MODELING AND SIMULATION

Following an evaluation of several expert system shells, we decided to construct a frame-based environment for modeling and simulation (FEMS) in-house [24]. The system is based on the assumption that many simulation users will, for the majority of the time, be working with a limited set of components from which various models can be constructed. The system is thus designed to make it easy to reconfigure existing models and to create new models based on an existing set of components or submodels. The creation of models for new components is considered to be a specialist task which will be carried out infrequently.

The robustness of this system is mainly due to an object-oriented knowledge representation scheme which allows complex knowledge to be expressed and organized into a single hierarchy of inheritance—the system. This is the frame-based approach which groups knowledge into generic and structured classes (or frames) and allows these frames to interact through inheritance mechanisms to generate a relatively complex and global model.

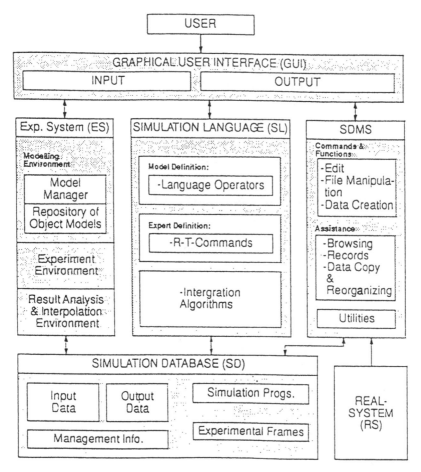

Figure 3.1 Functional specification for a knowledge-based modeling and simulation environment.

3.2.1 The Component Frames

The core of the expert system is the knowledge base of physical components which make up the repertoire of the systems engineer. The knowledge elements to be represented include the physical attributes which describe the physics of each component, the mathematical and simulation models, the structure of interconnected systems, and a system as a hierarchy of components. Hence, these basic knowledge elements need to be organized at two levels—a local level of representation which allows all the information describing a particular component to be grouped together, and a global level which allows information to be shared or "inherited." The modeling procedure for valves and pipelines is presented to illustrate the frame-based concept.

A valve knowledge base may be described by the following data structures:

<valve> = <instance>
          <class>
          <valve sizing specifications>
          <math model>
          <simulation model>

<valve sizing specifications> = <design specs>
                                <fluid specs>
                                <linespecs>

<design specs> = <order of magnitude of CD>
                 <noise limit>
                 <rangeability>
                 <corrosion resistance>

<fluid specs> =   <phase>
                   <name of fluid>
                   <flow rate>
                   <pressure drop>

<line specs> =   <reducers status>
                   <line diameter>

<valve specs> =   <body shape>
                   <trim>
                   <rated capacity>
                   <operating capacity>

<math model> = <equation>*
               <variable>*

<simulation model> = <block>*
                     *

where * denotes a list, and "block" and "parameter" are simulation operators.

The first slot in the valve frame ("instance") is a text string which distinguishes a unique valve so that a system may contain several valves which can be modeled unambiguously. The second slot ("class") specifies whether the valve is a servo, control, or process valve and this allows information to be inherited from more specific subclasses. A specification of the valve class and its geometry is usually sufficient to generate a model, but some process control applications require the choice of a valve to suit user-specified operating conditions. Hence, the "valve sizing" slot incorporates a knowledge base which allows a process valve to be selected to match certain nominal flow conditions. The CV curve corresponding to the geometry of the sized valve is then retrieved from a database to instantiate the model slot of the valve. The "simulation model" slot expresses the mathematical model in terms of the operators used by the simulation model. The valve frame presented above has been implemented in Prolog.

As a further illustration of the frame concept, consider a pipeline frame which can be translated into the following data structures:

<pipe> = <instance>
                    <configuration>
                    <line properties>
                    <math model>
                    <simulation model>

where configuration = 1 if inlet pressure and outlet flow are the output variables

                    = 2 if inlet flow and outlet flow are output variables

                    = 3 if inlet pressure, outlet and inlet flows are output variables

line properties   = lumped or distributed.

The other data structures have been defined. Basically, a pipeline model can either be lumped or distributed depending on the line and fluid properties. A lumped model will have one or more linear elements in series, depending on the predominant physics of the pipeline such as heat transfer mode, pipe geometry, and friction. A distributed model is expressed in terms of partial differential equations which can be converted into an equivalent partial difference form expressed in terms of time delays and impedances. The interconnection of the model elements depends on the configuration of the pipeline which is considered as a four-terminal device. The pipeline frame has also been implemented in Prolog.

Other components which have been implemented include controllers, pumps, servomechanisms, and actuators. These can be used to construct a hierarchical system of reasonable complexity.

A system is implemented as

system (name, structure, math model, simulation model)
structure (connectors, subsystems)
connectors = connector*
subsystems = system*
connector (NODE, upstream subsystems, downstream subsystems)
NODE, NAME = string

Hence, the hierarchical nature of a system is implicit in the definition of structure.

The frame concept illustrated above has at least two advantages: compactness and organization. The compactness derives from the fact that each frame represents only a generic class of objects which can be instantiated to more specific data. The organization is achieved through recursive inheritance of information between closely related objects and the fact that each frame can be regarded as a stand-alone unit of knowledge.

3.3 Knowledge-based Environment for Modeling and Simulation

Having established the knowledge representation style via FEMS, the next phase was the construction of good graphical interfaces and the incorporation of standard simulation languages [18]. The basic strategy has been to use unmodified commercially available packages as much as possible and to use a specifically written manager to establish links between the packages. The prototype uses three commercial packages: EASE+ for the user interface and PSI or ESL as the simulator. The overall structure of KEMS is shown in Figure 3.2.

KEMS encompasses a wide range of different tasks within a single application, and different parts of the system are coded in different languages. Because of this and the complexity of the application itself, the quality of the user interface is particularly important. We, therefore, decided to construct the user interface using a specialized frontend builder. The package chosen eventually for this task was Expert-EASE Systems' EASE+. EASE stands for Engineering Applications Software Environment, and it is essentially a flexible graphics database management system which contains an interface development toolbox. EASE+ was designed for use by the process control, aerospace, telecommunications, manufacturing, and electric power industries for the construction of user interfaces and simulation, control, monitoring, diagnostics, and design applications. EASE+ can be mounted on either PCs or workstations, and as well as being a development tool, it is capable of producing run-time software modules.

EASE+ contains a relational database and an interface toolkit which contains an icon editor, forms generator, menu generator, window facility, plot facility, help facility, and external program calls. The program developer uses this toolkit to create the building blocks of the interface and puts the building blocks together using the EASE+ procedural code.

KEMS was created originally for possible use in the gas industry. The primary user is considered to be the area engineer who would utilize KEMS to construct graphical models of possible system configurations and run simulations from which the performance of the corresponding real system could be predicted. These models rely, for their construction, on a library of preprepared icons which represent system components. The physical specifications for these components

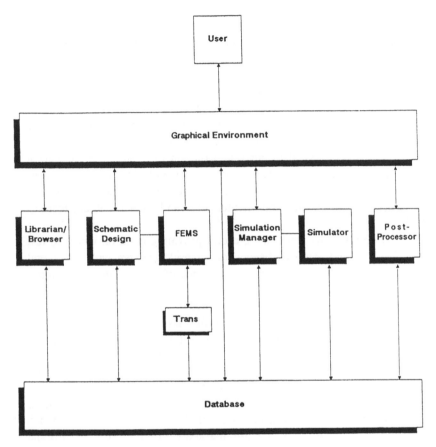

Figure 3.2 Schematic structure of KEMS.

must be stored within the EASE+ database prior to use by the area engineer. This is the function of the second type of KEMS user: the expert simulationist.

The area engineer's simulation interface is fairly typical of the type of application for which EASE+ was principally designed. It contains three parts: the user interface itself (written in EASE+), an expert system called FEMS (frame-based environment for modeling and simulation) written in Prolog, and a commercial simulation language called ESL. From an ergonomic viewpoint, KEMS was designed to simplify the process of modeling and simulation, both by integrating the different functions into a single environment, and also by removing the need for users to write a simulation code [3].

The system assumes that users will generally be creating new models or reconfiguring existing ones from a limited set of components or submodels. The user interface facilitates this by allowing the user to select and position component icons on the screen to construct the topology of the system which is to be simulated. A form associated with each icon is then presented to the user. Each form contains component-specific information such as the component's name, its parameters, valid data ranges, and so on. The on-screen representation of the system is then automatically translated into an output text file which lists the components and their interconnections. This file is then read by FEMS which constructs the model by instantiating each component and generating the appropriate simulation code. Control then returns to the EASE+ interface where the user is invited to enter information to control the simulation (run-time settings, length of run, and so on). At the end of the simulation run, control returns to the EASE+ interface and a plot of up to four selected variables is displayed. Subsequent experimental runs can be made without reentering FEMS.

The KEMS interface has provided a useful proving ground on which to test EASE+ . The Sheffield group has been using the package extensively for several years and, therefore, has considerable experience of its strengths and limitations. As a tool for building dynamic "mimic" diagrams of physical systems, EASE+ is excellent, possessing a good graphical representations system and a useful database. In addition, the interface toolkit provides facilities for generating customized data-entry forms which can be linked to icons. This facility permits validation checks to be performed on input data. These features make EASE+ very suitable for the KEMS simulation application. However, because many of the interface routines are "hard-coded," it is harder to use EASE+ to develop sophisticated mouse-driven interfaces up to the Apple-Mac standard. This presents problems for applications such as KAM (see Section 3.5) which require interface capabilities which are markedly different from traditional graphical simulation techniques. The group, therefore, concluded that, at present, EASE+ is a good system for graphical modeling but has only limited applications outside this field. Because of this, other rapid-prototyping interface packages may, eventually, prove to be more appropriate for a stand-alone version of KAM. Systems such as Expert Object Corporation's "Autocode" have been considered, but the most promising alternatives to EASE+ appear to be the various hypertext systems, such as OWL Guide Hypertext.

3.4   AN INTEGRATED SIMULATION DATABASE

An integrated simulation database provides a rich set of data structures to integrate components in a simulation environment and simplifies the process of adding future tools to the system. It also provides services required to manage the different kinds of data (graphics, text, etc.) and the large number of files involved in a simulation study. An integrated simulation database provides the means to the following:

1. *Structure the model and data hierarchically.* Hierarchy and regularity are well-known tactics for reducing the complexity of a large model. Modeling proceeds by a top-down decomposition of systems into submodels and a bottom-up synthesis of components from more primitive building blocks. The modeling is complete when all submodels can be implemented with existing components and primitives.

2. *Support multiple-component model representation.* Models are described in several representations. For example, a component may be represented as a simple or a detailed model. It may also have different mathematical representations (e.g., lumped versus distributed models). Each representation is appropriate for different purposes.

3. *Maintain model versions and alternatives.* Since modeling and simulation is an iterative and evolutionary process, the database must support model versions and alternatives. *Versions* are improvements or corrections to a model or object (component), whereas *alternatives* are different implementations of the same model with varying performance characteristics. Versions are needed for documentation. Alternatives, especially within model libraries, enable users to experiment with different implementations of the same model.

4. *Help users co-operate and interact as a team.* Large systems are usually simulated by a team. To help team members to work effectively, the models and components are made self-documenting by placing interface descriptions, dependency information, and model responsibility in the database. This helps a modeler to understand *how* an object is used, *where* it is used, and *who* is responsible for its design.

5. *Maintain the consistency of the model.* This involves automatic checking of the data and gives access to lower levels only to expert users.

The requirements of an integrated database for a simulation environment are similar to that of CAD, CACSD, and other engineering systems. Since simulation is used in many disciplines, the data representation has to be very flexible to map the data from various disciplines to its database. So, an important requirement of an integrated database is to provide complex as well as simple data structures. Here, we are concerned with simulating engineering systems and, therefore, concentrate on developing the data structures needed to describe such systems.

Before dealing with the methodologies for constructing data models for describing engineering systems in a simulation environment, let us look first at some of the characteristics of engineering data:

First, the large number of data types and complex relationships between data elements: The description of engineering components involves hundreds of pieces of data linked together through complex relationships. The database designer has to be familiar with engineering disciplines to grasp the full range of engineering design procedures and their underlying principles.

Second, the dynamic nature of the engineering data structure: Since modeling and simulation is an iterative process which evolves with time, a static data structure is not acceptable as it is impossible to envisage and analyze all the information to be stored in advance.

Third, representation for complex abstraction hierarchies: To solve complex engineering problems, we often recursively divide them into a hierarchy of subproblems whose solutions are conceptually simpler and can be assembled to provide a solution to the engineering problem. The integrated database must, therefore, support mechanisms for top-down design.

3.4.1 Database Management

No suitable package was found for the database, and the basic graphical data-storage facilities of the EASE+ tools were augmented in-house [3]. An integrated database not only provides a global repository system from which each component in an environment stores and retrieves information, it also provides support tools for browsing, documentation, version control, and backups. These are essential parts of a simulation environment to ease the process of model development, organization, and maintenance. KEMS is interfaced to a simulation browser and a librarian program to perform these tasks.

3.4.2 The Librarian

The librarian generates documentation entry forms for each stage of a simulation study. Once proper documentation is generated, the archiver stores the models, experiments, and the results (Figure 3.3). Since modeling and simulation is an iterative process, several versions and alternative models of a process may be developed and a number of experiments are usually performed on each model.

The basic role of the archiver is to control changes to source codes, object codes, graphic pages, etc. It stores the history of changes made to each model and allows the user to revert easily to a prior version of a model if necessary. It not only documents the changes but also documents who made each alteration and the nature and the time of the change. We use a commercial software package called PVCS [16], which provides functions for storage and retrieval of multiple versions of a model. To economize on the use of storage space, PVCS uses the "delta storage technique" to store the versions of models, experiments, etc. A "delta" is the set of differences between one version and its immediate ancestor. Only changes made to a model are, therefore, stored rather than the whole model.

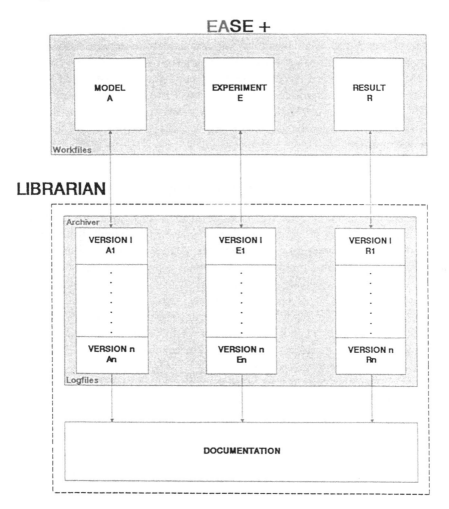

Figure 3.3 The librarian structure in KEMS.

3.4.3 The Browser

Ideally, a user-friendly intelligent simulation environment automatically selects the appropriate generic model from related knowledge bases which contain the code for user goals of a particular system. In a conventional simulation approach, we manually define a problem to be studied. Several versions of a model are generated, producing a large number of files with little or no documentation. After a period of time, users forget where a particular model is, who created it, and how it was developed.

The browser implemented in KEMS aims to solve this problem. On-line documentation is provided. The user can browse through the documents associated with models and select the one which satisfies his requirements. Once a particular version of a model and its corresponding experiment is tagged, the browser retrieves the model automatically and loads it into KEMS, ready for inspection or execution. Figure 3.4 depicts the browser structure. It shows that the model PRS.MOD has four experiments. As the cursor moves over each file, the description is updated.

3.5   KNOWLEDGE ACQUISITION MODULE

To construct models in the version of FEMS embedded in KEMS, one needs to be familiar with Prolog programming. This would clearly inhibit dynamicists who wished to enter frames for new components. A knowledge acquisition module, KAM, addresses this problem. KAM defines a pseudo-natural language, FKRL, which is used to construct a knowledge base of rules and frames in a simple syntax. The FKRL objects are then parsed into a form suitable for conversion into Prolog code. KAM also incorporates a range of other utilities to which allusion has been made. These utilities reinforce the basic functions of parsing and Prolog generation and enable KAM to function as a stand-alone program. Details of KAM are given [25].

The basic elements of KAM are shown in Figure 3.5. FEMS, KEMS, and KAM have been integrated into a set of tools which can be used by either novice or expert users of simulation.

3.5.1   Design and Implementation of KAM

The basic elements of KAM include a parser, Prolog-generator, tree-generator, file-handler, query language, and in-built editor.

The Parser

A preliminary step in describing the parser requires a definition of the FKRL language which forms the input to the parser. This is then followed by a description of the processes which constitute the parsing cycle. These involve the scanner, syntax-analyzer, and error-handler.

Structure of FKRL.   The syntax of FKRL frames is designed to accommodate all the desirable features of frame-based knowledge representation. These include such concepts as object hierarchies, data abstraction, semantic relationships between objects, the distinction between conceptual and instance frames, and the assignment of values to attributes.

There are two types of semantic link: the "IS A" and "INSTANCE OF" links. The "IS A" link delares a relationship between a prototypal class and a

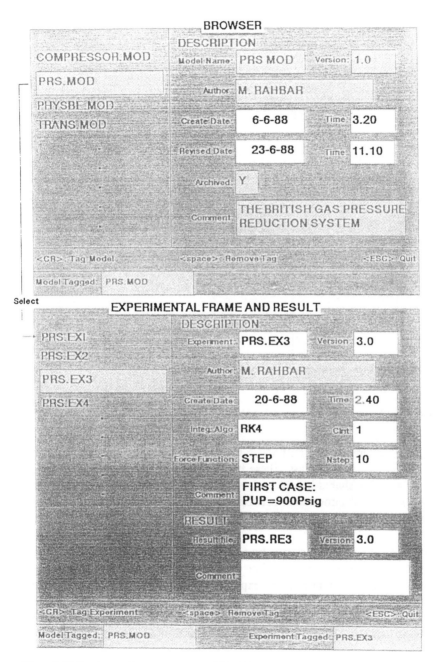

Figure 3.4 The hierarchical browser in KEMS.

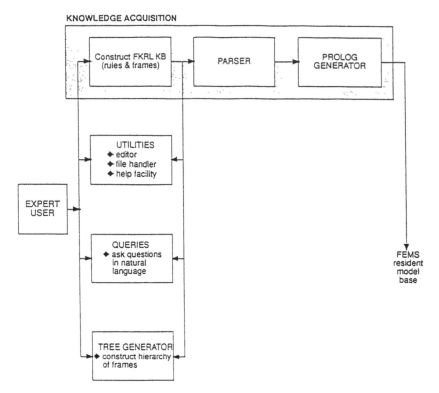

Figure 3.5 Schematic structure of KAM.

higher class, whereas the "INSTANCE OF" link assigns membership of a class to specific objects of that class. These two semantic links define paths through which data can be shared between related frames; a process called inheritance. It is this sharing of information between templates representing various levels of knowledge abstraction which makes frame-based knowledge representation efficient.

The range of values which can be assigned to an attribute of a frame must reflect the vast range of information used in systems modeling. Hence, attributes can either be numbers, text, expressions, or equations. The use of equations and expressions requires secondary considerations such as the attachment of symbols to attributes. Type checking is also required to impose sensible limits on various parameters used in modeling. The basic device used to implement type checking is called a "facet."

These considerations are embodied in the following FKRL frame construct:

FRAME pipe IS A flow component WITH diameter,D; length, L; area,A = 3.14*D*D/4; volume,V[real]; inletpressure; inlet-

flow; outletpressure; outletflow; pipe-friction[0,1]; equations = [C*dPin/dt = Qin-Qout, Qout = R(Pin-Pout)]; and INSTANCES pipe1, pipe2.

The keyword "FRAME" precedes the name of the object of which the frame is a template. This may (or may not) be followed by the "IS A" semantic link depending on whether the object has a higher class or not. The next subconstruct is a list of attributes preceded by the keyword "WITH." Attributes within this list are separated by semicolons. The "attribute" construct has several variants. The simplest of these simply states the name of an attribute (e.g., inletpressure). The second variant associates an attribute with a symbol. Hence "diameter,D" assigns the symbol, "D," to the attribute diameter. Both the attribute name and symbol can be used to refer to the attribute. The third variant attaches a data-type (facet) to an attribute. Hence "volume,V [real]" specifies that the attribute "volume" can only be assigned real values. Similarly, "pipe-friction [0,1]" specifies that the attribute "pipefriction" can only take values between 0 and 1. Attributes are assigned values via the equality operator " = ," with the value occurring after the operator. Hence, "area,A = 3.14*D*D/4" assigns an expression to the attribute "area." This expression can be used to compute the area of a pipe once its diameter has been specified.

Similarly, the syntax of rules is designed to reflect the antecedent–consequent structure of heuristic knowledge. The rule syntax is illustrated by the following FKRL items:

```
IF C1 is a component with upstreamnode = N1
AND C2 is a component with downstreamnode = N2
AND N1 = N2
THEN C1 connects-to C2.

IF P is a pipe with length = L
AND L < 20 ft
THEN pipe-friction of P = 0.
```

The first rule is a general connectivity rule which states that two components are connected if the upstream node of one is the same as the downstream node of the other. The second rule assigns a value of 0 to the pipe-friction attribute of any pipe shorter than 20 ft. These two rules illustrate the close integration between rules and frames. Both the antecedent and consequent parts of a rule can refer to attributes of a frame. The conclusions reached by rules can be used to infer values of frame attributes.

FKRL thus consists of a hierarchy of constructs which includes FRAME, RULE, ATTRIBUTES, INSTANCES, FACETS, EQUATIONS, EXPRESSIONS, ANTECEDENTS, and CONSEQUENTS. These constructs have various degrees of complexity with the low-level ones being relatively simple, whereas the high-level ones are quite complex. Taken together, this set of constructs amounts to a specification of FKRL.

The Scanner. The main function of the scanner is to define a set of legitimate tokens which can appear in FKRL text. These tokens includes all keywords, delimiters, numbers, and alpha-numeric literals. If an illegal token is encountered during scanning, parsing is temporarily halted. The error-handler is activated to generate an appropriate error message. The editor is automatically invoked, the illegal token highlighted, and error message displayed in a status-line which appears at the bottom of the screen. Parsing is again resumed when the user has corrected the illegal token. This interactive cycle continues until the scanning cycle is completed. The output from the scanner is a list of valid FKRL tokens which form the input to the syntax analyzer.

The Syntax Analyzer. This program module uses a look-ahead scheme to match groups of tokens to appropriate FKRL constructs. Successful syntax analysis results in a match between an FKRL object and a group of tokens. Syntax analysis fails when no match is found, in which case the error-handler is again activated to display an appropriate error message in the editor. The cursor position where syntax failure has occurred is also highlighted. Parsing is resumed when the syntax error has been corrected. This interactive process continues until all the FKRL text has been grouped into a list of valid syntax objects.

During syntax analysis, the user can invoke a help file which contains details of the FKRL syntax. The last activity in the syntax analysis cycle is "type checking," which consists of comparing the values of various attributes with any facets associated with them. Attributes with no facets can take any value, be it numeric, textual, an expression, or equations. Type checking is also an interactive process and results in suspension of the parsing process when an invalid data value is encountered. The error-handler is again activated to display an appropriate message in the editor. Parsing is resumed when a correct data value is entered.

The Error-Handler. This program module is closely integrated with all stages of parsing. The function of the error-handler is to assert various types of error messages into the working memory. Each such message contains the cursor position at which the error is found.

The Code Generator

The output of the syntax analyzer is a list of valid FKRL objects. The essence of code generation is to convert this list into Prolog structures.

The Tree-Generator

The basis of the tree-generator is a recursive data structure defined as:

```
TREE     = tree(NODE,TREELIST)
TREELIST = TREE*
```

The asterisk (*) denotes a list. This definition implies that a tree is a hierarchical structure of nodes and links. The nodes represent the various frames described in the knowledge base, whereas the links represent semantic relations between frames. It is, therefore, possible to generate a tree of arbitrary depth, depending on the complexity of the underlying knowledge base. As an example, consider a knowledge base which contains the following frames:

COMPONENT,VALVE,PIPE,COMPRESSOR,PIPE1,VALVE1, and COMPRESSOR1.

Suppose, the following semantic relationships exist between these frames:

pipe	IS A component
valve	IS A component
compressor	IS A component
pipe1	INSTANCE OF pipe
valve1	INSTANCE OF valve
compressor1	INSTANCE OF compressor

The corresponding tree structure for this hierarchy is

```
tree(component,[tree(valve,[tree,valve1,[ ])],
      tree(pipe,[tree(pipe1[ ])],
      tree(compressor,[tree(compressor1,[ ])]]]).
```

The square brackets denote the null list and define the extremities of the tree. Once the tree structure has been generated, the next activity is to sketch it on the screen. This involves two operations: scaling and sketching. Scaling involves the calculation of the relative sizes and screen positions of each node. The scaled data is then used by a graphical routine which sketches the tree as a set of boxes and lines.

The Query Language

The query language is a logical extension of FKRL which facilitates interaction with the knowledge base. Several types of query are available. These include queries of the form

SHOW AREA OF PIPE1
SHOW EQUATION OF VALVE

These are used to retrieve the values of various frame attributes. Queries of the form

FIND A PIPE of length 20
FIND A PIPE WITH length 20

are used to find a frame which has a prescribed attribute value.

Instances of a particular frame can be listed via queries of the form

```
SHOW PIPES
SHOW VALVES
SHOW COMPRESSORS
```

The template describing a given frame can be displayed through queries of the form

```
SHOW PIPE
SHOW VALVE.
```

The Help Facility

The help file contains text which is organized into headers which correspond to various FKRL syntax elements. Hence, there are headers for FRAME, RULE, ATTRIBUTES, INSTANCES, and QUERIES. The headers are used to retrieve the text which is most relevant to the concept being investigated. A help menu is displayed from which the user interactively selects the type of help required. Any number of headers can be selected from this menu. The help facility is only terminated when the user exits the help menu.

The File-Handler

The file-handler implements operations which enable files to be loaded, saved, browsed, and renamed.

The In-Built Editor

The in-built editor is invoked during the creation of FKRL files and during the parsing cycle whenever an error is encountered.

Testing the Knowledge Base

After constructing the FKRL knowledge base, the next step is to test it. To do this, a Prolog code is generated by selecting the PROLOG option from the KAM boxmenu. The QUERY option is then selected to interact with the knowledge base.

3.5.2 Graphics Interfacing for KAM

In contrast to simulation, the knowledge acquisition function is a significant deviation from typical EASE+ applications, and, as such, it represents a challenge to the versatility of EASE+. KAM was originally designed to enable the expert simulationist to enter new components into the KEMS database to expand the variety of models which can be generated, but there is also interest in developing KAM into a stand-alone knowledge acquisition system in its own right.

KAM was originally written in Turbo Prolog, a language with limited interface-building facilities. Despite this, a basic line-editor-style user interface was constructed. This enabled the user to create a knowledge base of components using a pseudo-natural language called FKRL (frame-based knowledge representation language). This language uses the concept of "frames," whereby individual components can be described hierarchically as specific instances of generic component "classes." Frames are, therefore, very efficient in this context because they avoid descriptive redundancy.

An ergonomic evaluation of this interface suggested an alternative interface which would enable the user to create knowledge bases graphically. Since Turbo Prolog has limited graphics, a specialized graphical interface package was required. Although it was known that EASE+ was not entirely suited to KAM, the need to maintain consistency with the rest of the KEMS environment required that EASE+ be used.

The new interface uses a hierarchical (tree-structure) network metaphor as a basis for building knowledge bases. Components and component classes are represented by rectangular icons called frames which the user can name and position anywhere on the screen. The user specifies the hierarchical inheritance relationships between the components and component classes by using the mouse to draw interconnecting lines between the icons. This approach not only saves the user from having to write inheritance relationships using FKRL syntax but it also enables the user to see the hierarchical structure of the knowledge base on the screen while it is being built.

Because KAM would probably be used only intermittently, it was felt that a form-filling procedure would be easier to use for knowledge acquisition than the line-editor used previously because it would remove the need for the user to recall the FKRL syntax before using the system. Therefore, to enter knowledge about a particular component instance or component class, the user clicks on the relevant icon. This opens up a window containing a data-entry form. This form contains columns of fields, where each column corresponds to a particular argument of FKRL syntax. The user then fills in the fields, with each field linked to a data-checking routine to ensure that the correct type of information is being entered. From this, an output file of FKRL text is generated. This is then read by the KAM application software which is written in Turbo Prolog.

3.5.3 Hierarchical Modeling

Hierarchical modeling, as implemented in KAM, hinges on three key concepts: subsystems, connectors, and viewpoints. A system is decomposed into subsystems which can themselves be systems. Systems are, therefore, hierarchical structures, whereas connectors define the links between subsystems and they contain five descriptors. The first descriptor defines the TYPE of connector (flow, signal, or more abstract types). The second and third define the subsystems linked by the connector. The fourth and fifth define the ports linked. Ports are also typed to facilitate consistency-checking of the system structure. "Views" define the different modalities in which a system can be decomposed.

The concept of several model types has also been implemented to allow for the possibility of mixed-mode modeling. The system *template* has, therefore, been extended to include a steady state model, linear model, detailed model, and linguistic model. A linear model can occur at a high level in the system hierarchy, whereas a detailed model can only be constructed from low-level subsystems. A linguistic model, in its simplest form, is a textual description of the system. It can also be interpreted as a functional specification of what the system does.

The idea of connectors and ports can be illustrated by considering a system, S, containing three subsystems (S1,S2,S3) in series. S can then be defined as

```
system      : S
subsystems : [S1,S2,S3]
connectors : [ct(flow,S1,S2,2,1), ct(signal,S2,S3,2,1)]

system      : S1
ports       : [flow,signal]

system      : S2
ports       : [signal,signal]

system      : S3
ports       : [flow,signal]
```

The first connector of S is a flow connector which links port 2 of S1 to port 1 of S2. The second is a signal connector which links port 2 of S2 to port 1 of S3. Port 2 of S1 is a signal port, and port 1 of S2 is also a signal port. This is, therefore, a valid connection. On the other hand, port 2 of S2 which is a signal port cannot be connected to port 1 of S3 which is a flow port.

This representation of connectors has the distinct advantage of compactness since both ports and connectors are defined within the system. An alternative approach is to explicitly declare each port and connector as a separate object. This would require very large knowledge bases making knowledge base construction tedious. Another advantage of the representation described here is the opportunity it provides for connections to be checked. This is the first step toward model validation. Eventually, the definition of a connector can be extended to include a variable. A flow connector would then have a flow variable, whereas a signal connector will have a voltage or current. Consistency-checking would then include ensuring that the variables at two connected ports are the same.

The ideas presented here are further illustrated in the next subsection.

3.5.4 An Example of Hierarchical Modeling—Aircraft Control Modeling

A generic aircraft control system, such as that shown in Figure 3.6a, can be decomposed according to two criteria or "viewpoints": the function performed by each subsystem and the system block to which each control subsystem is

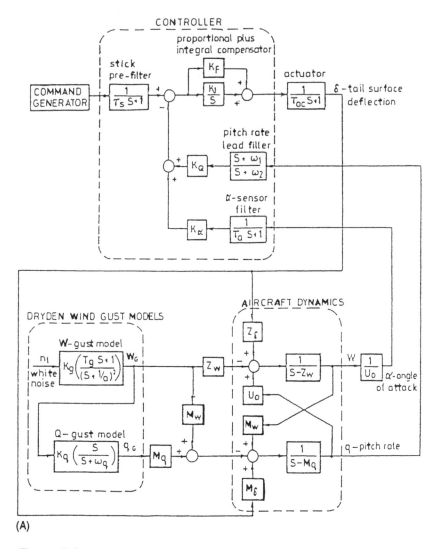

(A)

Figure 3.6 (A) Wind-gust aircraft control system. (B) KAM knowledge base for aircraft control. (C) Function-based decomposition of aircraft control. (D) System-based decomposition of aircraft control.

attached. A decomposition according to function would result in subsystems such as altitude, wind-gust, fuel flow, and engine temperature control subsystems. Each subsystem can then be subdivided into more specific subsystems.

A decomposition according to the system blocks served by the control system would result in subsystems such as spacecraft and engine. The spacecraft control system can then be divided into subsystems such as wind-gust, altitude, etc., whereas the engine can be divided into temperature, fuel, and vibration control subsystems.

It is interesting to note that both modes of decomposition include the wind-gust subsystem. The main difference is the hierarchical level at which the wind-gust subsystem is found. The wind-gust control system is modeled in greater

```
frame flight_control is a system with views=[control_function,control_blocks].
frame flight_control_functions instance of flight_control with
view=control_function;subsystems=[pitch_control,alpha_control].
frame flight_control_blocks instance of flight_control with
view=control_blocks;subsystems=[wind_gust,aircraft_dynamics,controller].
frame pitch_control instance of system with subsystems=
[q_gust,pitch_dynamics,pitch_filter,prefilter,compensator,actuator].
frame alpha_control instance of system with subsystems=
[w_gust,alpha_dynamics,alpha_filter,prefilter,compensator,actuator].
frame wind_gust instance of system with subsystems=
[w_gust,q_gust].
frame aircraft_dynamics instance of system with subsystems=
[rate_dynamics,pitch_dynamics].
frame controller instance of system with subsystems=
[alpha_filter,rate_filter,prefilter,compensator,actuator].
```

(B)

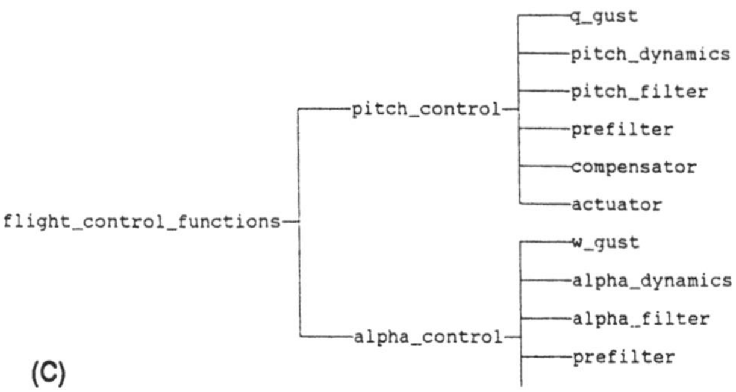

(C)

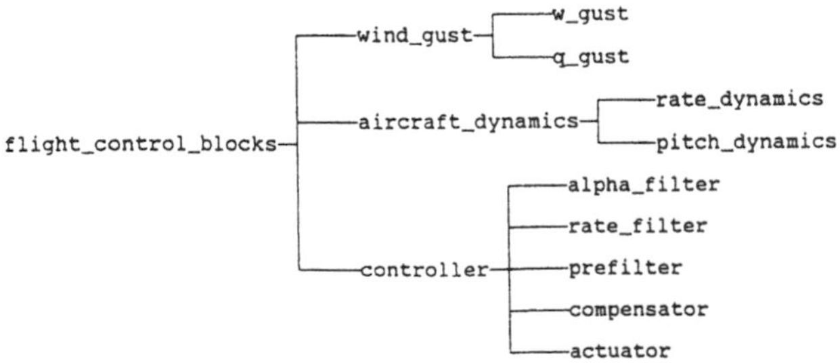

(D)

Figure 3.6 Continued.

detail in the knowledge base of Figure 3.6b. Figures 3.6c and 3.6d show the function-based and system-based hierarchies, respectively.

A further example of hierarchical modeling via KAM is given by Tanyi et al. [26], where a complex gas compressor station is modeled from process and functional "viewpoints."

3.6 SIMULATOR DRIVERS FOR KEMS

In this section, a Prolog-based driver, ESLGEN, for the ESL simulation language is described. Designed by the ISIM simulation group, ESL was developed by the Business Services group of Salford University. The ESL driver automates the various stages of simulation code generation so that the full potential of ESL can be realized by all users, even novices. The only requirement is that a user should specify the system to be simulated correctly and completely. The model construction interface in ESLGEN is a knowledge acquisition module (KAM). Previously designed as part of a knowledge-based simulation environment, KAM has proved to be a versatile tool with a wide range of applications. A knowledge base created in KAM consists of objects such as systems, subsystems, components, state variables, time constants, function generators, and experiments. ESLGEN converts these objects into ESL code.

The ESL driver has been tested on a wide range of different models and by several different users. It has enabled users having no knowledge of the ESL language to produce successful simulations.

3.6.1 Design and Implementation of ESLGEN

ESLGEN consists of five main parts: the model definition interface, equation solver, submodel generator, model generator, and archiver. A brief description of each of these facilities follows:

The Model Definition Interface
During the system definition stage, ESLGEN invokes KAM which allows the user to build a model of the system in the pseudo-natural language called FKRL. This language is based on the AI frame construct and allows the creation of prototypal classes of objects and their instances, as described in Section 3.5. It can be seen that a frame consists of the name of the object it describes. A KAM knowledge base consists of several objects which are then translated into Prolog.

The Equation Solver
The equation solver converts the equations in the Prolog database into ESL format. This is done by rearranging all differential equations so that highest-derivative term appears on the left-hand side. If this term has a nonunity

coefficient, the equation is normalized by dividing all other terms by this coefficient. As an example, the equation

$$K \frac{d^2Y}{dt^2} + T \frac{d^3Y}{dt^3} + \frac{dY}{dt} + Y = X$$

is rearranged into the form

$$\frac{d^3Y}{dt^3} = \frac{X - Y - dY/dt - K\,d^2Y/dt^2}{T}$$

The equation solver also generates subsidiary state equations for high-order equations. Hence, for the equation mentioned above, subsidiary state equations are required for the first and second derivatives.

The Submodel Generator

During submodel generation, ESLGEN collects all the components in the Prolog database into a list. The user is then prompted to select a component from this list. After a component has been selected, a corresponding submodel is generated by mapping the name of the component to the name of the submodel, output variables of the component to those of the submodel, parameters of the component to the data section of the submodel, and equations of the component to the dynamic section of the submodel. The template description of KAM component is

```
component      : name
inputvariables  :
outputvariables :
equations      :
parameters     :
```

This maps onto the following template of a generic submodel

```
submodel name(OUTPUTVARIABLES = INPUTVARIABLES);
              DATA                 ;
              INITIAL CONDITIONS;
              DYNAMIC EQUATIONS;
     END name;
```

The conversion from a KAM component to ESL submodel is, therefore, fairly straightforward since the two templates have a similar structure. The data section of a submodel can also contain arrays. These might, for instance, be function generators. Once a submodel has been generated, it can be stored in the submodel library via the archiver. There is also a submodel "importation" facility which allows the archiver to pick up standard ESL submodels from the library and "import" them to the ESLGEN environment. Such an imported submodel can then be used in the same way as locally generated ones.

Model Generation

During model generation, the name of a system is mapped to the name of an ESL model. All the constituent submodels are "included" in the submodel section. A list of output variables is formed by combining all the output variables of the constituent components of the system. The dynamic section of the model is generated by making calls to the various submodels. All output variables which are differential state variables are initialized. Finally, the experiment section is constructed by retrieving information from the Prolog database about integration method, integration step, experimental variables, and communication step. If no integration method is specified, Runge–Kutta 5 is used by default.

The model generation cycle is quite detailed and, as it progresses, the user is updated with appropriate messages from the error-handler. If no errors are found, the error-handler declares that the cycle has been successfully executed. If an error occurs, model generation is halted and the user is informed of the cause of the error. Errors can occur at three levels: during calls to submodels, when a variable is not found, or when data is either unavailable or inconsistent with a declared type. Errors encountered during calls to submodels arise when a component has no corresponding submodel in the library, or when the user attempts to import a nonexistent submodel. Errors associated with variables occur when the output variables of a component do not correspond to the definition of the submodel output. Data errors can occur when a function generator has the wrong number of points.

The Archiver

The archiver is used to store submodels and models created by the driver. It is also used to import standard ESL submodels from the ESL library to prevent the user from redefining submodels which already exist.

3.6.2 Case Study—Code Generation for a Satellite Roll Axis Control System

The use of ESLGEN is best illustrated by an example. In this section, we will generate ESL code for the roll axis system of Figure 3.7.

Describing the Roll Axis System in KAM

The following components are readily identifiable in the roll axis system–rate gyro, demodulator, compensator, torque motor, and roll axis. The KAM Knowledge base for these components and their instances is presented in Figure 3.8.

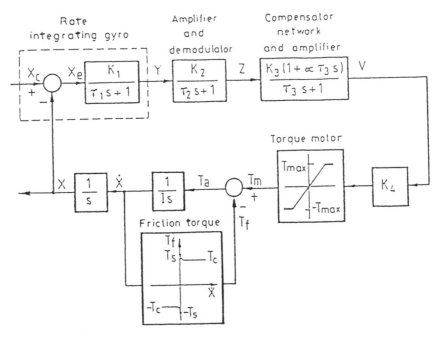

Figure 3.7 Satellite roll axis control system.

Generating the ESL Code

Code generation is done in two stages. First, submodels are generated for all the components by selecting the SUBMODEL option from the ESLGEN boxmenu. All submodels so generated are archived, and some of these submodels are shown in Figure 3.9. After submodel generation, the MODEL option is selected to generate the model shown in Figure 3.10. This model was tested in ESL and it ran successfully. Other models have been generated and tested in the same way to validate the driver module.

3.7 MODEL VALIDATION

This section investigates the addition of model validation aspects to the KEMS environment and it discusses briefly the specification and development of a knowledge-based adviser for model validation.

3.7.1 The Model Development Cycle

Model validation should not be regarded as a confirmation exercise carried out at the end of a modeling process. It is an iterative process that is an integral part of the model development cycle (Figures 3.11 and 3.12). KEMS exhibits some

```
frame rate_gyro is a component with
command_angle,Xc;roll_angle,X;
output_voltage,Y;gain,K1;time_constant,T1;equations=[T1*dY/dt+Y=K
1*(Xc-X)];
outputvariables=[Y];parameters=[T1,K1];inputvariables=[Xc,X].

frame demodulator is a component with derivative_constant,alpha;
gain,K2;time_constant,T2;equations=[T2*dZ/dt+Z=K2*Y,Zdot=dZ];
output_voltage,Z;output_derivative,Zdot;input_voltage,Y;
outputvariables=[Z,Zdot];parameters=[T2,K2];inputvariables=[Y].

frame roll_axis is a component with inertia,I;motor_torque,Tm;
friction_torque,Tf;roll_angle,X;roll_position,P;roll_velocity,V;
equations=[I*d2X/dt2=Tm-Tf,V=dX,P=X*180.0*3600.0/(4*ATAN(1.0))];
outputvariables=[X,V,P];inputvariables=[Tm,Tf];parameters=[I].

frame gain is a component with input,X;output,Y;gain,G;
inputvariables=[X];outputvariables=[Y];parameters=[G];
equations=[Y=G*X].
frame compensator is an imported submodel with.
frame limit is an imported submodel with.
frame coulomb is an imported submodel with.

frame racont instance of system with
components=[rg1,d1,g1,comp1,lim1,coul1,ra1].
frame g1 instance of gain with X=Vc1;Y=Y1;G=1.356.
frame rg1 instance of rate_gyro with Xc=0.009;X=Xr;K1=15.0;
T1=0.006;Y=Yg;initial_conditions=[0.0].

frame d1 instance of demodulator with
K2=1000.0;T2=0.01;alpha=10.0;
Y=Yg;Z=Zd;Zdot=Z1;initial_conditions=[0.0].

frame comp1 instance of compensator with outputvariables=[Vc1];
parameters=[0.0,0.1,0.055,Zd,Z1].

frame lim1 instance of limit with outputvariables=[Tm1];
parameters=[-1.356,1.356,5.424].

frame coul1 instance of coulomb with outputvariables =[Tf1];
parameters=[0.2712,0.1356,dXr,Tm1].

frame ra1 instance of roll_axis with
Tm=Tm1;Tf=Tf1;X=Xr;P=Pr;I=271.2;

V=dXr;initial_conditions=[0.0,0.0].
```

Figure 3.8 KAM knowledge base for the roll axis system.

basic conceptual validation procedures. During entry of the schematic diagram presenting the model structure, KEMS performs an automatic validation check on the integrity of component interconnections. Similarly, following the next stage of model building—the entry of parametric data—KEMS again validates the data entered. Automatic generation of simulation code from the conceptual model by the expert system in KEMS reduces the likelihood of coding errors. This dramatically reduces model development time and greatly enhances the

```
SUBMODEL roll_axis(REAL:X,V,P:= REAL: ICdX,ICX,I,Tm,Tf);

 REAL :dX;
 INITIAL
dX:=ICdX;
X:=ICX;
DYNAMIC

P:=X*180*3600/4*ATAN(1);
V:=X';
dX':=(Tm-Tf)/I*1;
X':=dX;
END roll_axis;

SUBMODEL demodulator(REAL:Z,Zdot:= REAL: ICZ,T2,K2,Y);

 INITIAL
Z:=ICZ;
DYNAMIC

Zdot:=Z';
Z':=(K2*Y-Z)/T2*1;
END demodulator;

SUBMODEL rate_gyro(REAL:Y:= REAL: ICY,T1,K1,Xc,X);

 INITIAL
Y:=ICY;
DYNAMIC

Y':=(K1*Xc-Y-X)/T1*1;
END rate_gyro;
```

Figure 3.9 Submodels for the roll axis system.

iterative nature of the KEMS environment. However, the general user has no real knowledge of the level of complexity of the model of a particular component: This is decided by the expert who enhances the expert system knowledge base. The completed simulation model must be tested to see if it is operationally valid. Its performance must have the accuracy required for its intended purpose, over the domain of its application.

3.7.2 Model Validation for KEMS

Analysis of KEMS suggests two levels where model validation techniques could be of greatest value. The first is to support the system modeler in operational validation at the postprocessing level of the environment. The system modeler, for operational validation, requires techniques for direct evaluation of performance using both real system data and simulation data. The second is to support the domain expert (environment developer) in adding new knowledge (components) to the environment, that is providing support at a more conceptual level. The domain expert uses KAM to describe and add new knowledge to the envi-

```
STUDY
include "roll_axis";
include "coulomb";
include "limit";
include "compensator";
include "demodulator";
include "rate_gyro";
MODEL racont;

REAL :
Pr,dXr,Xr,Tf1,Y1,Tm1,Vc1,Z1,Zd,Yg;
DYNAMIC
     Xr,dXr,Pr:=roll_axis(0.0,0.0,271.2,Tm1,Tf1);
     Tf1:=coulomb(0.2712,0.1356,dXr,Tm1);
     Y1:=gain(1.356,Vc1);
     Tm1:=limit(-1.356,1.356,Y1);
     Vc1:=compensator(0.0,0.1,0.055,Zd,Z1);
     Zd,Z1:=demodulator(0.0,0.01,1000.0,Yg);
     Yg:=rate_gyro(0.0,0.006,15.0,0.009,Xr);

STEP prepare " racontsystem",Pr,dXr,Xr,Tf1,Tm1,Vc1,Z1,Zd,Yg;
communication
  Tabulate t,Pr,dXr,Xr,Tf1,Tm1,Vc1,Z1,Zd,Yg;END racont;
ALGO:=RK4;CINT:=0.1;NSTEP:=1;
TSTART:=0; TFIN:=20;
racont;
END_STUDY
```

Figure 3.10 ESL model code for the roll axis system.

ronment. Because of the diverse range of model structures, it is difficult to specify the precise requirement for an expert tool to perform conceptual validation.

Validation Techniques

Validation techniques typically used in modeling and simulation fall into two broad categories: qualitative and quantitative. The qualitative approach concentrates on the cause–effect relationships within a model, whereas quantitative techniques are used either to investigate the parametric behavior of a model or to statistically analyze its performance against that of the real system. A particular criticism of many quantitative techniques is that they concentrate solely on identifying the existence of errors within a model; few attempt to locate the source of these errors.

Quantitative and qualitative techniques developed for fault detection and diagnosis can also be applied to the problem of error location in simulation models. For example, Leary [8] has developed a qualitative approach concerned with isolating faults in a real process. It places emphasis on hierarchical modeling and component connectivity. These similarities to the KEMS style of modeling may provide the solution to a qualitative expert validation tool for KEMS.

The quantitative methodology is best suited to identifying and locating potential sources of modeling errors, where the error is parametric rather than structural. The technique relies on a standard statistical test between system and

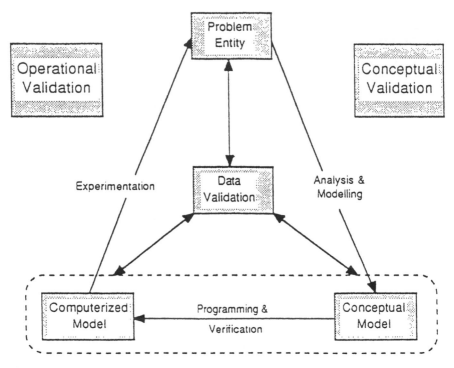

Figure 3.11 Validation in the model development cycle.

model parameters. The nature of the parameters is such that they are unmeasurable and an estimation method is applicable. A gray-box estimation method is implemented which uses the structure of the theoretical model. Running the system through selected operating conditions will yield solutions for the dynamic parameters in the estimated model. Statistical deviation between theoretical and estimated coefficients will locate errors in the model.

There are difficulties in developing an expert system for model validation directly [21]. It was decided, as a first step, to develop an advisory system which would recommend appropriate validation techniques based on information supplied by a user [23].

Design and Implementation

A software specification was drawn up from consideration of the Human Computer Interface (HCI) aspects and knowledge representation issues. Several expert system shells were reviewed and rejected along with Prolog, C, and KAM. The most common reason for rejection was a poor interface capability. Nexpert [15] was selected as it supports an object-based knowledge representation scheme with a well-established interface facility to the graphics package, EASE+.

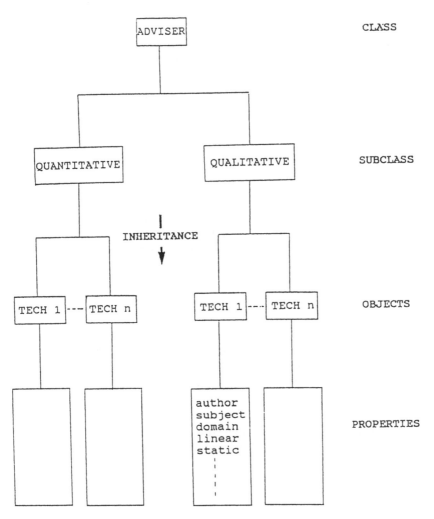

Figure 3.12   Class hierarchy for the model validation advisor.

EASE+ already provides the host environment for KEMS. The knowledge is represented as a hierarchical framework. Each technique is regarded as an object. The particular constraints of each technique are described as properties of the object. Objects are grouped into related classes and superclasses (Figure 3.12).

The user interacts with the adviser via a mouse-driven graphical interface. The option is available to reference the knowledge base by author, subject, or attributes. For the attributes search, the user completes questionnaire data forms on his own expertise, the system under investigation, and the model of that system. These descriptions are held in the EASE+ database. During knowledge processing, this knowledge can be retrieved by a manager/translator module. The required search strategy is then configured from the user description. The inference engine pattern-matches through the knowledge base to locate validation techniques best suited to the descriptive attributes. The inference process normally runs transparently to the user; however, the user can select a transcript mode and, hence, see a trace of the inference process. Advice on the selected techniques is presented to the user.

3.8 CONCLUSIONS

3.8.1 Architecture and Implementation

The prototype environment for dynamic systems simulation has been constructed using a mixture of commercial packages and in-house software generation. The aim is to obtain an "open" architecture into which new concepts and tools can be integrated easily. This parallels the approach of the much larger MEAD project [27]. Whereas MEAD uses ADA for the supervisor language, we used Modula-2 to provide the Simulation Manager. In both cases, the manager is used to reformat files for interpackage communication.

The KEMS hybrid toolkit runs on PC286 or PC386 systems under MS-DOS, using "medium coupling" between software segments. A mixture of languages was involved in the early KEMS environment. The decision to use a PC-based implementation platform was influenced by the industrial motivation to make simulation design tools available to a range of users at widely differing sites. At various phases in the project, the limitations of MS-DOS have been particularly apparent. It is not a good operating system with which to achieve "medium coupling" between a mixture of commercial and in-house developed software.

3.8.2 Database Integration and Management

In agreement with the MEAD project, no commercial database package was found to be suitable for this CADCS application. The question of knowledge representation for control systems has been addressed by Rimvall [20], who noted the need for a more flexible information-retrieval system than the classical help-function. A core data model has been proposed by Maciejowski [13], who also noted the need for more flexible data types than those provided by conventional databases. These themes have also been considered by Åström and Kreutzer [1], using the underlying notion of system representation. Many features in commercial database systems are unnecessary, and other requirements are lacking. Instead, we chose to use the graphics database facilities of EASE + as the core format and to add the necessary additional database support tools, written in-house in Pascal. EASE+ is far from being an ideal UIMS (User Interface Management System) and could not be claimed to be that [22]. However, its high-level facilities have been very useful in implementing graphics scenarios which illustrate the modeling modalities which we wished to embed in the environment.

Color was particularly important and the PC-based platform offered, at the time (and still does), a cost-effective basis for the project. We knew of no other software package which would give us the necessary prototyping graphics and databasing facilities required, and only recently has EASE+ been made available for SUN systems. EASE+, however, has been an awkward language to utilize, and its database mechanisms for linkage to other languages and programmes is far from ideal for our concept of a "medium coupled" environment.

A browser was required for fast retrieval of models and experiment frames. In addition, a librarian was needed for on-line documentation and archiving. For efficient storage and retrieval, version control provided model data compression via the Polytron Version Control System. The need for protection of validated modeling components is provided for by multiple security levels in the EASE+ toolkit. Further facilities involve a consistent naming convention and encouragement in documentation for archiving both models and experimental results.

The early version of FEMS included rudimentary "help" facilities for performing simulations. Basically, it was a summary of the user manual. It is recognized that simulation is a skill-based discipline, with very few adequately trained practitioners in industry. Thus, it is intended to provide an expert-adviser, similar to the Validation Adviser, which will assist the novice user in model experimentation, simulation analysis, and run-time advice. Further phases in the modeling design cycle include problem formulation and specification. Thus, the future aim is to produce an integrated simulation environment (ISE) which uses multiple cooperating Expert Systems, together with co-ordinating database management tools [19]. Utilizing an "open" architecture, it should be relatively simple to incorporate other features such as computer algebra, and modeling methodologies like that of Bond Graphics [11], to widen access to include workers in the life sciences as well as engineering. The bond-graph modality is being incorporated into KEMS at the "medium coupling" level since it should be utilized throughout the modeling and simulation cycle. However, because of unfamiliarity and difficulty of notation, it is also intended to produce an ITS (Intelligent Tutoring System) for Bond Graphs. This would be "loosely coupled" into the Environment, in a similar manner to the Model Validation Adviser. In addition to the hierarchical concepts incorporated in the database structure, similar principles have recently been built into the modeling infrastructure. Such hierarchical modeling is deemed to be essential in the design and simulation of complex systems. The embedding of further IT-related concepts into the KEMS environment will make the decision-making aspects of design for complex systems more tractable. Thus, the object-oriented paradigm of message passing has already been incorporated into the KEMS framework and illustrated via some initial extensions to on-line fault diagnosis. It is also being used to commence studies into linguistic forms of conceptual modeling.

ACKNOWLEDGMENTS

Throughout the development of KEMS, industrial cooperation and advice has been obtained from the Simulation Group at British Gas Engineering Research Station, Newcastle. In particular, A. S. Clark and B. Thompson have provided very valuable assistance, enabling us to use industrially relevant case studies for validating the concepts within KEMS. Acknowledgments are also due to S. Bennett and a number of researchers including M. Rahbar, E. Tanyi, M. Smith, A. Scott, and S. Xia, and also SERC who provided financial support for this research.

REFERENCES

1. Åström, K. J., and Kreutzer, W., System representations, in *IEEE 3rd Symp. on CACSD*, Arlington, Virginia, 1986, pp. 24–26.
2. Araki, M., Industrial applications in Japan of CAD packages for control systems, *IFAC Symp. CADCS*, Lyngby, Denmark, 1985, pp. 71–76.
3. Bennett, S., Rahbar, M. T., Linkens, D. A., Tanyi, E., and Smith, M., A knowledge-based environment for modeling and simulation (KEMS), *Proc. SCS Western Multi Conference*, San Diego, January 1989.
4. Borchadt, G. C., STAR: A computer language for hybrid AI applications, in *Coupling Symbolic and Numerical Computing in Expert Systems* (Kowalik, ed.), Elsevier, Amsterdam, 1986, pp. 169–177.
5. Fellheim, R. A., A knowledge-based interface to process simulation, in *AI Applied to Simulation* (Kerckhoffs, Vansteenkiste, and Ziegler, eds.), Simulation Series, Vol 18, No 1, SCS, California, 1986, pp. 97–102.
6. Gladd, N. T., and Krall, N. A., AI methods for facilitating large-scale numerical computations, in *Coupling Symbolic and Numerical Computing in Expert Systems* (Kowalik, ed.), Elsevier, Amsterdam, 1986, pp. 123–136.
7. Langen, P. A., Application of AI techniques to simulation, in *Simulation and AI* (Luker and Birtwistle, eds.), Simulation Series, Vol 18, No 3, SCS, California, pp. 49–57.
8. Leary, J., Process fault detection using constraint suspension, *Proc IEE*, 134, 1987 pp. 264–271.
9. Lehmann, A., Expert systems for interactive simulation of computer system dynamics, in *Simulation and AI* (Luker and Birtwistle eds.), Simulation Series, Vol 18, No 3, SCS, California, 1987, pp. 21–26.
10. Linkens, D. A., Tanyi, E., Rahbar, M. T., and Bennett, S. Artificial intelligence techniques applied to simulation, *IEE Conf Control '88*, Oxford.
11. Linkens, D. A., Bond graphs for an improved modelling environment in the life sciences, *IEE Colloquium, Bond Graphs in Control*, April 1990, pp. 3/1–3/4.
12. Linkens, D. A., AI in control systems engineering, *Knowledge Eng. Rev.*, 5(3), 1990, pp. 181–214.
13. Maciejowski, J. M., *A Core Data Model for Computer-Aided Control Engineering*, Engineering Dept. Technical Report CUED/F-GAMS/TR25, Cambridge University, 1985.

14. Neilson, N. R., The impact of using AI-based techniques in a control system simulator, in *Simulation of AI* (Luker and Birtwistle, eds.), Simulation Series, Vol 18, No 3, SCS, California, 1987, pp. 72–77.
15. *NEXPERT manual*, Neuron Data Inc., Palo Alto, California, 1987.
16. *POLYTRON Version Control System (PVCS)*, Version 2.0, Polytron Corporation, 1988.
17. Rahbar, M. T., Bennett, S., and Linkens, D. A., Functional specifications for an Intelligent Simulation Environment, *Proc UKSC Conf 'Computer Simulation'*, Bangor, 1987. pp. 182–197.
18. Rahbar, M. T., Tanyi, E., Bennett, S., and Linkens, D. A., A framework for a knowledge-based modeling and simulation environment (KEMSE), *Proc 12th IMACS World Congress*, Paris, 1988, Vol. 4.
19. Rahbar, M. T., Bennett, S., and Linkens, D. A., Strategies for designing a knowledge-based environment for modelling and simulation, *UKSC '90*, Brighton, September 1990, pp. 146–149.
20. Rimvall, M., Interactive environments for CACSD software, *IFAC Symp on CACSD*, Beijing, 1988.
21. Sargent, R. G., An exploration of possibilities for expert aids in model validation, in *Modelling and Simulation Aides in the AI Era*, 1985.
22. Scott, A., Linkens, D. A., Bennett, S., and Smith, M., The use of Expert-EASE's 'EASE+' package to develop user-interfaces for mathematical modelling and simulation, *IEE Colloq. Software Tools for Interface Design*, November 1990.
23. Smith, M., Linkens, D. A., and Bennett, S., "Model validation for a knowledge-based environment for modelling and simulation (KEMS), *UKSC '90*, Brighton, September 1990, pp. 50–53.
24. Tanyi, E., and Linkens, D. A., A frame-based modelling and simulation environment, *Proc. UKSC Conf., Computer Simulation*, Bangor, 1987, pp. 215–219.
25. Tanyi, E., and Linkens, D. A., Addition of a knowledge acquisition facility to a knowledge-based environment for modelling and simulation (KEMS), *Proc. of ESC '89*, Edinburgh, Scotland, September 1989.
26. Tanyi, E., Linkens, D. A., and Bennett, S., Knowledge acquisition and hierarchical structures for modelling and simulation, *13th IMACS Congress on 'Computation and Applied Maths'*, Dublin, 1991.
27. Taylor, J. H., Frederick, D. F., Rimvall, M. C., and Sutherland, H. A., Computer-aided control engineering environments: architecture, user interface, data-base management, and expert aiding, *Proc. IFAC World Congress*, Tallinn, USSR, August 13–17, 1990.
28. Van den Bosch, P. P. J., and Van den Boom, A. J. W., Industrial applications in the Netherlands for CAD packages in control systems, *IFAC Symp. CADCS*, Lyngby, Denmark, 1985, pp. 58–61.
29. Ziegler, B. P., and de Wael, L., Towards a knowledge-based implementation of multi-faceted modelling methodology, in *AI Applied to Simulation* (Kerckhoffs, Vansteenkiste, and Ziegler, eds.). Simulation Series, Vol 18, No 1, SCS, California, 1986, pp. 42–51.

4

Qualitative Modeling in Control

Roy Leitch

Heriot-Watt University
Edinburgh, Scotland

4.1 INTRODUCTION

These days the word "model" is a heavily overworked term. In its most general form, it can be used to mean any description of an entity. However, what is crucial is to clearly understand the *role* of the model. Within engineering, models have long been used to predict the temporal evolution of the attributes of a physical system, often now called the behavior of the system. However, recently, mainly stemming from the AI community, modeling techniques for reasoning about the topological properties or *spatial* position of objects and methods for representing and reasoning about the *function* of systems have also been developed. Although these latter developments are interesting, they have not yet impacted on control engineering. We will, therefore, restrict the subsequent discussion to models for the purpose of predicting behavior, sometimes called *behavioral* models and descriptions.

Further, we must also consider the *purpose* (or task) for which the model is being developed. For example, it has long been recognized that models for open-loop and feedback control require different amounts of detail to achieve a similar performance. Now, with control engineering expanding its horizons to include other tasks, e.g., fault diagnosis, process monitoring, planning, training, etc., we must carefully consider the relationship between task and model requirements. There will be no one model that is best suited to all tasks. This "no best model'' is fundamental to engineering, whereas in science, where the task of modeling is almost exclusively analytic—to describe the physical world as accurately as possible—the notion of best model may be valid. In engineering, concerned with synthesis as well as analysis,

> *a model is correct if it satisfies its purpose.*

Also, synthesis is usually expressed as a set of performance specifications for the system. So, even a best or optimal model can be difficult, and sometimes impossible to obtain. We are normally faced with a tradeoff between some of the specifications. For example, accuracy of predictions and generality of the model can sometimes be conflicting requirements. Further, AI-based approaches emphasize the need for "understandability" or *perspicuity* of models as an important specification requirement. In fact, many of the existing AI approaches and those under development explicitly address this issue of enhancing "perspicuity," sometimes at the notional expense of accuracy, so that the system can be more easily modified or extended. Therefore, in developing a model, we have to consider the role (behavioral prediction), the task (control, diagnosis, training, etc.), and the performance specifications (accuracy, flexibility, generality, verifiability, perspicuity, and honesty).

Honesty! What has honesty got to do with modeling? Well, what has been under development within AI-based approaches are techniques that allow the modeler to represent the available knowledge in a model at the degree of precision and certainty that is confidently known—no less and no more. That is, if the knowledge is uncertain, and perhaps even incomplete, we should provide representation and reasoning mechanisms to explicitly support such knowledge, and not require the modeler to make "guesses" or estimates that he may not believe in, in order for the model to become tractable. This last insight is particularly important and has resulted in an enormous interest in using AI techniques to develop alternative (qualitative or non-numeric) modeling approaches to cope with such issues [1–4]. However, what has resulted is a whole *plethora* of techniques based on a wide range of assumptions and normally developed for specific tasks. It is important that we now try to understand the relationships between such models and most crucially identify a *methodology* for selecting the most appropriate technique for a given purpose, task, specification, and characteristics of the available knowledge.

4.2 QUALITATIVE MODELING

The development of qualitative models of physical systems is currently a worldwide research community. It consists of a set of eclectic techniques designed to generate a qualitative description of the behavior of physical systems from a description of its structure, a set of initial conditions, and some initial "disturbance." The term "qualitative" has been used in many ways and generally to mean "non-numerical" models. However, the whole field of Artificial Intelligence utilizes non-numeric symbolic models. We will, therefore, use the term "qualitative modeling" to mean reasoning about systems characterized by continuously changing variables (of time) by discrete abstractions of the value of the associated variables. Most of the work on qualitative modeling concerns identifying appropriate "abstractions" that allows the important distinctions or landmarks in the behavior to be computed. This requires quantization of the real-number line into a finite set of distinctions, known as the *quantity space*; one of the main goals of qualitative modeling research is to identify useful discretizations for particular generic purposes.

Abstract descriptions of state make it possible to have more concise representations of behavior. However, the generation of the behavior from such descriptions tends not to produce a *unique* solution. This, of course, is to be expected, as the information required to produce a unique description has been eliminated in the intentional abstraction. Therefore, with respect to the real-valued description, qualitative models may produce ambiguous descriptions of the behavior. However, for some purposes, such ambiguous behavior can still contain useful information. For example, if it is required to predict whether the current state can lead to a critical or faulty condition, it may be sufficient to show that *none* of the possible behaviors leads to a critical situation. It is important to show, therefore, that the set of possible behaviors includes the "actual" behavior of the system, i.e., that the inferences made are sound. In this way, a purpose can be satisfied even with incomplete descriptions. In traditional methods, all of the information needs to be available, and it needs to be precisely and uniquely characterized, before any inference can be made.

The motivation for the initial work in qualitative reasoning came from the development of an Intelligent Tutoring System for trouble-shooting electronic circuits, called Sophie III. There the need was to represent "simpler" or more abstract models of electronic circuits such that a student could obtain an understanding of the function of the device without requiring a precise description of the detailed behavior. In this case, a detailed description in the form of a conventional mathematical model is available and reliable. However, a qualitative description should support a simpler computational mechanism than the detailed model (in practice, this has not yet been achieved) and also permit a conceptual understanding of the operation to be gained without the clutter of unnecessary detail.

In other cases, however, a reliable mathematical model, usually in terms of numerical differential or difference equations, is either not available or the expense, in time and effort, of generating one is not justified for the particular application for which the model is to be used. For these applications, a qualitative model may be all that can be reliably used to generate a behavioral description.

We can, therefore, identify four (nondistinct) motivations for developing qualitative models of physical systems:

(i)   To provide simpler computational mechanisms than those already existing, implying that a conventional model already exits and is well-posed, but that the computation on this model is complex and prohibitive. This is usually the motivation of people working in the domain of astronomy, for example.

(ii)   To provide a description for systems where traditional methods are ineffective, either due to lack of knowledge of the system description or the development costs are nonviable. This is often the case in the chemical or process industries, where chemical reactors, flow processes, and thermal systems are more difficult to characterize than, say, mechanical structures typical of the aerospace industry or robotics.

(iii)   To provide modeling paradigms that accord more closely with our "common sense" intuition of the operation of physical systems. Such descriptions do not require that the modeler, or the person utilizing the model, have a knowledge of physics to understand the way a system behaves. This motivation is based on the premise that people interact with the physical world based on their experience and intuition and not on an understanding of the "principles" of physical laws.

(iv)   To develop modeling methods based on the principles of knowledge-based systems. This includes representing the model in a declarative format such that the same description can be used for a number of different purposes, and reasoning with partial, uncertain, or incomplete information. Additionally, this motivation includes the goal of providing effective explanation facilities. This aspect is, of course, vitally important for real-time applications where the need to be able to "justify" a given decision is crucial to its acceptance and verification.

These motivations are not independent and any application domain will exhibit a mixture of these. However, the developers of the qualitative modeling techniques usually have one of the above motivations as their primary goal.

The last 10 years have seen many significant developments in the basis for formal methods for qualitative reasoning [2]. However, as yet, these developments are still short of real industrial applications. The main reason for this "applications gap" is the inherent ambiguity, usually manifested in the prediction of spurious behaviors, of the original techniques. Recently, however, a number of theoretical advances have been made that substantially reduce the ambiguity, although they can never completely eliminate it. Such developments include the use of more descriptive quantity spaces and improvements to the associated simulation algorithms. Each of these topics is discussed in the rest of this section, concluding with a brief overview of one of these techniques.

4.2.1 Extending the Quantity Space

The quantity space describes the set of values that the variables may take. In many ways, this is the distinctive feature of qualitative reasoning in that the variables take their values from an underlying finite *support set* that represents a symbolic or qualitative description. Early work concentrated on the simple three-valued quantity space $\{+, 0, -\}$, interpreted as either positive, zero, negative, or increasing, steady, decreasing, respectively. While intuitively appealing, this simple quantity space has resulted in many computational problems due to the inherent ambiguity in the associated calculus. In fact, the computational approaches to overcoming the ambiguity, for example, the use of Truth Maintenance Systems [5] has, itself, resulted in very complex simulation mechanisms, hence negating one of the original motivations for qualitative reasoning. However, it is now clear that in all but the simplest systems, this quantity space is impractical. A contemporary quantity space, consisting of open intervals and real-valued landmarks, was developed as part of the QSim algorithm [6]. This has received the greatest attention and has come nearest to practical application. The landmark values are associated with physical parameters, e.g., top of the tank, and define the operating range for the variables. Landmarks can also be used to change the underlying modeling assumptions. In QSim, the quantity space consists of a qualitative magnitude taken from the landmark quantity space and a "qualitative derivative" taken from the three-valued space interpreted as increasing, steady, or decreasing. This permits a reasonable representation of the value of a variable; however, its use in qualitative reasoning still results in ambiguity or, in the case of QSim, in the generation of spurious behaviors.

Since the late 1980s, a number of proposals have been made to extend the quantity space to allow more detailed descriptions of variables but without requiring the use of real numbers; these are shown in Figure 4.1. The most evident of these is the use of mathematics of non-standard analysis to perform order-of-magnitude reasoning [7]. This allows the quantity space {inf, fin, negl, fin, inf} to be used where inf represents infinite values, fin, finite values, and negl, negligible values. This supports a calculus reflecting *order-of-magnitude* axioms such that "second-order" effects in models can be neglected. This represents an automation of everyday pragmatic assumptions in building realistic models. These have been used in static systems by Raiman [8] and in dynamic simulation by Weld [7]. However, there are difficulties in interpreting this quantity space with respect to the standard real line [9]. The use of these values, and its associated calculus, distorts the simulation so that it is no longer a faithful qualitative abstraction of the solution to an underlying differential equation. However, supporters argue that it is just this distortion that is required to produce caricature or exaggerated behaviors that are used to evaluate designs at their extrenum. Recently, the present authors have proposed the use of fuzzy sets to extend the

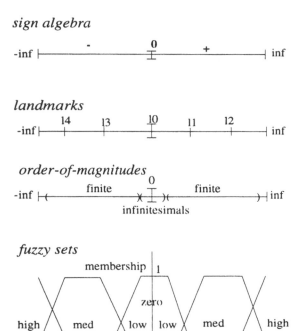

Figure 4.1 Extending the quantity space.

quantity space. In particular, we have proposed a fuzzy quantity space [10] that permits an arbitrary discretion of the real-number line and allows the subjective element of assigning values to be incorporated through the use of the graded membership afforded by the membership function of fuzzy sets. Further, as discussed later, both the qualitative magnitude and its derivative are represented in the same quantity space, bringing significant advantages to the simulation algorithms.

And so, the industry of generating alternative quantity spaces continues. The trend is, of course, to come closer and closer to the real line; then why not give up and just use real numbers? The overriding principle is, however, to utilize the most abstract quantity space, i.e., with least cardinality, that is consistent with the task for which the model will be utilized. If the task requires real numbers and they are available, accurate, and at a justifiable cost, then they should be used. If, however, a more abstract description will suffice and may also provide additional benefits, its terms of understanding and explanation, then it should be used. The relationship between quantity space, task, and available knowledge has still to be properly understood.

4.2.2 Approaches to Simulation

The most basic distinction here is whether the equations represent a static or dynamic model of the system. One of the early misconceptions was in misguided attempts to generate dynamic behavior from algebraic models of a system in equilibrium. This led to strange notions such as mythical causality and mythical time being the (false) interpretation of the steps in the solution procedure as the temporal evolution of the system. Notwithstanding these misconceptions, these approaches identified two basic attributes of simulation, depending on whether the relations or constraints used in the model are directional or not. In the case of the former, an efficient algorithm can be developed for propagating values through the directional constraints. Such a representation can be used to support a casual interpretation that is supplied by the model builder. Alternatively, bidirectional constraints do not require that an explicit ordering, or casual interpretation, be ascribed to the model. In cases where casual directions may change, this can be very important, e.g., in fault diagnosis. The price paid for this generality is in a more complex inference mechanism. Constraint propagation based on generate and test search mechanisms, and supported by Truth Maintenance Systems [5] to attempt to manage the ambiguity, has been developed for such simulation systems. Here the basic tradeoff is between explicit representation of causality and general search mechanisms. However, a technique for the automatic generation of causally ordered equations from a bidirectional representation has been developed by Iwasaki and Simon [11]. Basically, it determines when a given exogeneous (external variable) set of equations is self-contained. When applied to dynamic systems, this approach gives intuitive causal orderings. The behavior of the system is then generated by propagating the values through these constraints. Actually, the behavior here refers to the determination of the existence of equilibrium states consistent with a perturbation of the system state. The term simulation should, in theory, be reserved for the behavior of dynamic systems.

Dynamic system simulation is an increasingly important form of qualitative reasoning. The pioneer of this approach has been the QSim algorithm of Kuipers [6,7]. This utilizes the constraint ontology with a landmark quantity space. The simulation algorithm is a particular form of the above generate-and-test procedure from the current state, assuming only that the variables are continuous. These successor states are then checked for consistency with the modeling constraints, and conflicting states are eliminated. Therefore, directionality of the constraint is not required and, hence, no causal interpretation of the behavior can be derived. The algorithm can, however, be shown to predict the "correct" behavior, interpreted as the qualitative abstraction of the real-valued solution of the associated differential equation. Unfortunately, the "nonconstructive" nature of the algorithm also leaves behind many spurious behaviors that tend to obscure the real behavior, often to the extent of making the simulation unusable. The spurious behaviors are the manifestation of the essential ambiguity in the underlying qualitative calculus. Because of the generate-and-test procedure, the ambiguity in the constraints does not allow the elimination of sufficient possible next-state behaviors. A further problem with the QSim algorithm is that it is state based. That is, the complete state is calculated every time that any system variable changes qualitative state, thereby forcing unnecessary distinctions to be made. Without ambiguity, this would only result in inefficient computation; however, with ambiguity, each distinction results in spurious behaviors being generated, further obscuring the "real" behavior. Attempts to reduce such spurious behavior have become a major preoccupation within this branch of qualitative reasoning. Global filters based on system properties of uniqueness of behaviors and energy conservation are used to remove some of the spurious behaviors. Additionally, external knowledge or heuristics may be used; however, this removes the generality of the method in generating behaviors directly from a structural model.

The behavior generated by such algorithms consists of a logical ordering in the evolution of system state through the quantity space. This allows a temporal ordering in the states, but no indication of the temporal extent of the *duration* between states. Strangely, this is not a fundamental limitation of the approach, but simply an effect of restricting the representation of the qualitative derivative to three values; in which case, no qualification on the rate of change, e.g., increasing slowly is possible and, hence, no calculations of the temporal duration may be made. More recently, durations have been derived from knowledge of the change in magnitude and the relative rate of change.

To date, these most recent developments have been realized as particular experimental systems only. Whether the combination of all of these techniques will result in practical qualitative simulation mechanisms is still unknown. However, it is clear that such approaches will significantly improve the performance of earlier systems.

4.3 FUZZY QUALITATIVE SIMULATION

As an example of recent work in qualitative simulation, we now present an overview of a system, developed by the authors, that combines the use of fuzzy sets with features of qualitative simulation to provide a number of significant advantages over other approaches.

Fuzzy qualitative simulation (FuSim) [10] adopts a constraint-centered ontology in that the model is derived from an underlying differential equation representation of the system. In FuSim, a qualitative value of a system variable is a fuzzy number chosen from a subset of normal convex fuzzy numbers. This subset, called the fuzzy quantity space, is generated by an arbitrary but finite discretization of the underlying numeric range of a variable. For computational efficiency, such values are characterized by a 4-tuple parametric representation of their membership functions; see Figure 4.2. The sets of possible values which system variables can take are restricted by either algebraic derivative of function relational constraints among the variables. More specifically, the algebraic operations performed within a fuzzy quantity space are those of the fuzzy

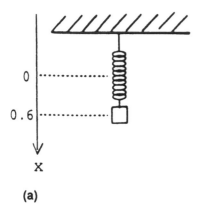

(a)

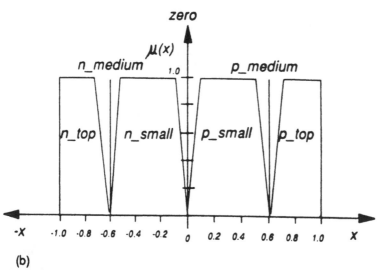

(b)

Figure 4.2 Example of fuzzy qualitative simulation. (a) Simple mass on a spring; (b) fuzzy quantity space; (c) fuzzy model: the matrix represents a fuzzy relation representation of Hooke's Law; (d) fuzzy description of the behavior; O indicates the fuzzy magnitude, → indicates the fuzzy rate of change.

Constraints

$$\text{deriv } x = v$$
$$\text{deriv } v = a$$

Functional relations

$$\begin{pmatrix}
a & x & -b & -0.6 & -s & 0 & s & 0.6 & b \\
-b & & 0 & 0 & 0 & 0 & 0 & 1 & 1 \\
-0.6 & & 0 & 0 & 0 & 0 & 1 & 1 & 0 \\
-s & & 0 & 0 & 0 & 0 & 1 & 1 & 0 \\
0 & & 0 & 0 & 0 & 1 & 0 & 0 & 0 \\
s & & 0 & 1 & 1 & 0 & 0 & 0 & 0 \\
0.6 & & 0 & 1 & 1 & 0 & 0 & 0 & 0 \\
b & & 1 & 1 & 0 & 0 & 0 & 0 & 0
\end{pmatrix}$$

(c)

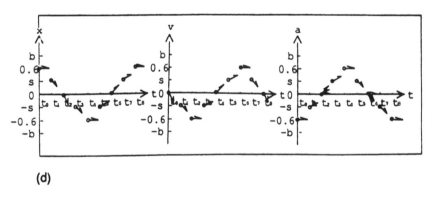

(d)

Figure 4.2 Continued.

numbers. A derivative constraint simply reflects that the qualitative value of a variable's magnitude must be the same as that of another variable's rate of change. Functional relationships within FuSim are represented by fuzzy relations, thereby allowing imprecise and/or partial numerical information on functional dependencies between variables to be exploited, if indeed such information is available.

Based on the above qualitative representation of values and constraints, FuSim takes as input a set of system variables, a set of constraints relating the variables, and a set of initial values for the variables, and produces a tree of states with each path representing a possible behavior of the system as output. In fact, FuSim first generates a set of transitions from one qualitative state description to its *possible* successor states by applying state transition rules, derived from the continuity of the system variables. Further restrictions on these possible successor states are then imposed by checking for consistency between constraints which share a variable, called constraint filtering, and information on the rates of change of the system variables held as part of the fuzzy qualitative state, called temporal filtering. Importantly, associated with each sequence of states, i.e., each path of the output tree, FuSim also generates a sequence of temporal intervals that provides durations for how long a variable will persist within a particular state. This a distinct advantage of FuSim over other qualitative simulation systems, especially when used for diagnosing faults on dynamic systems [14]. Figure 4.5, shows a simple example of FuSim applied to the simulation of a simple second-order system of a mass on a spring. It shows the fuzzy quantity space, the fuzzy constraints (model), and the resulting (unique) qualitative prediction of the oscillatory behavior. Complete details of FuSim are available in [10,14].

4.3.1 Multiple Models

An exciting theme in the utilization of qualitative reasoning techniques is in the development of *multiple models* of the system that can be used for a variety of purposes. They can be used to improve the efficiently of the computation by first searching on a simpler model. They can be used to improve the effectiveness of the system by utilizing more abstract models in preference to more detailed ones. Also, they can be used to provide further constraints on which to reduce ambiguity or spurious behaviors of another, more detailed model. Such multiple models require a clear understanding of the process by which models can be developed from one another and the consequent relationship between them.

Three important modeling *dimensions* can be identified and visualized as in Figure 4.3. Although the terminology of these dimensions is still unstable, there appears to be convergence on some of the basic concepts. The first dimension, called *abstraction* is, in fact, the choice of quantity space discussed in previous sections. In this sense, more abstract models are those with weaker quantity spaces that support less distinctions between values. Qualitative reasoning may, therefore, be seen as a process of abstraction from an underlying real-valued representation. Similarly, three-valued algebra systems are more abstract than, say, order-of-magnitude or fuzzy quantity spaces.

A second dimension is one of *abstraction* which represents the accuracy of a model with respect to a given reference model. This is the classical choice that the system modeler must make and is, of course, represented in conventional modeling by the system order. The process of moving from one model to another

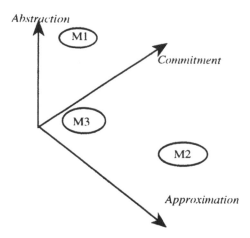

Figure 4.3 Dimensions for multiple models.

of lower resolution is of aggregating system variables to generate simpler models that will *approximate* the behavior of the more detailed model. Several approaches to this aggregation of variables have been proposed depending on the metric used for simplification. One approach is the use of the physical structure to determine the prime variables, as those that interconnect the identifiable subsystems. In this way, internal variables can be neglected in the simpler models. This approach is also closely related to the use of function to decompose the system. Functional units are often those with explicit terminal interconnections. However, it is also possible to have a given functionality distributed across a number of physically separate subsystems. Metrics on system variables can also be used as aggregation criteria. Specifically, the magnitude of the variable determines the strength of interaction and may also be used to neglect those with weak influences on other variables. Similarly, the time response can also be used to replace a variable within a more detailed model with a constant parameter representing the steady-state value, thereby reducing the number of variables. This simpler model then only accurately represents the system at a coarser temporal granularity. In general, this aggregation process results in an *approximation*.

The third dimension with which to classify model variations is in explicitly quantifying the belief or *commitment* associated with a given model or particular modeling primitives. This, of course, depends on the type of uncertainty being represented within the models. In particular, vagueness or imprecision is captured in fuzzy qualitative simulation through the graded set membership. Also, nondeterminism in the form of randomness can be represented by probability approaches. This provides a dimension along which commitment is increased as uncertainty is reduced in the model.

It can be seen, therefore, that multiple models can be classified along (at least) these three dimensions: abstractions, approximation, and commitment. The goal in the utilization of these models is to start with the most abstract, most approximate, and least commitment and "pilot" within this model space until the task is satisfactorily achieved. This approach is still in its infancy and model switching on one dimension only has been reported so far. However, the general approach would appear to have significant potential, particularly for diagnostic and training systems.

In summary, the "first generation" approaches to qualitative reasoning, while not resulting in effective systems, have been inspirational in generating further approaches which greatly improve and extend the prospects. Such approaches include less-abstract quantity spaces, constructive algorithms, and the principled use of multiple models. These more recent developments are currently restricted to experimental systems. However, the vibrant activity within the community suggests that qualitative reasoning is alive and well. It is hoped that these "second generation" approaches will result in practical systems that will realize the ambitions of the early goals of qualitative reasoning.

4.4 APPROACHES TO CONTROL

Although modeling, in all its various guises, is intrinsically important, the real advantage comes when using these techniques for control applications. The last decade has seen a rapid expansion in the techniques available to develop solutions. However, so far there has not been a significant attempt to identify the "best" approach for a given application. Such a methodological approach is now becoming crucial as both the approaches to modeling and the range of applications continually expand. What is required is a set of relations that will identify the most appropriate model, and the corresponding solution technique, for a given class of application, the class being determined by the characteristics of the domain. In this regard, the model classifications presented in [15] begin to form a basis from which such a methodological approach can be generated. In this section, we begin the process of attempting to identify particular solution techniques with generic problems appearing within the control engineering literature.

4.4.1 Expert Control

One of the obvious approaches to the utilization of Artificial Intelligence techniques within Control attempts to use Expert System techniques [12] as an adjunct to conventional methods. By placing an Expert System, usually using rule-based technology "on top of an existing numerically based control system" (see Figure 4.4), the range of applicability of the controller can be extended by encoding into the Expert System rules for the adjustment of the control, either by

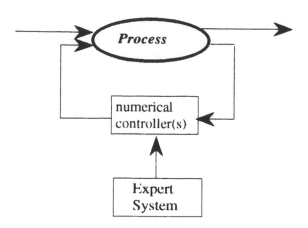

Figure 4.4 Expert Control.

modifying the control algorithm or by replacing it with another approach altogether. This essentially puts the Control Engineer on-line so that knowledge normally only used during the (off-line) design process is available during the actual operation. This approach is now being called *Expert Control* and represents a major activity for control engineers wishing to become involved in Artificial Intelligence approaches. It is attractive in that it utilizes existing techniques, and hence skills; we believe that the majority of control engineers have adopted this route.

In interpreting the structure of Figure 4.4 we have to be careful in understanding exactly what kind of knowledge is being automated within the general Expert Control framework. This requires that we clearly identify the content of the knowledge, sometimes called *knowledge level,* and where the knowledge comes from, sometimes termed the *knowledge source* [15].

By knowledge level, we mean that there are basically two options to representing Control Engineering knowledge. The second-class dimension determines the subject of the knowledge. In control engineering terms, we can have two options. We can represent the knowledge of the process itself, i.e., model-based approaches, or of the control algorithm; we term this *object-level* knowledge. Alternatively, we may choose to represent knowledge about the control design methods so that they can be modified on-line. We term this *meta-level* knowledge, as it reflects knowledge about the knowledge used to control rather than the modeling knowledge itself. Both approaches are actively being developed, both with AI-based techniques and "traditional" control methods. In the case of conventional control techniques, examples of object-level control are classical three-term controllers or indeed an Linear Quadratic Gausian-derived state feedback control system. Adaptive control systems are a common form of meta-level control as the performance of the system is monitored on-line, usually in the form of some performance index, and used to modify the (design) of the object-level controller. This clear separation of meta- and object-level knowledge allows different techniques (models) to be used at each level, thereby greatly expanding the range of applicability of the control techniques. This distinction is fundamental in control applications; however, it is also valid within other domains, e.g., diagnosis.

By *knowledge source,* we mean from where the knowledge that is used to build the model comes. Two major sources of such knowledge have been identified [1] as empirical and theoretical. Empirical knowledge relates to that which is obtained directly from first-hand experience. It attempts to capture knowledge that has been induced from direct observation of a particular system. As such, it can be highly effective but is limited in its generality. Empirical knowledge has traditionally been omitted from control systems design, sometimes resulting in reduced performance, and hence requiring subsequent empirical tuning. However, the development of Expert Systems techniques has brought such knowledge to the fore and emphasized its importance and, more recently its limitations. On the other hand, theoretical knowledge, i.e, knowledge of scientific laws and principles, has long been the basis of control system design. However, the use of such knowledge has, until fairly recently, been almost exclusively restricted to numerical descriptions, usually in the form of differential or difference equations. And, as discussed in Section 4.1, often there does not exist adequate knowledge to make use of the powerful methods associated with real-valued equations. This is exactly the motivation and role of the approaches to Qualitative Reasoning discussed in the previous section. Theoretical knowledge is, of course, general and transferable from one application to another and, in fact, removes much of the knowledge acquisition problem associated with empirical knowledge. However, it can also be inefficient, and its very generality may mean that it is less effective. Clearly, theoretical and empirical knowledge are complementary; the best solution is obtained by a symbolic combination of the two. However, such combinations are, by necessity, specific to a given application, and care has to be taken to ensure that performance is indeed improved.

In the above context, Expert or Intelligent Control can be described as a meta-level approach, usually with empirical knowledge at the meta-level and a conventional numerical controller(s) at the object level. In contrast, Fuzzy Logic Controllers can be regarded as object-level empirical knowledge (with uncertainty).

4.4.2 Qualitative Control

The second approach uses AI techniques directly to model the system at a level of detail consistent with the available modeling knowledge and the task to be executed. Such approaches, sometimes called *Qualitative Control*, can be regarded as directly "closing the feedback loop" by using AI methods, or loosely, an Expert System in the feedback path as shown in Figure 4.5. In this way, qualitative representation of the control policy is used to compute the value of the control variable. This exposes a major shortcoming of qualitative methods for control applications: Practical controllers still must output a numerical value. This requires that the qualitative value be "approximated" by a numerical value, a symbol-to-signal transformation that is highly subjective. Fuzzy Logic Controllers are prime examples of this approach. The use implicit empirical models at the object level, with fuzzy sets to represent the inherent vagueness or uncertainty in human knowledge. However, many other AI techniques may also be used. This approach is appropriate whenever there is some inherent difficulty with conventional modeling (numerically based) techniques. Further, in many cases, it has been shown that equivalent control performance can be achieved; however, qualitative methods have a distinct advantage of *perspicuity*, thereby supporting a much easier interaction and not demanding highly trained (educated) staff. Meta-level control can also be used with qualitative controllers. A good example of this is Self-Organizing Fuzzy Logic Controllers, where self-organizing rules are used to modify the fuzzy rule-base to improve the overall performance.

A further distinction has to be made, and that is whether the knowledge represents an *explicit* model of the physical world to be reasoned about or whether it represents our procedures for controlling or diagnosing the world. In the latter case, the model would be termed *implicit*. Explicit models relate system inputs to outputs in the same way as the real system. They can, therefore, have a causal interpretation associated with the structure of the representation. Conversely, implicit models effectively relate outputs (symptoms in the case of diagnosis)

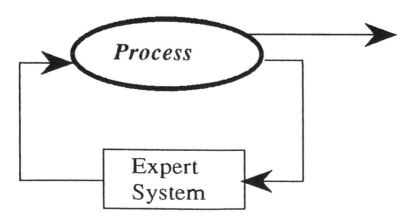

Figure 4.5 Qualitative Control.

to inputs and are inherently acausal. Almost all of Expert Systems work has been based on implicit empirical models, sometimes now called classification-based approaches.

4.4.3 Qualitative Model-Based Control

Awaiting development is the qualitative counterpart of Model Reference Control (see Figure 4.6). In this case, the techniques of Qualitative Simulation can be used to represent the "reference model," i.e., the ideal model, and the control adjusted such that the observed behavior approaches the predicted response. This is directly the qualitative counterpart to "conventional" model-reference adaptive control [16]. The development of such an approach will allow "weaker" specification of the reference behavior, thereby explicitly recognizing that often there is no unique ideal behavior, but that the specifications can be achieved by a *set* of behaviors. This set is, of course, captured in the qualitative description of the behavior. The key issue is, therefore, to find the most appropriate quantity space to capture the specification at the correct level of abstraction.

The comparison between behaviors will also require a form of symbol-to-signal transformation to identify real-valued adjustments to the controller. Such a discrepancy detector has been developed by the authors for use in their Fuzzy qualitative simulation-based approach to model-based diagnosis [14]. Interestingly, the duality property between control and identification can also be used to develop qualitative model-based controllers from model-based diagnostic systems. We are not aware of this work being reported or even pursued.

4.5 CONCLUSION

We have attempted to show that underlying a great part of what is currently being developed, both within the AI community and the control engineering

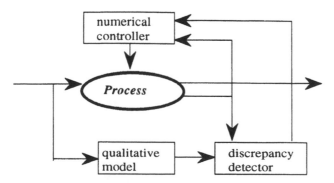

Figure 4.6 Qualitative Model Based Control.

community, is fundamentally concerned with generating a range of approaches to the modeling of physical systems. It is here that AI is having a very profound impact on approaches to Control, if not yet on theory. By extending the formal methods of reasoning to include the prediction of the qualitative evolution of a dynamic system, as in qualitative simulation, we now have a much greater diversity of tools with which to more effectively tackle complex applications. These methods are not *ad hoc*. They can be analytically rigorous as methods based on the differential calculus for example, perhaps even more so.

And so, the scene is set for an exciting future for Control Engineering based on a vastly expanded repertoire of techniques; some AI based, some numerically based, and some as symbiotic combinations of the two. We hope that the above discussion will contribute to the development of an understanding of the relative merits of the different techniques currently existing and also under development. Only with such an understanding can we begin to find the most appropriate technique for a given application.

REFERENCES

1. Leitch, R. R., The modeling of complex dynamic systems, *Proc. IEE, Part D, 134*(3), 245–250 (1987).
2. Weld, D., and de Kleer, J. (eds.), *Readings in Qualitative Reasoning about Physical Systems*, Morgan Kaufmann, San Mateo, CA, 1989.
3. Davis, E., *Representations in Common Sense Knowledge*, Morgan Kaufmann, San Mateo, CA, 1990.
4. Leitch, R. R., A review of the approaches to qualitative reasoning of complex physical systems, in *Knowledge Based Systems for Process Control*, J. McGhee, M. J. Grimble, and P. Mowforth (eds.), Peter Peregrinus, 1990.
5. de Kleer, J., An assumption-based Truth Maintenance System, *Artificial Intelligence*, 28, 127–162 (1986).
6. Kuipers, B. J., Qualitative simulation, *Artificial Intelligence*, 29, 289–338 (1986).
7. Weld, D. S., Exaggeration, *Artificial Intelligence*, 43 311–316, (1990).
8. Raiman, O., Order of magnitude reasoning, *Proc. 5th National Conference on Artificial Intelligence*, 1986, pp. 100–104.
9. Shen, Q., and Leitch, R. R., On extending the quantity space in qualitative reasoning, *Intern. J. Artificial Intelligence* Eng.
10. Shen, Q., and Leitch, R. R., Combining qualitative simulation and fuzzy sets, in *Recent Advances in Qualitative Physics* (B. Faltings and P. Struss, (eds.), MIT Press, Boston, MA, 1991.
11. Iwasaki, Y., and Simon, H. A., Causality in device behavior, *Artificial Intelligence*, 29, 3–23 (1986).
12. Astrom, K. J., Anton, J. J., and Arzen, K. E., Expert Control, *Automatica*, 22(3), 277–286 (1986).
13. Linkens, D. A., and Hasnain, S. B., Self organising Fuzzy Logic Control and its application to muscle relaxant anaethesia, *Proc. IEEE, Part D, 138*(3), 57–95 (1991).

14. Leitch, R. R., and Shen, Q., Finding faults with model based diagnosis, *Proc. 2nd. Int. Workshop on the Principles of Diagnosis*, Milan, 1991.

15. Leitch, R. R., Artificial Intelligence in Control: some myths, some fears, but plenty prospects, *IEE Comput. Contr. J. 3*(4), 153–164 (1992).

16. Landau, Y., *Adaptive Control: The Model Reference Approach*, Marcel Dekker, Inc., New York, 1979.

5

Symbolic Modeling in Control

P. J. Gawthrop

University of Glasgow
Glasgow, Scotland

5.1 INTRODUCTION

As with many branches of engineering, system modeling can be aided by suitable computer-based tools. One such set of tools is the simulation packages, for example, ACSL and SimuLab, which have very successfully built on the *numerical* processing capabilities of computers. Perhaps due to this success, system modeling has (unfortunately) almost become synonymous with system simulation; system models are often represented as simulation codes. This chapter argues that a simulation code is a poor system representation.

However, computers are also good at *symbolic* processing, but this aspect has been less used by system modelers in the control engineering field. A purpose of this chapter is to suggest that symbolic computing can augment numerical computing in giving the modeler insight into system structure and behavior.

This chapter presents a philosophy of system modeling as a set of transformations between representations, and it is argued that *bond graphs* form a good representation for this purpose. An experimental toolbox (MTT) for implementing such model transformations, using a bond-graph representation, is described. The toolbox is implemented as a set of Unix tools making use of xfig as a graphical frontend and Prolog, Reduce, and Matlab for transforming between representations. The toolbox is evolving and this chapter describes the current situation. Three examples are given to illustrate the main features of our approach.

The chapter is organized as follows. Section 5.2 presents a philosophy of system modeling [1] of which MTT is an implementation. Section 5.3 discusses how systems are represented by MTT; Section 5.4 discusses how representations are transformed by MTT. Section 5.6 gives a number of illustrative examples. Section 5.8 concludes the chapter and indicates possible research directions.

5.2 SYSTEM MODELING

As discussed in a recent paper [1], a system model has a *use*: for example, control design, system design, system simulation, or system understanding. A detailed discussion can be found in [2]–[6].

System modeling, the procedure for arriving at an appropriate (for its use) model, can be viewed as a sequence of *transformations* between system *representations* as indicated in Figure 5.1. The start of this chain of transformations is the physical system; an intermediate representation is the *core* representation; the final representation is the system model in an appropriate form.

With reference to Figure 5.1, the following considerations are important.

1. There is a unique "core" representation of any system. There are many routes *from* this core representation, each leading to an appropriate model via intermediate representations and corresponding transformations. There are many possible routes *to* this core representation, each leading from a particular physical system.

- Physical system

- $Transformation_1 \Rightarrow Representation_1$

- $Transformation_2 \Rightarrow Representation_2$

- ...

- $Transformation_N \Rightarrow$ Core representation

- $Transformation_{N+1} \Rightarrow Representation_{N+1}$

- $Transformation_{N+2} \Rightarrow Representation_{N+2}$

- ...

- $Transformation_{N+M} \Rightarrow$ Model

Figure 5.1 System modeling: transformations.

2. Because the core representation is unique, it is easy to expand the toolbox to include additional transformations from the physical system to the core representation and additional transformations from the core representation to the model.
3. *Transformation*₁ probably cannot, and certainly should not, be completely automated. Engineering insight, knowledge, and experience is essential to capture the essence (with respect to the particular use) of the physical system while discarding the irrelevant form.
4. In the light of consideration 3, *Representation*₁ should be "close" in some sense to the physical system.
5. The core representation, and hence the representations leading to it, must contain enough information to generate all of the required models.
6. Representations must be easily *extensible*: It must be possible to add extra components or attributes without restructuring the representation.
7. Representations must be *hierarchical* in two senses.
   (a) It should be possible to describe a complex system in a modular fashion: subsystems must be reusable.
   (b) Subsystems with fast time constants (relative to the level above) should be replaceable by steady-state representations. Thus, automatic *approximation* of complex systems should be possible.

There is clearly some tension between considerations 4 and 5: the former giving a tendency to simplification, the latter toward complexity.

In the context of this framework, we have adopted bond graphs [2]–[6] to provide the core representation. Some of the reasons for this follow.

1. Bond graphs provide a cross-disciplinary representation and modeling methodology. They have been applied to a range of systems with components in a number of domains, including mechanical, electrical, hydraulic, chemical, and biological.
2. Bond graphs provide a precise and unambiguous modeling representation; many domains (with the notable exception of electrical circuits) do not have such a representation.
3. Bond graphs can be constructed without reference to computational causality. This is in distinction to other representations such as block diagrams, signal flow graphs, and a simulation code such as ACSL.
4. Bond graphs provide a clear diagrammatic picture of system structure. This is in distinction to equation-based approaches.
5. Bond graphs can be constructed without identifying the system states, which may not be unique anyway. The system states may be deduced from the bond graph. This is in distinction to describing systems in state-space form.
6. Bond graphs make a clear distinction between system structure and component behavior.

7. Bond graphs of individual components can be combined directly to form systems. This is in distinction to system transfer functions which cannot be directly combined unless the systems which they represent do not interact.
8. The individual components within the bond graph structure may be linear or nonlinear.

The purpose of this chapter is to present, and illustrate, a series of software tools to implement transformations helpful to the systems modeler. At the moment, we have not implemented the hierarchical ideas [7], so the presentation is restricted to simple models. The examples are chosen to illustrate some of the points listed above.

5.3 REPRESENTATIONS

MTT makes an explicit distinction between two aspects of system representation:

- the system structure
- the component constitutive relationships

This is important for a number of reasons:

1. The system designer needs to make a clear distinction between the two aspects.
2. Causality is usually determined by structure except for cases where some components have causality restricted to one type.
3. Once structure has been determined, the effect of constitutive relationships determines system behavior.

Some representations employed by MTT appear in Table 5.1 The core representation (Figure 5.1) that we use at the moment encapsulates *structure* in a Prolog textual representation of an acausal bond graph (code pab in Table 5.1) and *constitutive relations* in a Reduce representation (code rcr in Table 5.1).

Parameters can be supplied in Reduce (code rpa in Table 5.1) and/or Matlab format (code mpa in Table 5.1).

At the moment, the easiest way to access MTT is via a graphical frontend based on the "fig" or "xfig" utility, a standard graphical editor which generates the bond graph (code fab in Table 5.1).

5.3.1 Components

There are two categories of components recognized by MTT:

- One-port components
- Two-port interjunction components

The former are associated with one junction, the latter with two. One-port components are listed in Table 5.2; two-port components are listed in Table 5.3.

Table 5.1 Model Representations

Code	Model type	Language
fab	Acausal bond graph	Fig
lbl	Labels for fab	
pab	Acausal bond graph	Prolog
rcr	Constitutive relations	Reduce
rpa	Parameters	Reduce
mpa	Parameters	Matlab
pcb	Causal bond graph	Prolog
tsu	Summary	LAT$_E$X
req	Equations	Reduce
rde	Definitions	Reduce
rse	State/nonstate equations	Reduce
fse	State/nonstate equations	Fortran
mse	State/nonstate equations	Matlab
ree	Explicit (state) equations	Reduce
rsm	State matrices—A, B, C, D	Reduce
fsm	State matrices—A, B, C, D	Fortran
msm	State matrices—A, B, C, D	Matlab
tsm	State matrices—A, B, C, D	LAT$_E$X
rtf	Transfer-function matrix	Reduce
ttf	Transfer-function matrix	LAT$_E$X

Table 5.2 One-Port Components

Name	Description
S	Either an effort or a flow source
C	Energy store with effort output
I	Energy store with flow output
R	Dissipator (R component)
M	Measurement

Table 5.3 Two-Port Components

Name	Description
TF	Power-conserving transformer
GY	Power-conserving gyrator

In addition to their energy port(s), these components can have any number of signal connections. Hence, for example,

- *n* one-port components can be interconnected to form arbitrary *n*-port components.
- Two-port components can, with the addition of one or more signal connections, form modulated transformers or gyrators.
- A source, with the addition of one or more signal connections, forms a modulated source representing, for example, a power amplifier.
- System *inputs* are indicated by sources with an *atomic* CR; that is, sources with no modulation.
- System *outputs* are indicated by measurement components.

Constitutive relations are represented in Reduce form (code rpa in Table 5.1). Some examples follow.

Linear CR

Consider the generic linear CR with name Lin. The purpose of this CR is to allow a component gain *k* to be defined corresponding to a default output causality. Thus, if *y* is the component output and *u* is the component input, there are two possibilities: If the component output causality is the same as that of the default causality, then

$$y = ku \tag{5.1}$$

Alternatively, if the component output causality is not the same as that of the default causality, then

$$y = \frac{1}{k} u \tag{5.2}$$

A Reduce implementation is

```
OPERATOR Lin;
FOR ALL
  Causality, DefaultCausality, Gain, Input
SUCH THAT
  Causality = DefaultCausality
LET Lin(Causality, DefaultCausality, Gain, Input)
= Gain*Input;

FOR ALL
  Causality, DefaultCausality, Gain, Input
SUCH THAT
  Causality NEQ DefaultCausality
LET Lin(Casuality, DefaultCausality, Gain, Input)
= (1/Gain)*Input;
```

Thermal Capacity

In the context of heated fluids, the thermal capacity CR involves an additional input from the hydraulic domain—the mass of the fluid involved. A Reduce representation is

```
%Thermal capacity
OPERATOR Thermal_CAPACITY:

FOR ALL cp, Density, Enthalpy, Volume
LET Thermal_Capacity(Causality, cp, Density, Input, Volume)
= lin(Causality, effort, 1/(cp*Volume*Density), Input);
```

5.4 TRANSFORMATIONS

The *representations* used by MTT (discussed in Section 5.3) are *transformed* using a set of UNIX tools. Each such tool is encapsulated in a UNIX Bourne shell script. Table 5.4 shows some of the transformations that have been implemented, together with the implementation language and the additional files required. The first three characters give the code for the input representation; the last three characters give the code for the output representation.

The dependency tree of the representations, together with the transformations between them, is another Bourne shell script invoking the UNIX "make" facility. This provides a simple and effective way of accessing the transformation tools. The user just asks for the representation required, and "make"

Table 5.4   Model Transformations

Code	Language	Additional files
fab2pab	C	lbl
pab2pcb	Prolog	
pab2tsu	Prolog	
pcb2req	Prolog	
req2rse	Reduce	rcr, rpa
rse2fse	Reduce	
fse2mse	sed	mpa
rse2ree	Reduce	rcr, rpa
ree2rsm	Reduce	rcr, rpa
rsm2fsm	Reduce	
fsm2msm	sed	mpa
rsm2tsm	Reduce	
rsm2rtf	Reduce	
rtf2ttf	Reduce	

automatically determines the transformations required given the revision history of the files containing the representations.

We have used appropriate languages at each stage; thus, for example, the essentially logical step of causal completion is implemented in Prolog, whereas symbolic algebra makes use of the Reduce package. Alternatives would, of course, be possible; for example, Lisp could have been used in place of Prolog, and Maple or Mathematica in place of Reduce.

5.5 GRAPHICAL INTERFACE

At the moment, the system bond graph is entered directly by the user using a standard graphics editor "fig" (available under suntools) or "xfig" (available under X11).

The bond graph is drawn in the conventional way using half arrows to represent bonds and full arrows to represent signals. Examples appear in Figures 5.2, 5.7, and 5.9. Casual strokes can be added to any bond, although a transformation is provided to complete causality where necessary. This is stored in the fab file of Table 5.1.

As the bond graph has to be automatically interpreted by computer, it must be precise and unambiguous. To this end, components and junctions are described as TYPE:LABEL, where TYPE is the code from Table 5.2 or Table 5.3, and LABEL is a string of characters. The LABEL string is looked up in a separate file (lbl of Table 5.1) and hence linked to the appropriate constitutive relation (linear or nonlinear) described in the corresponding rcr file of Table 5.1. An example appears in Figure 5.3. Within the lbl file, the first string on each line is the label, the second the name used in the derived description (pab of Table 5.1), the third the name of the constitutive relation (in the rcr file of Table 5.1), and the fourth supplies appropriate (symbolic) parameters.

The diagram must also be complete, so some additional notation has been provided. Measurements are indicated with a full arrow, modulating signals with a dotted arrow.

5.6 EXAMPLES

This section provides three illustrative examples of the use of MTT in system modeling, the first two examples are a linear dc motor, the third is a (nonlinear) process system. More substantial robotic examples appear in another paper [7].

5.6.1 Example 1: dc Motor

Description

A dc motor converts power from the electrical to the mechanical domain (or vice versa). This simple model includes armature inductance and resistance as well

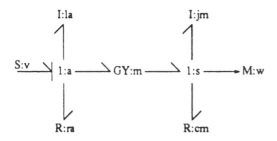

Figure 5.2 dc motor: bond graph.

as inertia and friction. In this example, the acausal bond graph is supplied by the user, via the xfig utility (Figure 5.2), together with the corresponding lbl file (Figure 5.3).

The bond graph (Figure 5.2) has two junctions: an electrical junction with common current (labeled armature) and a mechanical junction with common velocity labeled shaft. A voltage source is assumed.

For example, consider the following two representations:

- Textual transfer function (ttf)
- Matlab state-space matrices (msm)

The bond graph was entered with xfig and the label file with xedit. The mtt commands on the first line of Figure 5.4 were issued. The resulting transformations appear on the remaining lines. Notice that the second mtt command only invokes transformations not already accomplished by the first command.

The pab representation of the dc motor is given in Figure 5.5.

Transfer Function: Textual Symbolic Form
The file resulting from the first mtt command is included; intermediate representations are also given for completeness.

State Equations: tse

$$x = \begin{pmatrix} \lambda \\ h_m \end{pmatrix}; \quad y = (\omega); \quad u = (v) \tag{5.3}$$

$$\dot{x}_1 = \frac{-(r_a j_m x_1 - j_m l_a u_1 + l_a k_m x_2)}{j_m l_a} \tag{5.4}$$

$$\dot{x}_2 = \frac{j_m k_m x_1 - l_a c_m x_2}{j_m l_a} \tag{5.5}$$

$$y_1 = \frac{x_2}{j_m} \tag{5.6}$$

State Matrices: tsm

```
dc         %System name

%%%%%%%%%%%%%%%%%%%%%%%%%%%%%%%%%%%%%%%%%%%%%%
%%%%% Model Transformation Tools: Example %%%%%
%%%%%%%%%%%%%%%%%%%%%%%%%%%%%%%%%%%%%%%%%%%%%%

% Labels for dc motor
% File dc.lbl

%System name

%Input voltage source
v         v

%The two junctions
a         armature
s         shaft

%Armature components
ra        resistance       lin        [effort,r_a]
la        lambda           lin        [state,l_a]

%Shaft components
cm        friction         lin        [effort,c_m]
jm        h_m              lin        [state,j_m]


%Motor gain
m                          lin        [effort,k_m]

%Output velocity
w         omega      unity           []
```

Figure 5.3 The file dc.lbl.

$$A = \begin{pmatrix} \dfrac{-r_a}{l_a} & \dfrac{-k_m}{j_m} \\ \dfrac{k_m}{l_a} & \dfrac{-c_m}{j_m} \end{pmatrix}$$

(5.7)

```
bill% mtt dc ttf ; mtt dc msm
fab2pab dc
pab2pcb dc
pcb2req dc
reduce_separator <dc.req >dc.req_
req2rse dc; tidy dc.rse
reduce_separator <dc.rse >dc.rse_
rse2rsa dc; tidy dc.rsa
reduce_separator <dc.rsa >dc.rsa_
rsa2ree dc; tidy dc.ree
reduce_separator <dc.ree >dc.ree_
rss2rsv dc; tidy dc.rsv
reduce_separator <dc.rsv >dc.rsv_
ree2rsm dc; tidy dc.rsm
reduce_separator <dc.rsm >dc.rsm_
rsm2rtf dc; tidy dc.rtf
reduce_separator <dc.rtf >dc.rtf_
rtf2ttf dc; latex_tidy <dc.ttf >dc_ttf.tex
rsm2fsm dc
fsm2msm dc; Matlab_tidy  <dc.msm >dc_msm.m;
Matlab_tidy  <dc.mpa >dc_mpa.m
```

Figure 5.4 Invoking mtt.

$$B = \begin{pmatrix} 1 \\ 0 \end{pmatrix} \tag{5.8}$$

$$C = \left(0 \, \frac{1}{j_m} \right) \tag{5.9}$$

$$D = (0) \tag{5.10}$$

Transfer Function: ttf

$$G(s) = \frac{k_m}{(k^2_m + r_a c_m) + (l_a c_m + j_m r_a)s + l_a j_m s^2} \tag{5.11}$$

Transfer Function: Textual Numerical Form

The following text was generated using Matlab in conjunction with mtt. Note that state matrices are used as the mtt/Matlab interface. The Matlab "echo" function is used to expose the contents of intermediate files: "dc_msm.m" is generated by mtt (see Figure 5.4); "dc_mpa.m" contains the numerical values.

```
system(dc_motor,
[
  junction(shaft, flow,
  [
    bond(friction,dissipator,out,_,[lin,[effort,cm],[]]),
    bond(h_m,fstore,out,_,[lin,[state,jm],[]]),
    bond(armature,gyrator,in,_,[lin,[effort,k_m],[]]),
    bond(omega,sensor,out,flow,[unity,[],[]])
  ]),

  junction(armature, flow,
  [
    bond(v,source,in,effort,[la,1_1,lin]),
    bond(l_1,fstore,out,_,[lin,[state,l_a],[]]),
    bond(resistor,dissipator,out,_,[lin,[effort,r_a],[]]),
    bond(shaft,gyrator,out,_,[lin,[effort,k_m],[]])
  ])
]).
```

Figure 5.5 The pab representation of the dc motor.

```
>> dc
!rm *.met
% Matlab script for dc motor example.
%File: dc.m

%Create state matrices
!mtt dc msm

%Set up some parameters
dc_mpa
%parameters
r_a = 0.1;
l_a = 0.01;
k_m = 0.1;
j_m = 1.0;
c_m = 0.01;
k_t = 1.0;

%Read in the A,B,C and D matrices
dc_msm
%state matrices for system dc;
%file dc_msm.m;
%generated by mtt;
A(1,1)=-c_m/j_m;
```

```
A(1,2) = k_m/l_a;
A(2,1) = -k_m/j_m;
A(2,2) = -r_a/l_a;
B(1,1) = 0;
B(2,1) = 1;
C(1,1) = 1/j_m;
C(1,2) = 0;
D(1,1) = 0;

%Find the system transfer function
[dc_num, dc_den] = ss2tf(A,B,C,D,1)

dc_num =
    0    0    10
dc_den =
    1.000    10.0100    1.1000
```

Transfer Function: Graphical Form

Figure 5.6 was generated using Metlab and based on the state-space matrices generated by MTT.

5.6.2 Example 2: dc Motor with Current Input

This example is identical to that of Section 6.1 except that the motor is driven from a current, not a voltage source. The corresponding bond graph is given in Figure 5.7 and has one difference from Figure 5.2: The causal stroke on the source is moved. In contrast, the corresponding block diagram, or signal flow graph, would be quite different due to the changed causality.

Transfer Function: Textual Symbolic Form

The following equations were again automatically generated using MTT. Intermediate representations are given for completeness. Notice that the system is now first order.

State Equations: tse

$$x = (h_m); \qquad z = (l_l); \qquad y = (\omega); \qquad u = (i) \tag{5.12}$$

$$\dot{x}_1 = \frac{-(c_m x_1 - j_m k_m u_1)}{j_m} \tag{5.13}$$

$$y_1 = \frac{x_1}{j_m} \tag{5.14}$$

$$z_1 = l_a u_1 \tag{5.15}$$

State Matrices: tsm

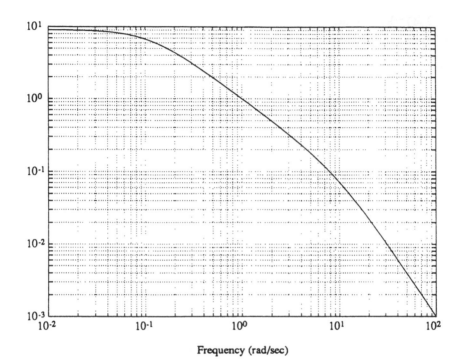

Figure 5.6 dc motor: frequency response magnitude.

$$A = \left(\frac{-c_m}{j_m}\right)$$ (5.16)

$$B = (k_m)$$ (5.17)

$$C = \left(\frac{1}{j_m}\right)$$ (5.18)

$$D = (0)$$ (5.19)

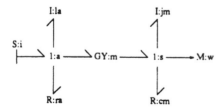

Figure 5.7 dc motor (current source): bond graph.

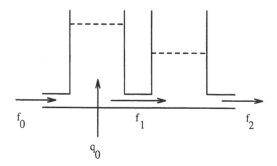

Figure 5.8 Stirred-tank heaters: physical system.

Transfer Function: ttf

$$G(s) = \frac{(k_m)}{sj_m + c_m}$$

(5.20)

5.6.3 Example 3: Two Stirred-Tank Heaters

Description

With reference to Figure 5.8, two uniform tanks of nonvolatile incompressible liquid are connected by a short pipe with significant resistance. The first tank is fed by a flow of liquid with flow rate $f_0 m^3 s^{-1}$ and temperature T_0 degrees Celsius above ambient, the liquid flows out of the second tank driven by the pressure in that tank.

Heat is put into the first tank via a heater at the rate of $q_o W$. For simplicity, the three flow resistances are taken to be linear and of equal value.

$$r_0 = r_1 = r_2 = r$$

(5.21)

This is represented using pseudo-bond-graphs [8,9]. The upper half of the bond graph (Figure 5.9) corresponds to the hydraulic properties (pressure and mass flow); the lower half correspond to the thermal properties (temperatures and enthalpy flow). There are implicit connections from the upper to the lower R and C components (described in the .lbl file). Temperature is a transport property (it is carried along by the liquid) and signal bonds are used to indicate the lack of interaction with downstream temperatures. The *modulated* thermal R components represent the nondynamic relation that

$$\hbar = (c_p \dot{m})T$$

That is, the enthalpy flow $\hbar$ is the product of the effort T and the "conductance" $c_p \dot{m}$. The mass flow $\dot{m}$ thus modulates this "conductance."

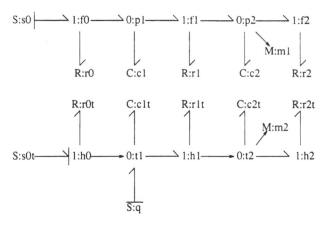

Figure 5.9 Stirred-tank heaters: bond graph.

State Equations: Textual Symbolic Form

The following equations were automatically generated using MTT.

$$x = \begin{pmatrix} m_1 \\ m_2 \\ h_1 \\ h_2 \end{pmatrix} ; \ y = \begin{pmatrix} p_2 \\ t_2 \end{pmatrix} ; \ u = \begin{pmatrix} f_0 \\ t_0 \\ q \end{pmatrix} \tag{5.23}$$

$$\dot{x}_1 = \frac{-((x_1 - x_2)g - ra u_1)}{ra} \tag{5.24}$$

$$\dot{x}_2 = \frac{(x_1 - 2x_2)g}{ra} \tag{5.25}$$

$$\dot{x}_3 = \frac{((c_p u_1 u_2 + u_3)ra - gx_3)x_1 + gx_2 x_3}{ra x_1} \tag{5.26}$$

$$\dot{x}_4 = \frac{((x_3 - x_4)x_1 - x_2 x_3)g}{ra x_1} \tag{5.27}$$

$$y_1 = \frac{gx_2}{a} \tag{5.28}$$

$$y_2 = \frac{x_4}{c_p x_2} \tag{5.29}$$

The system has four states: the mass and enthalpy of the liquid in each tank. The first two state equations correspond to the hydraulics and are linear; the third and fourth state equations, reflecting the thermal properties, are nonlinear due to the interaction with hydraulics.

Linearized Equations: Textual Symbolic Form

The steady state was computed for constant inputs f_0, q_0, and $t_0 = 0$. Linearizing the nonlinear equations about this steady state, the corresponding A, B, C, and D matrices are:

$$A = \begin{pmatrix} \dfrac{-g}{ra} & \dfrac{g}{ra} & 0 & 0 \\[2ex] \dfrac{g}{ra} & \dfrac{-2g}{ra} & 0 & 0 \\[2ex] \dfrac{-(gq_0)}{2raf_0} & \dfrac{gq_0}{raf_0} & \dfrac{-g}{2ra} & 0 \\[2ex] \dfrac{gq_0}{2raf_0} & \dfrac{-gq_0}{raf_0} & \dfrac{g}{2ra} & \dfrac{-g}{ra} \end{pmatrix} \tag{5.30}$$

$$B = \begin{pmatrix} 1 & 0 & 0 \\ 0 & 0 & 0 \\ 0 & f_0 c_p & 1 \\ 0 & 0 & 0 \end{pmatrix} \tag{5.31}$$

$$C = \begin{pmatrix} 0 & \dfrac{g}{a} & 0 & 0 \\[2ex] 0 & \dfrac{-gq_0}{raf_0^2 c_p} & 0 & \dfrac{g}{raf_0 c_p} \end{pmatrix} \tag{5.32}$$

$$D = \begin{pmatrix} 0 & 0 & 0 \\ 0 & 0 & 0 \end{pmatrix} \tag{5.33}$$

Note that f_o and q_o remain as symbols at this stage, and the transfer functions are

$$G_{11}(s) = \frac{g^2 r}{g^2 + 3agrs + a^2 r^2 s^2} \tag{5.34}$$

$$G_{12}(s) = 0 \tag{5.35}$$

$$G_{13}(s) = 0 \tag{5.36}$$

$$G_{21}(s) = \frac{-g^2 q_0 (c_p r_s f_0 + 1)}{g^2 c_p f_0^2 (c_p r_s f_0 + 1) + (3agrc_p f_0^2 (c_p r_s f_0 + 1))s + 2a^2 r^2 c_p^2 r_s f_0^3 s^2} \tag{5.37}$$

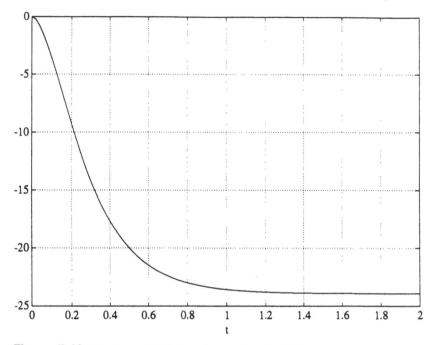

Figure 5.10 Nonlinear simulation: temperature deviation.

$$G_{22}(s) = \frac{g^2(c_p r_t f_0 + 1)}{g^2(c_p r_t f_0 + 1) + (3agr(c_p r_t f_0 + 1))s + 2a^2 r^2 c_p r_t f_0 s^2} \quad (5.38)$$

$$G_{23}(s) = \frac{g^2(c_p r_t f_0 + 1)}{g^2 c_p f_0 (c_p r_t f_0 + 1) + (3agr c_p f_0 (c_p r_t f_0 + 1))s + 2a^2 r^2 c_p^2 r_t f_0^2 s^2} \quad (5.39)$$

Simulation

Using files automatically generated by MTT, the full nonlinear system was simulated using the metlab "ode23" integrator and starting at the steady-state states and with a 10% increase in flow. Figure 5.10 shows the corresponding *deviation* in temperature. In addition, the linearized system was also simulated; the solid line represents the nonlinear response, and the dotted line, the linearized response.

5.7 CONTROL APPLICATIONS

One of the many uses of system modeling is an aid to control system design. At one level, the generation of frequency responses, such as that displayed in Figure 5.6, is such an aid.

However, automated symbolic system modeling has more power than this. Some examples of this currently under investigation are as follows:

- Automatic generation of inverse systems [10]
- Automatic generation of system-specific controller structures [11]
- Automatic generation of code for partially known system identification [12,13]
- Automatic generation of code for partially known system adaptive control [14]

5.8 CONCLUSION

An approach to modeling viewed as a set of transformations between representation has been advocated, and its implementation illustrated.

An experimental implementation of this approach, the model transformation toolbox (MTT) has been described and it use illustrated by three examples. Our experience is that bond graphs do provide a good core representation for describing dynamical systems.

ACKNOWLEDGMENTS

This work is supported by SERC through the Engineering Design Research Center, 1 Todd Campus, West of Scotland Science Park, Maryhill Road, Glasgow, G20 OXA.

Dr. Donald Ballance suggested the use of UNIX tools as a basis for MTT and, along with other members of the Control Group, contributed to the genesis of ideas incorporated in this chapter.

The chapter is based on a paper [15] appearing in an IFAC workshop.

REFERENCES

1. Gawthrop, P. J., and Smith, L., An environment for Specification, Design, Operation, Maintenance, and Revision of Manufacturing Control Systems, *Proceedings of UKIT90*, 1990.
2. Wellstead, P. E., *Introduction to Physical System Modeling*, Academic Press, New York, 1979.
3. Shearer, J. L., Murphy, A. T., and Richardson, H. H., *Introduction to System Dynamics*, Addison-Wesley, Reading, MA, 1971.
4. Karnopp, D. C., and Rosenberg, R. C., *System Dynamics: A Unified Approach*, John Wiley, New York, 1975.
5. Thoma, J., *Introduction to Bond Graphs and their Applications*, Pergamon Press, New York, 1975.

6. Rosenberg, R. C., and Karnopp, D. C., *Introduction to Physical System Dynamics*, McGraw-Hill, New York, 1983.
7. Gawthrop, P. J., Bond graphs: A representation for Mechatronic Systems, *Mechatronics*, *1*(2), 127–156 (1991).
8. Karnopp, D. C., Pseudo bond graphs for thermal energy transport, *J. Dynamic Syst.*, *Measurement and Control*, *100*, 165–169 (1978).
9. Karnopp, D. C., State variables and pseudo bond graphs for compressible thermo-fluid systems, *J. Dynam. Syst. Meas. Contr.*, *101*, 201–204 (1979).
10. Gawthrop, P. J., and Smith, L., Inverse systems: Bond graph and descriptor representations, *J. Franklin Inst.* (submitted).
11. Gawthrop, P. J., Jones, R. W., and MacKenzie, S. A., Bond graph based control: A process engineering example, *American Control Conference*, 1992.
12. Gawthrop, P. J., Jones, R. W., and MacKenzie, S. A., Identification of partially-known systems, *9th IFAC/IFORS symposium on identification and system parameter estimation, Budapest, Hungary*, 1991, pp. 1347–1352.
13. Gawthrop, P. J., Jones, R. W., and MacKenzie, S. A., Identification of partially-known systems, *Automation*, (to appear).
14. Gawthrop, P. J., Bond-graph-based adaptive control, *4th IFAC Symposium on Adaptive Systems in Control and Signal Processing*, 1992.
15. Gawthrop, P. J., Marrison, N. A., and Smith, L., MTT: A bond graph toolbox, *Proceedings of the 5th IFAC/IMACS Symposium on Computer-aided Design of Control Systems: CADCS91*, 1991.

6

Identification of Linear Systems

Lennart Ljung

Linköping University
Linköping, Sweden

6.1 INTRODUCTION

System identification is the art and science of building mathematical models of dynamical systems using observed data. The theory and methodology is by now well-established and in its mainstream edition it is well-founded in the methods of mathematical statistics. See, e.g., [8,16]. Current theoretical development also includes several optional routes that do not originate from statistics; see, e.g., [5,10,17,21].

In any case, the development of a model from data is actually quite a demanding numerical and computational task. It is, thus, natural that early on in the history of system identification, substantial efforts went into development of algorithms and software. One can rather easily distinguish between three generations of such software.

The first generation: "The batch routine period." During the 1960s, many (Fortran) subroutines were written for the estimation of parameters in different model structures. This development took place at many universities and research institutes in parallel, e.g., [2]. The subroutines were used in a batch mode, and the obtained models had to be subjected to analysis using other batch programs.

The second generation: "The interactive identification package period." The next step started around 1970 when the different subroutines for model estimation and model analysis were collected together with a better user interface and direct graphical facilities (much like the way the LINPAC and EISPAC Fortran subroutines were linked together into mainframe MATLAB). One of the first packages of this character was IDPAC [1,18–20], which was commercialized around 1974. Among several other packages of this character, we may mention The System Identification Toolbox (SITB) [9], MATRIX$_x$ [11], and PIM [6]. The help facilities in this generation are syntax oriented. One might say that the step from the first generation to the second generation meant a tremendous increase in efficiency for the user, but from a conceptional point of view, it was only just that: More user-friendly and efficient than the first generation numerical routines.

The third generation: "Identification packages with decision support and advanced help." The development of new software techniques has opened up the possibilities to offer more than just passive numerical computations and plotting from an identification package. The third generation will be characterized by knowledge-based-system (KBS) techniques of various kinds. Although this generation has not yet really been commercialized, substantial experimentation in the area has been reported, e.g., [3,4,7,12–15]. A survey is given in [22].

It is the purpose of this chapter to describe and discuss features, algorithms, and user interface primarily for the second-generation identification software. We shall also discuss desired features of the third generation software and report on some experimentation in that area.

6.2   SYSTEM IDENTIFICATION

The problem to build a mathematical model of a process based on measured data is characterized by three main ingredients:

- The *data* collected from the process
- A *model structure*, i.e., a set of possible mathematical descriptions of the process
- A *selection criterion*, by which a particular member of the set of candidate models is selected, guided by the data.

Once these three items are at hand, we have, at least implicitly, defined the *model* as that member in the model structure that best describes the data, according to the selection criterion. The actual computation of the model may be quite laborious.

The procedure is not finished when the model has been computed. A fourth ingredient in system identification is most important:

- *Model validation*, in which we twist and turn the model to examine and evaluate its properties to see if it will be useful for our intended application.

Often the result is that the model does not meet the requirements, and we must revise our choice of model structure and/or work further on the data.

In the next two sections we shall discuss model structures and selection criteria in more detail. There are, however, two specific techniques which deal with the broad model structure:

Model structure = all linear systems (6.1)

A general, linear, time-invariant system is uniquely described by its *impulse response* $g(k)$;

$$y(t) = \sum_{k=1}^{\infty} g(k)u(t - k)$$ (6.2)

Here $y(t)$ is the output at time t and $u(t)$ is the input at time t. We assume that the observed data have been sampled at equally spaced sampling times and that the sampling interval is one time unit.

A common way to estimate the coefficients $g(k)$ directly from u and y is *correlation analysis*. This means that u and y are first filtered through the same filter giving u_F and y_F. The filter is chosen so that u_F becomes as white as possible. Then the cross-covariance function for the filtered data $Ry_Fu_F(k) = Ey_F(t + k)u_F(t)$ is estimated in a straightforward way. After proper scaling, this value is the estimate of the impulse response coefficient $g(k)$.

The system (6.2) can be represented equally well by its *frequency function* $G(e_{i\omega})$ which is the Fourier transform of the impulse response:

$$G(e^{i\omega}) = \sum_{k=1}^{\infty} g_k e^{-ik\omega}$$ (6.3)

Spectral analysis is a common and useful method to directly estimate the frequency function G. This estimate is essentially formed as the ratio between the cross spectrum estimate between y and u and the auto-spectrum estimate for u. See, e.g., Chapter 6 in [8].

6.3 MODEL STRUCTURES

The model structures come in two different styles. They are:

Ready-made model structures. These are blackbox parameterizations, such as difference equations or canonical form state-space models, whose parameters are just vehicles to obtain flexible model descriptions, without their own physical interpretations.

Tailor-made model structures. These are parameterizations based on physical modeling of the object, where the parameters correspond to unknown values of physical constants that appear in the descriptions. Most often, these models are given in state-space form, linear or nonlinear.

We shall generically denote a model structure by

$$\hat{y}(t \mid \theta); \quad \theta \in \mathcal{D}_\mathcal{M} \tag{6.4}$$

indicating that $\hat{y}(t|\theta)$ is the predicted, or "guessed," value of the output vector $y(t)$, according to the model with parameter θ and based on available information about inputs and outputs up to time $t - 1$. See, e.g., [8] for many examples of model structures.

6.3.1 Ready-made Structures

A typical ready-made structure is the ARX structure

$$
\begin{aligned}
y(t) &+ a_1 y(t - 1) + \cdots + a_{n_a} y(t - n_a) \\
&= b_1 u(t - n_k) + \cdots + b_{n_b} u(t - n_k - n_b + 1) + e(t)
\end{aligned} \tag{6.5}
$$

where $u(t)$ and $y(t)$ denote the input and output, respectively, at time t. The natural predictor for this model is

$$
\begin{aligned}
\hat{y}(t|\theta) &= -a_1 y(t - 1) - \cdots - a_{n_a} y(t - n_a) \\
&\quad + b_1 u(t - n_k) + \cdots + b_{n_b} u(t - n_k - n_b + 1)
\end{aligned} \tag{6.6}
$$

where the parameters are

$$\theta = (a_1 \cdots a_{n_a} b_1 \cdots b_{n_b})^T \tag{6.7}$$

Another common structure is the ARMAX structure

$$
\begin{aligned}
y(t) &+ a_1 y(t - 1) + \cdots + a_{n_a} y(t - n_a) \\
&= b_1 u(t - n_k) + \cdots + b_{n_b} u(t - n_k - n_b + 1) \\
&\quad + e(t) + c_1 e(t - 1) + \cdots + c_{n_c} e(t - n_c)
\end{aligned} \tag{6.8}
$$

where $\{e(t)\}$ denotes disturbances that are modeled as white noise. The corresponding predictor $\hat{y}(t|\theta)$ is then obtained from

$$
\begin{aligned}
\hat{y}(t|\theta) &+ c_1 \hat{y}(t - 1|\theta) + \cdots + c_{n_c} \hat{y}(t - n_c|\theta) \\
&= b_1 u(t - n_k) + \cdots + b_{n_b} u(t - n_k - n_b + 1) \\
&\quad + (a_1 - c_1) y(t - 1) + \cdots + (a_{n^*} - c_{n^*}) y(t - n^*)
\end{aligned} \tag{6.9}
$$

(n^* is the maximum of n_a and n_c. "Missing" a_i and c_i are replaced by zero.) A third common model structure is the *output error* (OE) structure

$$y(t) = \eta(t) + e(t) \tag{6.10}$$

$$\eta(t) + f_1\eta(t - 1) + \cdots + f_{n_f}\eta(t - n_f)$$
$$= b_1 u(t - n_k) + \cdots + b_{n_b} u(t - n_k + n_b + 1) \tag{6.11}$$

The predictor for this structure is

$$\hat{y}(t|\theta) = \eta(t,\theta) \tag{6.12}$$

$$\theta = (f_1 \cdots f_{n_f} b_1 \cdots b_{n_b})^T$$

where η is computed from the input u as

$$\eta(t,\theta) + f_1\eta(t - 1,\theta) + \cdots + f_{n_f}\eta(t - n_f, \theta)$$
$$= b_1 u(t - n_k) + \cdots + b_{n_b} u(t - n_k - n_b + 1) \tag{6.13}$$

There are several other variants of model structures within this family of difference equations, parameterized in terms of difference-equation coefficients, such as the Box–Jenkins (BJ) structure, the ARARMAX structure, etc. See, e.g., [8], Chapter 4, for some examples. They can all be summarized by the generic model

$$A(q)y(t) = \frac{B(q)}{F(q)} u(t) + \frac{C(q)}{D(q)} e(t) \tag{6.14}$$

where $A(q)$, etc., are polynomials in the delay operator $q^{-1}(q^{-1}u(t) = u(t - 1))$:

$$A(q) = 1 + a_1 q^{-1} + \cdots + a_{n_a} q^{-na} \tag{6.15}$$

6.3.2 Tailor-made Model Structures

Tailor-made model structures are most often of the form

$$\dot{x}(t) = A(\theta)x(t) + B(\theta)u(t) + w(t)$$
$$y(t) = C(\theta)x(t) + D(\theta)u(t) + e(t) \tag{6.16}$$
$$x(0) = x_0(\theta)$$

where $\{w(t)\}$ and $\{e(t)\}$ are noise sequences with certain (parameterized) covariance matrices. In these cases, the parameter vector θ will contain unknown physical constants, and the components of the state vector $x(t)$ will have explicit physical significance. The user will arrive at Eqs. (6.16) after some phase of physical modeling. In this case, the predictor $\hat{y}(t|\theta)$ is found from Eqs. (6.16) by applying the Kalman filter with sampled observations of $y(t)$ and $u(t)$. A useful alternative is to parameterize directly the system in innovations form

$$\dot{x}(t) = A(\theta)x(t) + B(\theta)u(t) + K(\theta)e(t)$$
$$y(t) = C(\theta)x(t) + D(\theta)u(t) + e(t) \tag{6.17}$$
$$x(0) = x_0(\theta)$$

For a discrete-time system description, the corresponding model is

$$x(t + 1) = F(\theta)x(t) + G(\theta)u(t) + \bar{K}(\theta)e(t)$$
$$y(t) = C(\theta)x(t) + D(\theta)u(t) + e(t) \qquad (6.18)$$

which gives a particularly simple form for the predictor based on sampled data:

$$\hat{y}(t|\theta) = C(\theta)\hat{x}(t,\theta) + D(\theta)u(t) \qquad (6.19)$$

where

$$\hat{x}(t + 1,\theta) = (F(\theta) - \bar{K}(\theta)C(\theta))x(t,\theta) + (G(\theta) - \bar{K}(\theta)D(\theta))u(t)$$
$$+ \bar{K}(\theta)y(t) \qquad (6.20)$$

In practice, the predictor for Eqs. (6.16) or (6.17) is usually implemented as Eqs. (6.19) and (6.20), where $F(\theta)$, $G(\theta)$, and $\bar{K}(\theta)$ are computed from $A(\theta)$, $B(\theta)$, and $K(\theta)$ [or the covariance matrices for $w(t)$ and $e(t)$] and the sampling interval in a well-defined but somewhat complicated manner.

6.4 THE SELECTION CRITERION

The criterion by which a particular model is a model structure is selected in the light of the data may come in several different shapes. "Mainstream" identification uses a scalar measure of fit between the predicted and measured outputs:

$$\theta_N = \arg \min_{\theta \in D_M} \frac{1}{N} \sum_{t=1}^{N} \ell(y(t) - \hat{y}(t|\theta)) \qquad (6.21)$$

where "*arg min*" means the minimizing argument and $\ell(\epsilon)$ is a positive, scalar-valued function, such as

$$\ell(\epsilon) = |\epsilon|^2 \qquad (6.22)$$

(but typically modified so as to increase slower than quadratically for large values of ϵ).

Behind the innocent-looking "arg min" in Eq. (6.21), substantial computation may hide. Most of the effort of the first-generation identification software went into algorithmitizing and programming this mapping. A leading idea in the second-generation software is to "trivialize" the mapping (6.21) for the user: that is, the user names a model structure and—snap!—the computer returns the best model, $\hat{\theta}_N$, in this structure. This can be achieved quite well for ready-made model structures, but poses a problem for tailor-made ones. The reason is that the minimizing procedure often needs human interaction due to somewhat "tricky" shapes of the surface to be minimized.

The typical way to perform the minimization of Eq. (6.21) is the *Gauss–Newton iterative procedure*. This is given by

$$\hat{\theta}_N^{(i+1)} = \hat{\theta}_N^{(i)} + \mu R_N^{-1} \frac{1}{N} \sum_{t=1}^{N} \psi(t, \hat{\theta}_N^{(i)}) \ell'(\varepsilon(t, \hat{\theta}_N^{(i)})) \tag{6.23}$$

$$R_N = \frac{1}{N} \sum_{t=1}^{N} + \psi(t, \hat{\theta}_{N^{(i)}}) \ell''(\varepsilon(t, \hat{\theta}_{N^{(i)}})) \psi^T(t, \hat{\theta}_{N^{(i)}})$$

Here

$$\varepsilon(t, \theta) = y(t) - \hat{y}(t|\theta)$$
$$\psi(t, \theta) = \frac{d}{d\theta} \hat{y}(t|\theta)$$

and the prime and double prime in ℓ', ℓ'' denote differentiation once and twice w.r.t. ε. See, e.g., [8], Chapter 10, for more details.

The step size μ in Eqs. (6.23) is chosen so that criterion (6.21) strictly decreases.

An important application is when the estimates have to be computed online, as the measurements are made. This means that at time $N + 1$, $\hat{\theta}_{N+1}$ should be computed using the information already condensed in $\hat{\theta}_N$. The archetypical algorithm for this is obtained from Eq. (6.23) as

$$\hat{\theta}(N + 1) = \hat{\theta}(N) + \mu_N R^{-1}(N) \psi(t) \ell'(\varepsilon(t))$$
$$R(N + 1) = R(N) + \mu_N(\psi(t) \ell''(\varepsilon(t)) \psi^T(t) - R(N)) \tag{6.24}$$

where $\varepsilon(t)$ is some approximation of $\varepsilon(t, \hat{\theta}(N))$ and $\psi(t)$ is some approximation of $\psi(t, \hat{\theta}(N))$. See [8], Chapter 11, for more details.

6.5 TYPICAL FEATURES OF THE SECOND-GENERATION IDENTIFICATION CAD PACKAGE

The identification process is characterized by a loop in which various model structures (and data-set variants) are tried out until an acceptable model is found. See Figure 6.1. The rectangular boxes in this figure denote typical duties for the software.

More specifically, existing software packages for system identification typically contain the following facilities:

(a) **Data handling**—plotting, filtering, detrending, "polishing"
(b) **Nonparametric identification methods**—special analysis, correlation analysis

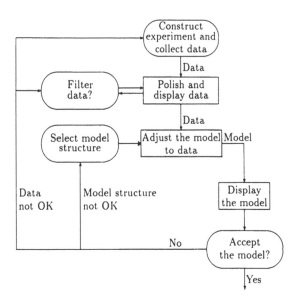

Figure 6.1   The system identification loop.

(c) **Defining model structures**
(d) **Parameter estimation**—computation of Eq. (6.21) for various parameter-
     ized model structures
(e) **Evaluation of models**—prediction, simulation, residual analysis, fre-
     quency response, zeros and poles, various transformations between differ-
     ent model representations
(f) **Support for comparisons of different models**

All commercial packages support all these items to various degrees. The main
differences between the packages lie in the command structure/user interface and
in what model structures are covered by (c) and (d). Virtually all packages support the ARX structure [6.5]; most also cover the ARMAX.

    Looking at Figure 6.1, we see that the packages support the user with the
tasks in square blocks, whereas for the tasks in ovals, the user is still on his/
her own. As we pointed out in Section 6.1, the value of this second-generation
software is primarily that the heavy numerical computational burdens (within
the squares of Figure 6.1) have been trivialized from the user's point of view.
Some monitoring and interaction of the minimization in Eq. (6.21) may still
be needed.

    The System Identification Toolbox (SITB) [9] supports, under (c) and (d)
above, all linear, time-invariant model structures of the kinds described in Section 6.3. How to define them in the user interface will be described in Section 6.7.

SITB also allows recursive (on-line) computation (6.24) of the parameters for ready-made model structures.

SITB also, under (e), supports all kinds of transformations of models between continuous time and discrete time, as well as to state space, transfer functions, zero pole, and frequency function representations.

Moreover, the SITB, under (f) above, allows efficient, simultaneous computation and comparison of many models of ARX type.

6.6 THE USER INTERFACE

The first interactive CAD system used to be *menu-driven*, i.e., the user had to select the action to be taken by answering a sequence of questions. Later, *command-driven* systems took over. In these, the user types the desired command directly, and often also decides upon a file-name for the result without any restrictions. This gives a considerable increase in freedom, but it also requires that the user has a good knowledge of the package. Lately, in window-oriented user interfaces with pull-down menus, there has been a renaissance of menu-driven systems in which the advantages of the two approaches can be combined.

The SITB is a command-driven package with a syntax that is inherited from MATLAB. The future may show a trend to complement it with pull-down menus (as in SIMULAB). This gives a fairly straightforward dialog as indicated by the following example, where y and u are column vectors containing the output and input data:

```
ze = [y(1:500) u(1:500)]       % the estimation data set
zv = [y(501:700) u(501:700)]   % the validation data set
m = arx(ze, [2,3,1])           % an arx model with
                               na = 2, nb = 3, nk = 1 (see (6.5))
compare (zv, m)                % simulate the model with
                               validation data and compare
                               model output with measured
                               output
g = th2ff(m)                   % the frequency function of the
                               arx model
gs = spa(ze)                   % the frequency estimation by
                               spectral analysis
bodeplot ([g,gs])              % comparing the Bode plots of g
                               and gs.
```

An important aspect of the user interface of the SITB is *"the theta-format."* This is an object into which all relevant model information is coded. All models produced by the parameter estimation routines (such as arx for [6.5] and pem for general structures) are returned in the theta format. This contains, in addition to the estimated parameters, the covariance matrix, also information about how and when the model was created. The actual, internal representation depend on the character of the model, but this is concealed from the user. This means that all model transformations, such as

th2ff	computing frequency functions
th2zp	zeros and poles
th2th	transfer functions
th2ss	state-space matrices
th2arx	arx parameters
th2thd, thd2thc	transformations between discrete-time and continuous-time representations

have identical syntaxes, like

   [num, den] = th2tf(m)

regardless of the type of model. The actual code, though, typically branches into different subroutines for state-space models, blackbox models and multioutput models.

The same is true for other model applications, such as simulation ("y = idsim(u,m)") and *k*-step ahead prediction ("yp = predict(z,m,k)").

6.7 DEALING WITH ARBITRARY LINEAR MODEL STRUCTURES

6.7.1 Defining Structures

The most general way of defining a model structure like (6.17) is to write an *m*-file (a macro) of the structure

   [Ad, Bd, C, D, Kd, X0] = myname(pars, T, aux)

Here, the output arguments are the six matrices in the sampled data description (6.18), pars is a vector that contains the current values of the parameters θ,T is the sampling interval, and aux are some auxiliary variables that the user may want to work with (to try out various alternative structures within the same *m*-file, for example). The structure is then packed into the "theta-format" of the SITB (used for all model structures) by

   th = mf2th ('myname','c',nompar)

Here mf2th is mnemonic for "*m*-file to theta." The argument c indicates that the underlying parameterization refers to a continuous-time model (d is the alternative) and nompar are the nominal/initial parameter values. Once the description th is created, all the SITB functions for model transformation, model presentation, simulation, prediction, and estimation can be applied in an entirely transparent way.

For model structures with parameters that do not enter in several matrix entries (i.e., the parameter in, e.g., entry *ij* of the A_c matrix does not depend on the parameters in any other matrix entry), there is a simpler way to define model structures. The method is simply to define the matrices in (Eq. (6.17) or (6.18) explicitly, with NaN (MATLAB for "Not a Number") in the entries for unknown parameters. For example,

```
A = [1, 0; NaN, NaN]
B = [ NaN; NaN]
C = [1, 0;0, 1]
D = [0;0]
K = zeros(2,2)
x0 = [0;0]
```

The structure is then packed as

```
ms = modstruc(A,B,C,D,K,X0)
th = ms2th(ms, 'c', nompar)
```

Here ms2th is mnemonic for "model structure to theta." Alternatively, canonical form parameterizations can be defined by the command canform.

6.7.2 Estimating Models

Once the model structure is defined by mf2th or ms2th and output-input data is collected in a matrix z, estimation by (6.21) is straightforward

```
the = pem(z,th)
```

Here pem is for "prediction error method". The estimate the can now be scrutinized by all the commands of the SITB, such as th2ff (frequency functions), resid (residual analysis) compare (model comparison), present, th2zp (zeros & poles), th2ss (state space matrices) etc.

6.7.3 Manipulating Model Structures

Two commands

```
fixpar and unfixpar
```

allow that certain parameters in a structure are fixed to given values, or unfixed (left free to be estimated). The command

```
thinit
```

allows the manipulation of initial parameter values.

For canonical form structures, the command

    canstart

gives a special start-up procedure for estimating its parameters, to be used as initial values for pem.

6.7.4 An Example

The use of the software is illustrated by the following demo from SITB, ver. 3.0:

In this demo we shall demonstrate how to use several of the m-files that are contained in the state-space package. We shall investigate data produced by a (simulated) dc-motor. We first load the data:

    load dcm-data

The matrix y contains the two outputs: yl is the angular position of the motor shaft and y2 is the angular velocity. There are 400 data points and the sampling interval is 0.1 seconds. The input is contained in the vector u. It is the input voltage to the motor.

    z=[y u];
    idplot(z,[ ],0.1,2)

We shall build a model of the dc-motor. The dynamics of the motor is well known. If we choose xl as the angular position and x2 as the angular velocity it is easy to set up a state-space model of the following character neglecting disturbances: [see page 84 in [8]]:

$$d/dt\ x = \begin{vmatrix} 0 & 0 \\ 0 & -th1 \end{vmatrix} x + \begin{vmatrix} 0 \\ th2 \end{vmatrix} u$$

$$y = \begin{vmatrix} 1 & 0 \\ 0 & 1 \end{vmatrix} x$$

The parameter th1 is here the inverse time-constant of the motor and th2 is such that th2/th1 is the static gain from input to the angular velocity. [See [8] for how th1 and th2 relate to the physical parameters of the motor.] We shall estimate these two parameters from the observed data.

1. Free Parameters

We first have to define the structure in terms of the general description

$$d/dt\ x = Ax + Bu + Ke$$
$$y = Cx + Du + e$$

Any parameter to be estimated is entered as NaN (Not a Number). Thus we have

A=[0 1; 0 NaN];
B=[0;NaN];
C=[1 0; 0 1];
D=[0; 0];
K=[0 0; 0 0];
X0=[0; 0];

X0 is the initial value of the state vector; it could also be entered as parameters to be identified.

The model structure is now defined by

ms = modstruc(A,B,C,D,K,X0);

We shall produce an initial guess for the parameters. Let us guess that the time constant is one second and that the static gain is 0.28. This gives:

th-guess = [-1 0.28];

We now collect all this information into the "theta-format" by

dcmodel = ms2th(ms,'c',th-guess,[ ],0.1);

'c' denotes that the parameterization refers to a continuous-time model. The last argument, 0.1, is the sampling interval for the collected data. The prediction error (maximum likelihood) estimate of the parameters is now computed by

dcmodel = pem(z,dcmodel);

We can display the result by the 'present' command. The imaginary parts denote standard deviation.

present(dcmodel)

The estimated values of the parameters are quite close to those used when the data were simulated (-4 and 1). To evaluate the model's quality we can simulate the model with the actual input and compare it with the actual output.

compare(z,dcmodel);

The result is shown in Figure 6.2. We can now, for example, plot zeros and poles and their uncertainty regions. We will draw the regions corresponding to 10 standard deviations, since the model is quite accurate. Note that the pole at the origin is absolutely certain, since it is part of the model structure; the integrator from angular velocity to position.

zpplot(th2zp(dcmodel),10)

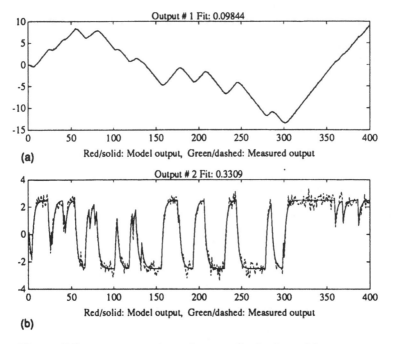

(a)

(b)

Figure 6.2 Simulated and actual outputs for the dc-model.

2. Coupled Parameters

Suppose that we accurately know the static gain of the dc-motor (from input voltage to angular velocity), e.g., from a previous step-response experiment. If the static gain is G, and the time constant of the motor is t, then the state-space model becomes

$$d/dt \, x = \begin{vmatrix} 0 & 1 \\ 0 & -1/t \end{vmatrix} x + \begin{vmatrix} 0 \\ G/t \end{vmatrix} u$$

$$y = \begin{vmatrix} 1 & 0 \\ 0 & 1 \end{vmatrix} x$$

With G known, there is a dependence between the entries in the different matrices. In order to describe that, the earlier used way with "NaN" will not be sufficient. We thus have to write an m-file which produces the A, B, C, D, K and X0 matrices as outputs, for each given parameter vector as input. It also takes auxiliary arguments as inputs, so that the user can change certain things in the model structure, without having to edit the m-file. In this case we let the known static gain G be entered as such an argument. The m-file that has been written has the name 'motor'.

```
type motor
function [A,B,C,D,K,X0] = motor(par,ts,aux)
```

This m-file describes the dc-motor with time constant t (= par) and known static gain G. The sampling interval is ts. The conversion to a sampled model is inhibited if ts is entered as a negative number. This is to allow for desirable features in the 'present' and 'th2ss' commands.

```
t = par;
G = aux(1);
A = [0 1;0 -1/t];
B = [0;G/t];
C = eye(2);
D = [0;0];
K = zeros(2,2);
X0 = [0;0];
if tx>0% Sample the model with sampling interval ts
s = expm ([[A B]*ts; zeros(1,3)]);
A = s(1:2,1:2);
B = s(1:2,3);
end
```

We now create a "THETA-structure" corresponding to this model structure: The assumed time constant will be

```
tc-guess = 1;
```

We also give the value 0.25 to the auxiliary variable G (gain) and sampling interval

```
aux = 0.25;
dcmm = mf2th ('motor','c',tc-guess,aux,[ ],0.1);
```

The time constant is now estimated by

```
dcmm = pem([y u], dcmm);
```

We have thus now estimated the time constant of the motor directly. Its value is in good agreement with the previous estimate.

```
present(dcmm)
```

With this model we may now proceed to test various aspects as before. The syntax of all the commands is identical to the previous case.

6.8   FUTURE ENHANCEMENTS

6.8.1   Experimentation into the Next-Generation Software for Identification

The rapid development in software techniques has opened up many new possibilities for better user support in all programming tasks. Let us see what features we could wish for when it comes to identification software! What are the shortcomings of the second generation?

First, let us stress that system identification/model building has a clear subjective component. Both the user's knowledge and intuitive insight about the system and the user's intended model application will have a strong influence on the modeling procedure and the resulting model. We could and should, therefore, not ask for completely automated procedures for system identification. It is *not* a goal to replace the user in this process, only to offer better support.

In Section 6.1, we mentioned several contributions that deal with the "third-generation software" for system identification. We shall here briefly comment on which problems they address.

One often distinguishes between the inexperienced and the experienced user and the different types of support these will need. The former's basic question is What shall I do next?, while the latter's question is What have I done so far? This calls for different approaches.

First, we shall point out three major problems in the practice of system identification. These are difficult both for the experienced and the inexperienced user.

1. *To select a suitable model structure*
2. *To select (and possibly polish) a suitable portion of the data for identification*
3. *To monitor and affect the minimization procedure [while effecting Eq. (6.21)].*

Problem (1) was the major concern of the ESPION project [4].

The experienced user will need support for the decision making in these three areas, primarily because there are many different indications to keep track of. The inexperienced user will, in addition, need some help in interpreting the results (like "What does it mean that the cross-correlation between residuals and inputs is above 0.6 at this value?). ID TOOL EXPERT [15] is intended to handle both these aspects of problem (1).

A fourth basic question is

4. *What shall I do next?*

The question contains a basic need for syntax support. This is available already in most second-generation packages. However, for the user that is inexperienced in the theories of identification and/or in the use of the particular software, more advanced support is desirable. It is important to distinguish between two branches of the question:

4a. *What shall I do next (based on what I have done so far)?*
4b. *What shall I do next (depending on the results obtained so far)?*

Both these branches are fairly tricky to handle. One main goal with the work of [7] is to deal with the first question, (4a). They use so-called *scripts* to match the sequence of applied commands to typical sequences used by experts. This support is primarily for the inexperienced user. However, the experienced user also will face question (4), mostly because he has lost track of all the implicit information in all the earlier produced result. The paper [14] reports on an approach to give support for question (4b).

A fifth question is

5. *What have I done so far?*

In fact, this rather trivial-looking question is often felt as the most pressing one in a long interactive computer session. There is a clear need for *executive summaries*. One main objective for ID TOOL [13] is to alleviate the bookkeeping burden from the user.

A sixth area where we should ask more of support from future generations of identification software is

6. *Closer ties to physical modeling.*

Physical modeling means that physical insight is used for the construction of mathematical models, most often having some parameters whose numerical values have to be determined from experiments. (See the discussion on tailor-made model structures in Section 6.2.) Since "modeling" and "identification" aim at solving the same problem, it is typically desirable to bring their respective tools closer together. BOND TOOL [15] is an approach to integrate bond-graph physical modeling with parameter estimation.

Finally, we should point out that the desired features described here must not hamper the user. As [7] stresses,

7. *The added software should be noninversive.*

That is, the user must feel that he is in full control of the process and will get directives only when asked for. It is also important, both for the experienced and the unexperienced user, that

8. *The software is transparent.*

That is, there must be explanations for the advice; it must be possible to inspect the knowledge of the system. It is also desirable that this knowledge could be expanded and maintained by the user since part of the relevant knowledge is application-dependent.

6.9 CONCLUSIONS

We have in this chapter outlined the system identification problem and its main algorithms. The area was an early experimentation ground for interactive software and there are now many packages available. We have used the System Identification Toolbox (SITB) [9] to illustrate a typical package with its basic functions and syntax.

A desired feature of such packages is that they should support as wide a class of model structures as possible; at the same time, the syntax for these different structures is uniform and transparent. This has been a main goal in the SITB.

The identification area is also an interesting proving ground for various ideas how to support the decision making and bookkeeping by the use of modern software tools.

REFERENCES

1. Åström, K. J., Computer-aided modeling, analysis and design of control system— a perspective, *Cont. Syst. Mag.*, *3*(2), 4–16 (1983).
2. Åström, K. J. and Bohlin, T., Numerical identification of linear dynamic systems, in *Theory of Self-Adaptive Control Systems*, P. H. Hammond, ed.), Plenum Press, New York, 1966.
3. Gentil, S., Barraud, A. Y., and Szafnicki, K., Sexi: An expert identification package, *Automatica*, *26*(4), 803–809 (1990).
4. Haest, M., Bastin, G., Gevers, M., and Wertz, V. Espion: An expert system for system identification, *Automatica*, *26*(1), 85–95 (1990).
5. Kalman, R. E., *Nine Lectures on Identification*. Springer-Verlag, New York, 1990.
6. Landau, I. D., *System Identification and Control Design Using P.I.M. + Software*, Prentice-Hall, Englewood Cliffs, NJ, 1990.
7. Larsson, J. E., and Persson, P., An expert system interface for an identification program, *Automatica*, *27*(6), 919–930 (1991).
8. Ljung, L., *System Identification, Theory for the User*, Prentice-Hall, Englewood Cliffs, 1987.
9. Ljung, L., *The System Identification Toolbox: The Manual, 3rd edition*, The Math-Works Inc., 1991.
10. Ljung, L., Issues in system identification, *IEEE Contr. Syst. Mag.*, *12*(1), 25–29 (1991).
11. *MATRIX$_X$ User's Guide*, Integrated Systems Inc., Santa Clara, CA, 1991.
12. Monsion, M., Bergeon, B., Khaddad, A., and Ermine, J. L., Sesim: A knowledge based system for identification of industrial processes, in *Proceedings of the First European Control Conference*, Grenoble, 1991, Vol. 1, pp. 115–120.

13. Nagy, P. A. J., and Ljung, L., An intelligent tool for system identification, *IEEE Control Systems Workshop on Computer-Aided Control System Design (CACSD)*, Tampa, Florida, 1989, pp. 58–63.

14. Nagy, P. A. J., and Ljung, L., An intelligent tool for system identification, *Proceedings of the 1991 IFAC Symposium on Identification and System Parameter Estimation*, Budapest, 1991, Vol. 2, pp. 918–923.

15. Nagy, P. A. J., and Ljung, L., System identification using bond graphs, *Proceedings of the European Control Conference*, Grenoble, 1991, Vol. 3, pp. 2564–2569.

16. Söderström, T., and Stoica, P., *System Identification*, Prentice-Hall, London, 1989.

17. Walter, E., and Piet-Lahanier, H., Exact and recursive description of the feasible parameter set for bounded error models, *Proceedings of the 26th IEEE Conference on Decision and Control*, Los Angeles, California, 1987, pp. 1921–1922.

18. Wieslander, J., *Idpac—user's guide*, TFRT-3099, Dept. of Automatic Control, Lund Institute of Technology, Lund, Sweden, 1976.

19. Wieslander, J., *Idpac commands—user's guide*, CODEN: LUTFD2(TFRT-3157)/ 1-108/(1980), Dept. of Automatic Control, Lund Institute of Technology, Lund, Sweden, 1980.

20. Wieslander, J., *Interactive programs—general guide*, CODEN:LUTFD2(TFRT-3156)/1-30/(1980), Dept. of Automatic Control, Lund Institute of Technology, Lund, Sweden, 1980.

21. Willems, J. C., Paradigms and puzzles in the theory of dynamical systems, *IEEE Trans. Auto. Contr.*, *36*(3); 259–294 (1991).

22. zu Farwig, H. M., and Unbehauen, H., Knowledge-based system identification, *Proc. IFAC Symp. on Identification and System Parameter Estimation*, Budapest, 1991, pp. 20–28.

7

Computer-Aided Analysis of Nonlinear Systems

S. A. Billings

University of Sheffield, Sheffield, England

B. R. Haynes

University of Leeds, Leeds, England

7.1 INTRODUCTION

In general, the method of analysis applied to any nonlinear system depends to a large extent on the structure of the model representation. To date, little has been done to study the more qualitative aspects of the type of nonlinear models produced as a result of control system modeling and system identification. Analytical methods meet some of the requirements for certain classes of models. Such methods however, tend to be valid for only one particular model structure or type. For instance, dynamical system theorists have studied the general nonlinear ordinary differential equation (ODE) as a dynamical system in its own right; see, for example, Mees [43], where the general form of system representation is taken to be

$$\dot{x}(t) = f(x(t), \mu), \qquad x \in R^{n_x}, \mu \in R^{n_\mu} \tag{1}$$

where the vector μ represents some quasi-static parameter set. Fundamental results, both analytical and geometric in nature, have been provided for Eq. (1) and these now form the basis for much of the work in this area; see, for example, Guckenheimer and Holmes [24]. On the other hand, the attention of the control theorists has, to a large extent, been constrained to less general model forms. This is largely as a result of a desire to solve specific problems in system theory. Fundamental results in this area were obtained by Sandberg [55] and Zames [64,65]. An algebraic theory for the representation, realization, and analysis of nonlinear control systems has since developed [42]. In this work, bilinear models have attracted a lot of attention largely because they are more amenable to analysis. Many concrete examples of bilinear systems do exist; see, for example, Mohler [45,46]. Subsequent work based around these systems spawned a more general class of system known as the linear analytic system. These take the form

$$\dot{x}(t) = f(x(t), t) + u(t)g(x(t)) + u(t), \qquad u \in R^{n_u}$$
$$y(t) = h(x(t), t) \tag{2}$$

Realization theory for these systems was provided by Brockett [7], D'Alessandro, Isidori, and Ruberti [15], Sussmann [60], Fliess [19], and Jakubczyk [37]. An algebraic approach to the analysis of the discrete-time equivalent of Eq. (2) has been pursued by Normand-Cyrot and Monaco [47]. Realization theory for discrete systems was supplied by Clancy and Rugh [12], Sontag [57], and Schwartz and Dickinson [56]. Additionally, extensions to the familiar concepts found in system theory and control have been provided by many authors; see, for example, Hermann and Krener [29], Brockett [8], Rugh [51], Isidori [35], Desoer and Lin [16], Byrnes and Lindquist [9], Fliess and Hazewinkel [20], Vidyasagar [62], and Isidori [36].

The discrete equivalent of Eq. (1) sometimes referred to as a *recurrence* or *mapping*, has been shown to exhibit a richer set of characteristics than its continuous counterpart and spawned at least an equal, if not greater, number of analysis techniques. Discrete maps have been extensively studied as dynamical systems in themselves, prompted initially by the work of Li and Yorke [39]. Here, the general model representation takes the form

$$x_{k+1} = g(x_k, \mu), \qquad x \in R^{n_x}, \mu \in R^{n_\mu} \tag{3}$$

where μ is once again some parameter vector. Extensive effort has gone into trying to classify the recurrent behavior of Eq. (7.3); see for instance, Bernussou [5], Guckenheimer [23], Iooss [34], Gumowski and Mira [25], Collet and Eckmann [13], Preston [48], Salvadori [54], or Holmes and Whitley [31].

From the control viewpoint, the main application of the dynamical systems approach has been the analysis of chaos in feedback systems; see, for instance, Baillieul, Brockett, and Washburn [4], Sparrow [58,59], Holmes [30], Cook [14], Ushio and Hirai [61], and Salam [52]. Success in this area has not, however, had such widespread coverage as other areas of exploration, in particular nonlinear circuit theory; see, for example, Chua and Lin [11]. This may be partly due to the flourishing interest by control theorists in the algebraic and geometric approach. Notable exceptions that make direct use of the qualitative approach include Mehra [44] and Sparrow [58,59] on bifurcation free control; Aeyels

[3] and Abed [1,2] on constructing stabilizing feedback control for continuous systems; Hahn [26], using describing functions; Chang and Chen [10] who considered PID control; and Salam and Bai [53], Ydstie and Golden [63], and Mareels and Bitmead [40] who have considered bifurcation in adaptive control systems.

In control studies, the augmentation of the system model with one or more parameters enables the study of the system's behavior over a specified range of parameter values. It is important to see how the behavior of a system or model changes if the equations that make up that representation change in some manner, if only because such models are seldom known accurately. In the general nonlinear setting, this problem is placed within the framework of structural stability and bifurcation theory. Detailed knowledge of the models solution structure is then required to classify the complete *unfolding* of the behavior to be expected [21,22]. In control, this problem appears under the title of robustness. The dynamical systems approach is attractive in that it provides information on the very type of qualitative behavior the nonlinear model was constructed to emulate. In addition, the method becomes of use precisely at the point where traditional linear control theory breaks down, that is, when one or more linearized eigenvalues becomes degenerate.

Unfortunately, these methods do have some drawbacks. First, being generally analytically based, they depend on a good deal of a priori knowledge of the solution structure of the system. In a general parameterized nonlinear model, such information will not be available and, indeed, may be difficult to obtain. Second, most of the analytical approaches available require the model structure to take a particular form.

One possible solution to this problem, which we outline in this work, is to suspend the parametric models in a discrete, or *cellular*, state space. This allows for the application of a particularly attractive and simple numerical algorithm which can be applied to a wide variety of model types while at the same time maintain the attractive qualitative aspects of the dynamical systems approach. Information on the system's stationary and periodic solution structure is provided along with both local and global stability characterization over a predefined region of the systems state/parameter space. This information can be presented in much the same way as a traditional bifurcation diagram. Although the method itself does not attempt to identify any particular bifurcation behavior, such characteristics can usually be identified by virtue of the changes detected in the systems solution structure.

As an added bonus, the analysis of an unknown system can proceed in an interactive manner whereby the parameter space of the problem is probed over specified parameter region for nonlinear characteristics. This probing approach has proved to be a powerful aid in the analysis of competing nonlinear model structures [27].

7.2 GLOBAL ANALYSIS OF NONLINEAR SYSTEMS

The stability of a model, both absolute and structural, is dependent on the location and distribution of degenerate singularities on the solution manifold of that system. Analytical methods meet some of the requirements for certain classes of models. Such methods, however, tend to be valid for only one particular model structure or type and tend to rely on constructing a reduced order system around a point where the *interesting* dynamics are deemed to be situated. In dynamical systems studies, most analytical approaches tend to rely on a reduction method to reduce the dimension of the problem under consideration. Such methods, when employed, not only rely heavily on a priori knowledge of the solution structure of the problem, but even if successfully applied, are only locally valid in a small region about the degenerate eigenvalue. In discrete dynamical systems, where a profusion of periodic behavior is all too common, constructing such a system is even more cumbersome. Obviously, for application to the type of discrete models encountered in sampled data systems, system modeling, and system identification, a flexible approach that does not suffer these drawbacks is essential.

Traditionally, numerical analysts have used *path following* or *continuation* methods to trace out branches or arcs in the solution manifold of a system of nonlinear equations dependent on one parameter. To illustrate the basic concept, consider an autonomous system, the stationary solutions of which satisfy

$$f(x(t), \mu) = 0, \qquad x \in R^n, \mu \in R \tag{4}$$

A smooth *branch* of solutions, $x(s)$, is made up of a one-parameter family of solutions to Eq. (1) or (3), where s is some arbitrary parameter. Figure 1 shows some typical solution branches. The points μ_a and μ_b represent simple *bifurcations*. The point μ_c represents a *turning* point or *limit* point. At the point μ_d a multiple bifurcation occurs where more than one eigenvalue or complex conjugate pair of eigenvalues become degenerate. The basic problem then is to compute large segments of the solution structure of Eq. (4), including the branches, as μ varies. A secondary problem also exists, that is, the assignment of stability and the estimation of the corresponding influence domain within R^n.

In the general case, the stationary solutions will be smooth functions of μ. For $\mu \in R$, this solution set forms a *path* in R^{n+1}, for a two-parameter family, $\mu \in R^2$, a *surface*, and for $\mu \in R^m$, where $m > 2$, a *hypersurface* of solutions. If an initial point, x_0, μ_0, is known, then a path following the algorithm can be used to trace out the branch [38]. Given the general one-parameter nonlinear equation

$$g(x, \mu) = 0, \qquad g : R^n \times R \rightarrow R^n, x \in R^n, \mu \in R \tag{5}$$

and a point x_0, μ_0 satisfying $g(x, \mu) = 0$, consider the problem of calculating the solution set of Eq. (5) near x_0, μ_0. The mathematical basis for the path fol-

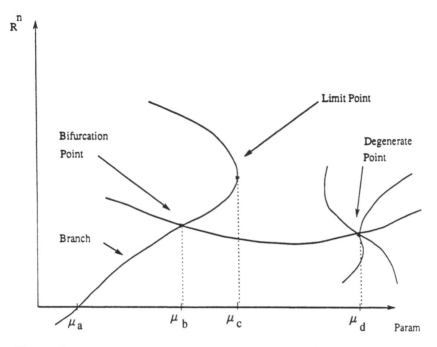

Figure 1 Typical solution branch structure.

lowing method is the Implicit Function theorem. This ensures the existence of a smooth path of solutions $x = x(\mu)$ near x_0, μ_0, provided $g_x(x_0,\mu_0)$ is nonsingular, that is, x_0, μ_0 is a regular point, where $g_x(x, \mu)$ is the usual $n \times n$ Jacobian matrix. Given a regular point x_0, μ_0, the problem is to compute a neighboring point corresponding to $\mu = \mu_0 + \Delta\mu$, where $\Delta\mu$ is a small increment in μ. A predictor-corrector method is typically used. Differentiation of $g(x(\mu), \mu)$ with respect to μ gives

$$g_x \frac{\partial x}{\partial \mu} + g_\mu = 0 \qquad (6)$$

where g_μ denotes the n vector with components

$$g_\mu(x, \mu) = \left[\frac{\partial g_1}{\partial \mu}, \cdots, \frac{\partial g_n}{\partial \mu} \right]^T \qquad (7)$$

Rearranging and evaluating about x_0,μ_0 gives

$$\frac{\partial x}{\partial \mu}(x_0, \mu_0) = -[g_x^0]^{-1}g_\mu^0, \qquad g_x^0 = g_x(x_0, \mu_0), \ g_\mu^0 = g_\mu(x_0, \mu_0) \qquad (8)$$

A Euler predictor approximation x_p to $x(\mu_0 + \Delta\mu)$ is then given by

$$x_p = x_0 + \Delta\mu \left[\frac{\partial x}{\partial \mu}\right]_0 \tag{9}$$

which is used as a starting value to solve the system

$$g(x, \mu + \Delta\mu) = 0 \tag{10}$$

This simple Euler–Newton method works well and forms the basis of many computer codes. However, the path following method runs into trouble at points where g_x^0 is singular; in particular, if the path $x(\mu)$ bends back upon itself, as in Figure 1 at μ_c, then the method fails. Addition of an auxiliary equation to Eq. (5) is often used to circumvent this difficulty.

Many implementations of the above algorithm are in existence [6,18,49]. Unfortunately, the available path following methods provide information on only the local stationary solutions. In addition, a priori knowledge is needed in the form of one or more points on a known solution branch to start tracing the solution arc. To detect any splitting at bifurcation points, an additional branch switching algorithm must be employed. Furthermore, continuation methods provide no information on the extent of the influence domains of the branches traced out. Indeed, no account at all is taken of the possibility of disjoint or isolated solutions existing.

Under natural conditions on f, the set of solutions of Eq. (5) constitutes a differentiable manifold in the product of state space and parameter space. The dimension of this manifold equals that of the parameter space. At present, the standard path following computational methods require the user to construct a picture of a *p*-dimensional manifold from information along a one-dimensional path [50].

In a true nonlinear control design, attention would center not so much on computing a few segments of this manifold, but rather in determining the form and special features of the entire solution manifold and designing control schemes or selecting system parameters to achieve some desired aim. Once a possible control has been identified, robustness or parameter sensitivity, and indeed structural stability, would be considered by allowing variations in the nominal parameter set. In general, this is not easy to achieve and can lead to misinterpretation of the solution structure.

In this work, a dual approach has been adopted specifically with this aim in mind. This combines the essentially qualitative ideas of Bifurcation Theory with a simple yet attractive numerical algorithm. The analysis of nonlinear systems using a *cell map* approximation was first carried out by Hsu; see, for example, Hus and Guttalu [33]. This method has been extended to the qualitative analysis of general nonlinear parameterized models such as Eqs. (1) and (3). The approach proves attractive for a number of reasons. First, it provides a method of enumerating both the stationary and periodic solution structure of the system over a given parameter range. Second, it has been shown to detect all of the typically found bifurcation phenomena [28]. Third, and more importantly, information of a global nature is provided on the extent of the systems stability domains. As a result, both local and global behavior can be studied with little need for a priori information.

7.3 CELL MAPPING SYSTEMS

The process of analyzing a system using cell map analysis comprises a number of steps. The first is the suspension of the nonlinear system, Eq. (1) or (3), in a *cell state space*, Z^n. This is an n-dimensional space whose elements are n-tuples of integers. Each element is called a *cell vector*, or simply a *cell*, and is denoted by z. There are many ways to obtain a cell structure over a given euclidean state space [32]. The simplest way, which we make use of here, is to construct a cell structure consisting of rectangular parallelepipeds of uniform size (squares, cubes, etc). Let x_i, $i = 1, \ldots, n$, be the state variables and let each coordinate axis of the state variable be divided into a number N_i of intervals of uniform size h_i. The interval z_i along the x_i axis is defined such that it covers all the x_i of interest and

$$(z_i - \tfrac{1}{2})h_i \leq x_i \leq (z_i + \tfrac{1}{2})h_i, \qquad z_i = 1, 2, \cdots, N_i \tag{11}$$

The n-tuple $z_i, i = 1, \ldots, n$, is then called the cell vector, denoted by z. A point x belongs to a cell z if x_i and z_i satisfy Eq. (11) $\forall i \in [1, n]$. Each cell z is considered as a *cell entity* and the entire collection of cells as the *cell state space*. Consider now the mapping between two cells $z(j)$ and $z(j + 1)$, where $j = 1, 2, \ldots$ is used to denote a sequence in the same manner as the iterates of a mapping. The *cell map*, C(z), is a mapping of a set of integers $\{N+\}$ such that

$$z(j + 1) = C(z(j)), \qquad z(j) \in Z^n \subset S \tag{12}$$

The *cell function*, F(z, C), is then defined as

$$F(z, C) = C(z) - z \tag{13}$$

A *singular cell* z* is a cell satisfying the relationship

$$F(z^*, C) = 0 \quad \text{or} \quad z^* = C(z^*) \tag{14}$$

A *periodic cell cycle*, given that $C^0(z)$ denotes the identity mapping, is a sequence of K distinct cells $z^*(j)$, $j = 1, \ldots, K$, K being the minimum value which satisfies

$$z^*(m + 1) = C^m(z^*(1)), \qquad m = 1, \ldots, K - 1, \, z^*(1) = C^K(z^*(1)) \tag{15}$$

Each element of the periodic cycle is a *periodic cell*. The complete cell cycle is labeled as P-K. Additionally, those cells eventually mapped onto the P-K cycle by Eq. (12) are defined as within the *domain of attraction*, or DOA, of the cell cycle and labeled as the DOA-K cells.

The size of the cell state space is determined by the system itself. For most practical systems, there are ranges of values of the state variable beyond which we are no longer interested. This means that there is only a finite region of the state space which is of concern. Similarly, for a dynamical system governed by a cell mapping, there is only a finite region of cell space of interest, and correspondingly a finite number, N_c, of cells, called *regular cells*. The *sink cell* is used to encompasses all possible cells outside the region of interest. If the mapped image of a regular cell lies outside the region of interest, it is then said to be mapped into the sink cell. The regular cells are labeled by positive integers $\{1, 2, \ldots, N_c\}$. The sink cell is labeled as $\{0\}$, the zero cell. This makes the total number of cells $N_c + 1$, such that $S = \{N_c+\}$. The set S is closed under the mapping described by

$$z(j + 1) = C(z(j)), \qquad z(j), z(j + 1) \in Z^n \subset S$$
$$C(0) = 0, \qquad S = \{N_c+\} \tag{16}$$

The sink cell, $C(0)$, is a P-1 cell. The set of regular cells within the influence domain of the sink cell, these being eventually mapped to $C(0)$, are in the domain of attraction of the sink cell, and labeled the DOA-Sink cells.

7.4 CELL MAPPING DISCRETIZATION

The cell map, $C(z)$, system may be considered as a discrete system similar to the point mapping

$$x_{k+1} = g(x_k), \qquad x \in R^{n_x} \tag{17}$$

To make use of the qualitative ideas provided by dynamical systems theory, it is necessary to extend the algorithm of Hsu to the analysis of a more general parameterized nonlinear system [28]:

$$x_{k+1} = g(x_k, \mu), \qquad x \in R^{n_x}, \mu \in R^{n_\mu} \tag{18}$$

Applying the center point method of discretization requires the division of R^{n_x} into a collection of cells according to Eq. (11) and the calculation of each cells center points $x^{(d)}(j)$ such that

$$x_i^{(d)}(j) = x_i^{(l)} + h_i z_i(j) - \frac{h_i}{2},$$
$$z_i(j) = 1, \ldots, N_i, j = 1, \ldots, N_i, i = 1, \ldots, n_x \tag{19}$$

where h_i is the cell size and $x_i^{(l)}$ the lower bound defining the region of interest such that $x^{(l)} \leq x \leq x^{(h)}$, where $x^{(l)} = (x_1^{(l)}, \cdots, x_{n_x}^{(l)})$ and $x^{(h)} = x^{(h)} = (x_1^{(h)}, \cdots, x_{n_x}^{(h)})$ Similarly R^{n_μ} is discretized using

$$\mu_i^{(d)}(j) = \mu_i^{(l)} + g_i z_i(j) - \frac{g_i}{2},$$

$$z_i(j) = 1, \ldots, N_i, j = 1, \ldots, N_i, i = n_x + 1, \ldots, n_x + n_\mu \qquad (20)$$

where g_i is the cell size over the region defined by $\mu^{(l)} \leq \mu \leq \mu^{(h)}$ where $\mu^{(l)} = (\mu_1^{(l)}, \cdots, \mu_{n_\mu}^{(l)})$ and $\mu^{(h)} = (\mu_1^{(h)}, \cdots, \mu_{n_\mu}^{(h)})$. The point mapping or image of the center point $x^{(d)}(j)$ is then calculated using Eq. (18) such that

$$x_{k+1}^{(d)} = g(x_k^{(d)}, \mu^{(d)}) \qquad (21)$$

and the cell map $C(z)$ can be constructed by determining the image of each cell within S using

$$C_i(z_i(j)) = z_i(j + 1) = \text{int}\left[\frac{x_i^{(d)}(k + 1) - x_i^{(l)}}{h_i} + 1\right],$$

$$i = 1, \ldots, n_x, j = 1, \ldots, N_i$$

$$C_i(z_i(j)) = z_i(j), \qquad i = n_x + 1, \ldots, n_x + n_\mu, j = 1, \ldots, N_i \qquad (22)$$

Note we are, in effect, constructing n_μ separate cell mappings of dimension n_x, each representing a *slice* within the parameter space $\mu = (\mu_1, \ldots, \mu_{n_\mu})^T$.

A similar approach to the construction of the cell map system may be taken for any continuous nonlinear model that can be written in the form

$$\dot{x}(t) = f(x(t), \mu), \qquad x \in R^{n_x}, \mu \in R^{m_\mu} \qquad (23)$$

It is straightforward to approximate the image of a cell's center point for Eq. (23). Take a point $x_k^{(d)}$, $\mu^{(d)}$ within $x \in R^n$, fix values for the parameters $\mu^{(d)} \in R^m$, choose a time interval τ, and calculate the resulting trajectory for $t = 0 - \tau$ to give $x_{k+1}^{(d)}$, $\mu^{(d)}$. All this requires is a suitable integration scheme to solve Eq. (21) given (23):

$$x_{k+1}^{(d)} = x(\tau)^{(d)} = x(0)^{(d)} + \int_o^\tau f(x^{(d)}(t), \mu^{(d)}) \, dt \qquad (24)$$

The only additional information needed is the time step τ and an integration step size. T is chosen as would be normal for the problem, bearing in mind the choice of integration method. The time step τ is dependent on the cell size and should be chosen such that image of the center point of a cell lies, on average, within a neighboring cell. This obviously depends on the strength of the local vector field and the cell size chosen. If the interior of the cell contains an attracting equilibrium, the trajectory described by Eq. (21) will remain within that cell. An interactive approach to the choice of τ is best adopted. Generally, choosing τ to be 3–15 times the integration interval T gives good results. Repeating the above calculations for each cell within S and applying Eqs. (22–24) approximates the cell map $C(z)$.

7.5 CELL MAP ANALYSIS

Having constructed $C(z)$, the classification of all cells within S is carried out using a modified version of the *unraveling algorithm* [33]. The algorithm involves calling up each cell in turn and processing it to determine its global characteristics. Each cell is then classified accordingly.

A cell may be a singular cell, P-1, or a periodic cell cycle, P-K, satisfying Eq. (5) or (6). The set of all such cells make up the invariant orbits within Z^n. When $K = 1$, a fixed point has been located; when $K > 1$, a periodic solution or limit cycle has been detected. Alternatively, a cell may simply be a regular cell in the DOA of a P-K cell. Each such cell is then said to be in the same *group* and have the same *periodicity* as that cell cycle and is labeled accordingly. Finally, a cell may be mapped by Eq. (21) outside the region of interest into the sink cell. Such a cell is then said to be in the domain of attraction of the sink cell, the DOA-Sink, and is labeled accordingly. For more detail on the unraveling algorithm and its variants, see Hsu [32] or Haynes and Billings [28].

The advantage of the parameterized cell map approach outlined above is that it combines the qualitative aspects of dynamical systems theory with the global aspect afforded by the unraveling algorithm. In implementing the above algorithm, many variations are possible. The version developed in this work focuses on the problems of enumerating the global characteristics of a wide variety of parameterized nonlinear systems. One aim throughout this work has been to maintain an interactive aspect to the algorithm. This enables the analyst to probe the dynamics of a problem by varying both the extent and form of the cell state space. To facilitate this, the algorithms outlined here have been packaged into a prototype CAD tool that enables the probing and analysis of a wide variety of nonlinear systems.

7.6 ANALYSIS OF A NONLINEAR FEEDBACK SYSTEM

In a nonlinear system, the type of bifurcation characteristic obtained can often depend on the bifurcational parameter selected. In real problems, where there may be a number of parameters of interest, the selection of the primary parameter may be critical. Two or more parameters may interact strongly, requiring greater powers of analysis to be applied to the problem. In the following exam-

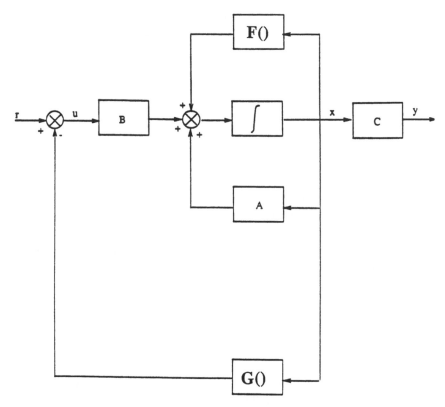

Figure 2 Nonlinear feedback system block diagram.

ple, a parameterized cell map system is constructed for the nonlinear feedback system depicted in Figure 2, described by

$$\dot{x}(t) = A(x(t)) + F(x(t)) + B(u(t)),$$

$$x \in R^{n_x}, u \in R^{n_u}, y \in R^{n_y}, F : R^{n_x} \rightarrow R^{n_x}$$

$$y(t) = C(x(t)), \qquad A : n_x \times n_x, B : n_x \times n_u, C : n_y \times n_x$$

$$u(t) = r - G(x(t)), \qquad r \in R^{n_u}, G : n_u \times n_x \tag{25}$$

Choosing $n_x = 2$, $n_u = 1$, and $n_y = 1$ gives a SISO system with second-order dynamics for the plant element. Note in this example we are dealing with a linear plant albeit as part of a nonlinear feedback loop. It should be noted that we could just as well be considering a nonlinear plant characteristic and are by no means limited to the model structure depicted in Figure 2.

In this problem, the main focus of our attention is on determining what influence the control matrix G and the set point vector r have on the overall system characteristic. To simplify this example, further assume that the feedback matrix comprises just a single parameter and the nonlinear element is a simple cubic polynomial

$$G = [k \ -k], \qquad F(x) = \begin{bmatrix} 2(x_1 - x_2)^3 \\ 0 \end{bmatrix} \qquad (26)$$

and

$$A = \begin{bmatrix} 0 & -1 \\ 1 & 0 \end{bmatrix}, \qquad B = \begin{bmatrix} 1 \\ 0 \end{bmatrix}, \qquad C = [0 \ 1] \qquad (27)$$

so that system set point, r, and feedback gain, k, become the chosen system parameters.

The characteristic of this system is dominated by the system gain k. This is easily demonstrated by performing a set of simple step tests. Large gain k, or large set point values r, result in an oscillatory response. As it stands, the system can be analyzed directly; both Hopf Bifurcation theory [41] and Describing Function methods [26] have been applied with some success.

7.6.1 Cell Map Analysis: $\mu = k, r = 0$

Consider the cell map analysis of the system, writing Eq. (25) in the form of Eq. (23) gives

$$\dot{x}(t) = F(x, r, k), \qquad x \in R^2, r \in R, k \in R \qquad (28)$$

Setting $r = 0$, reduces the dimension of the parameter vector to 1, further simplifying the problem, and leaving the primary parameter, $\mu = k$, the system gain. Thus,

$$\dot{x}(t) = F(x, \mu), \qquad \mu = k, r = 0.0, x \in R^2, \mu \in R \qquad (29)$$

It is now necessary to define the region of interest, that is, the range over which to construct the cell map (12). Choosing $x_1 \in [-1, 1]$, $x_2 \in [-1, 1]$, $\mu \in [-0, 2]$, and the number of cells to be $175 \times 175 \times 30$ cells allows the application of Eqs. (19–24) to give C(z). Subsequent application of the unraveling algorithm gives rise to the cell entities depicted in Figure 3. This diagram provides almost complete characterization of the system's behavior over the parameter range $0.0 \le k \le 2.0$ for $r = 0.0$. Within this range, the shaded areas represent the stable DOAs of a complex sequence of equilibria indicated by the dotted curves. These curves are made up of either P-1 singular cells (fixed points) or P-K cell cycles (limited cycles) that have been detected by the unraveling algorithm. The entire DOA comprises the set of all cells that are ultimately attracted to either a stable P-1 or P-K asymptote within this region.

For $0.0 \le k \le 1.0$, a number of P-1 cells are detected along the x axis of the diagram. This implies a stable attractor exists over this parameter range which

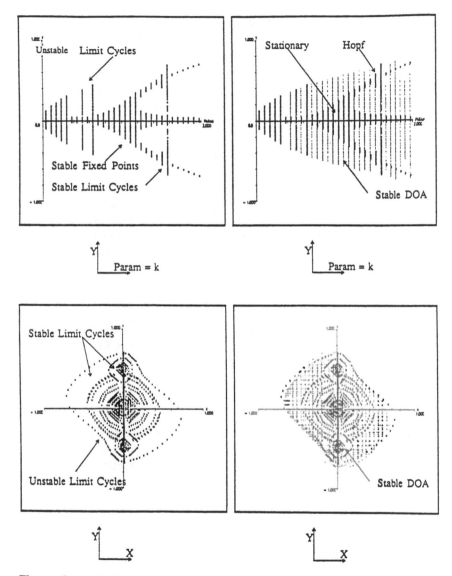

Figure 3. Cell diagram system: $\mu = k$, $r = 0.0$.

exerts an influence domain, the size of which is defined by the extent of the shaded DOA. The DOA forms a cone around the x axis about which a number of longer-period P-K cycles can be detected. These unstable cell cycles mark the separatrix between the stable DOA and the unstable sink cell, their form being more clearly depicted in the lower plots of Figure 3.

At $k = 1$, a split in the solution structure is evident leading to the formation of two new stable asymptotes located symmetrically either side of the x axis with a further, unstable, asymptote existing along the x axis for $1.0 \leq k \leq 1.5$. This qualitative change in the system characteristic implies a pitchfork bifurcation has occurred around $k = 1.0$ [26]. At $k = 1.5$, a further split in the solution structure occurs, resulting in the formation of a stable P-K limit cycle for $1.5 \leq k \leq 2.0$. This qualitative change implies a Hopf-type bifurcation has occurred. The lower diagrams of Figure 3 emphasize this essentially three-dimensional characteristic which we christen the *feedback induced* behavior for the system.

A number of points are worth emphasizing when considering this analysis. First, very little a priori knowledge of the system characteristic is necessary to carry out the analysis. Yet, a broad and complete picture of the system characteristic can be quickly obtained. Second, it is quite feasible to alter the coarse cell size chosen so that more quantitative information can be extracted from the cell diagrams at any particular point.

7.6.2 Cell Map Analysis: $\mu = r, k = 0.5, 1.0, 1.25$

To determine the influence of the set point r on the overall system characteristic, now choose the primary parameter $\mu = r$ and fix the value of feedback gain to $k = 0.5$, 1.0, and 1.25. The newly parameterized problem is suspended in the cell map framework such that

$$\dot{x}(t) = F(x, \mu), \quad \mu = r, k = 0.5\text{--}1.25, x \in R^2, \mu \in R \tag{30}$$

First, fixing $k = 0.5$ and defining the cell state space over $x_1 \in [-1, 1], x_2 \in [-1, 1]$, $\mu \in [-0.2, 0.2]$, for $175 \times 175 \times 30$ cells, allows the construction of a new cell map $C(z)$. Subsequent application of the unraveling algorithm results in the pattern of cell entities in Figure 4.

Once again, the upper plots show the distribution of P-K cycles, but this time the shaded area represents the DOA of just the P-1 cycles, the DOA P-1. That is, just those cells that are asymptotic to a fixed point. These vary with the input set point over the range $-0.2 \leq r \leq 0.2$. About the point $r \approx \pm 0.2$, a sequence of periodic, Hopf-type cycles appear, marking the transition to instability at the edge of the DOA. The lower phase plane plots show this *input induced* behavior more clearly. For $k = 0.5$ and a small input, the system is attracted to the stable equilibrium. For larger input set points, unstable limit cycles surrounding the origin induce a more oscillatory response.

By considering the problem as essentially two-dimensional, the analysis shows periodic behavior to occur at lower values of gain k than expected when $r = 0.0$; that is, we have detected limit cycles for $k = 0.5$. In essence, of

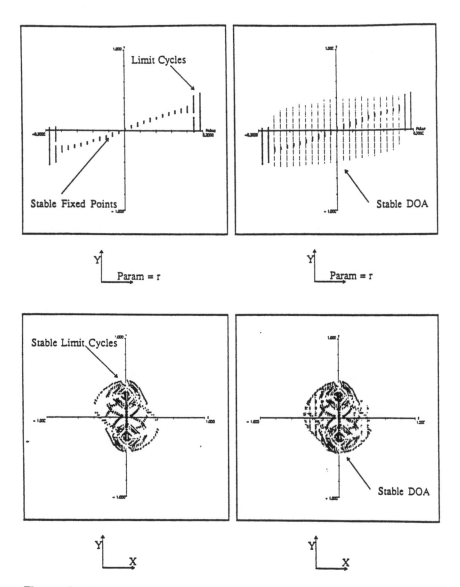

Figure 4   Cell diagram system: $\mu = r$, $k = 0.5$.

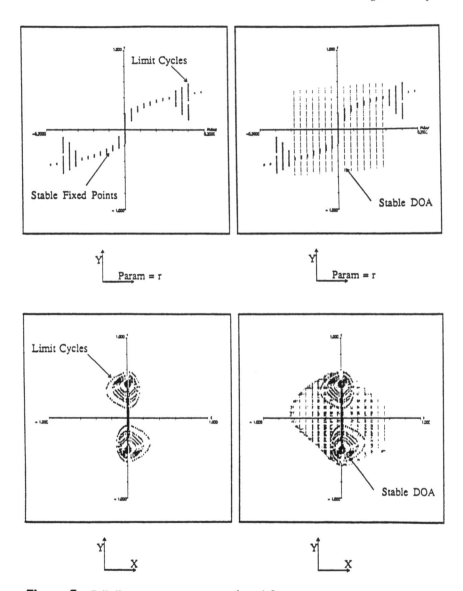

Figure 5 Cell diagram system: $\mu = r$, $k = 1.0$.

course, we have a two-parameter problem in which the limit cycle characteristic is both feedback and input induced.

Taking $k = 1.0$ and repeating the above analysis results in the diagram Figure 5. A similar scenario exists: the stable origin undergoes a pitchfork-type bifurcation. Note the system now has a reduced domain of stability with respect to

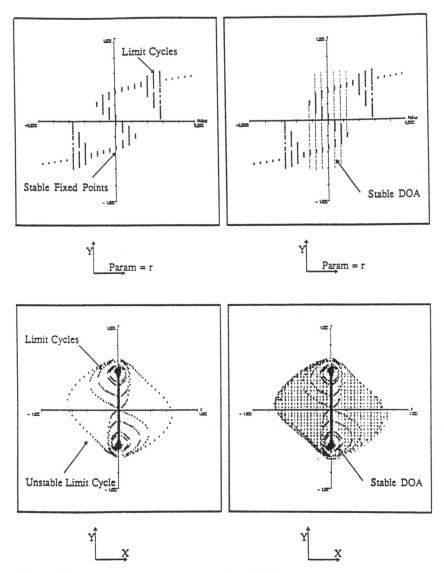

Figure 6 Cell diagram system: $\mu = r$, $k = 1.25$.

the set point, roughly $|r| < 0.1$, but increased internal stability with respect to the state variable x (the DOA in the lower plots being increased in size). For values outside this range, a sequence of periodic bifurcations again leads to instability.

Taking $k = 1.25$ results in the diagram Figure 6, which further confirms this trend. Oscillatory behavior increases further as the equilibria move closer to the unstable limit cycles marking the division between the stable and un-stable domains.

7.7 ANALYSIS OF A NONLINEAR SAMPLED DATA SYSTEM

To further illustrate the utility of our approach, consider now the analysis of the sampled data system depicted in Figure 7. If the feedforward element or plant model is linear, then

$$\dot{x}(t) = Ax(t) + Bu(t)$$

$$y(t) = Cx(t) \tag{31}$$

and, given digital control with sampling period T and an implied zero-order-hold (ZOH) element, then, within the $(k + 1)$th sampling interval Eq. (31) becomes

$$\dot{x}(t) = Ax(t) + Bu_k \qquad u(t_k) = u_k$$

$$y(t) = Cx(t) \qquad t \in [t_k, t_{k+1}] \tag{32}$$

and, under the usual assumption that the plant is time invariant,

$$x_{k+1} = e^{AT}x_k + \int_0^T e^{A(t-\tau)}Bu_k \, d\tau \tag{33}$$

Thus, an exact recursive relationship has been obtained in the form

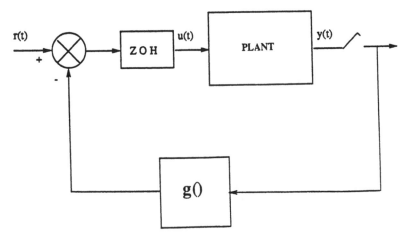

Figure 7 Nonlinear sampled data system.

$$x_{k+1} = \Phi x_k + \Delta u_k$$

$$y_k = C x_k \tag{34}$$

where Φ and Δ are constants dependent on T. Assume now that feedback is applied such that

$$u_k = r_k - g(y_k) \tag{35}$$

where $g(\cdot)$ is either a linear or nonlinear function of the output. If $g(\cdot)$ is linear, the system (32–35) can be analyzed using well-known techniques. If $g(\cdot)$ is nonlinear, Eq. (32) can be rewritten as

$$x_{k+1} = F(x_k, \mu), \qquad x \in R^{n_x}, \mu \in R^{n_\mu} \tag{36}$$

where the parameter μ is say the set point $\mu = r_k$. If the plant is, instead, a general nonlinear system, the model (31) is no longer valid and the description

$$\dot{x}(t) = f(x(t), u(t)), \qquad f : R^{n_x} \times R^{n_u} \rightarrow R^{n_x}$$

$$y(t) = h(x(t)), \qquad h : R^{n_x} \rightarrow R^{n_y} \tag{37}$$

must be used. Note in this case an exact discrete model of the plant cannot be formulated. However, all that is necessary for the system to be analyzed using a parameterized cell state space is that the feedforward and feedback elements can be combined into the form

$$x_{k+1} = F(x_k, \mu) \tag{38}$$

or

$$\dot{x}(t) = F(x(t), \mu) \tag{39}$$

In the first case, the plant has been transformed directly into the nonlinear map $F(\ )$. Normally, this requires that the feedforward element in Figure 7 be linear. The construction of the cell state-space system, $C(z)$, then proceeds using Eqs. (18–22). In the second case, the continuous plant exhibits a nonlinear continuous characteristic and use must be made of Eq. (24).

7.1.1 Cell Map Analysis

Consider now the effect of, say, the sampling interval on the qualitative behavior of a nonlinear sampled data system. Assume in this case that the plant is a linear system and the feedback $g(\cdot)$ is nonlinear. Setting $n = 2$, $l = 1$, and $m = 1$, and A, B, C, and $g(x)$ such that

$$A = \begin{bmatrix} \lambda_1 & 0 \\ 0 & \lambda_2 \end{bmatrix}, \qquad B = \begin{bmatrix} b_1 \\ b_2 \end{bmatrix}, \qquad C = [c_1 \quad c_1], \qquad g(x) = x^3 \tag{40}$$

and taking $\lambda_1 = 1$, $\lambda_2 = -1$, $b_1 = b_2 = 1$, and $c_1 = c_2 = 1$ completes the system definition.

The adoption of this particular block structure may seem restrictive, bearing in mind the flexibility of the cell map approach. However, this system was originally considered by Ushio and Hirai [61] to demonstrate chaos using an analytical approach. There the linear form of the feedforward element played an important role in the analysis. The same structure is adopted here purely as a basis for comparison.

Suspending the system in the cell map framework, the input set point r is selected as the primary parameter. Next define the cell state space such that $x_1 \in [-2, 2]$, $x_2 \in [-2, 2]$, and $\mu \in [-2, 2]$ over a mesh of $150 \times 150 \times 30$ cells. Then apply Eqs. (18–22) to construct C(z), over the region of interest for the three separate cases: $T = 0.1$, $T = 0.2$, and $T = 0.3$. Applying the unraveling algorithm results in the cell diagrams in Figure 8.

The diagrams produced exhibit a number of interesting features. For $T = 0.1$, the system exhibits an asymptotically stable fixed point over the range of input set point considered, $\mu = r \in [-2,2]$. This is indicated by the curve of P-1 periodic cells on the graph. For $T = 0.2$, a stable P-2 cycle is produced at $r = \pm 0.4$ which is in agreement with the behavior predicted by Ushio and Hirai [61]. For values of r outside this range, a sequence of periodic P-K cycles are detected which grow in amplitude as r increases. Note that the coarse cell size used in this analysis and the corresponding plots lead to some gaps that distort the conelike characteristic. This situation can easily be improved by adjusting the number of cells used; however, we choose to leave the results in this first-cut form, preferring to emphasize the ease of analysis of the method as a quick route to obtaining a broad global picture.

For $T = 0.3$, this trend increases with P-K cycles evident over the full range of set point values considered. Indeed, this periodic behavior is so pronounced that it deserves further attention. Note that the cell diagrams show the origin, at $r = 0$, as unstable for this system. This is due to two unstable manifolds emanating from the *fold* bifurcations located asymmetrically on either side of the origin.

Figure 9 shows the response for a step input $r = 0.4$ applied to the system for the cases $T = 0.1$, $T = 0.2$, and $T = 0.3$. The resulting outputs confirm the P-1, P-2, and P-K behaviors predicted above. A similar pattern of cyclic behavior is displayed if the sampling interval is fixed and the set point r is varied. In the lower plot in Figure 9, the behavior for $T = 0.3$ at $r = 4$, appears as P-K cycles of significantly larger period. Figure 10 shows the phase plane plot of this response plotted over 2000 points. This plot displays the characteristic typical of an aperiodic or chaotic system, the outline of the invariant strange attractor being clearly visible in this graph. Of course due to the finite number of cells used in this analysis, the unraveling algorithm cannot detect aperiodic chaotic phe-

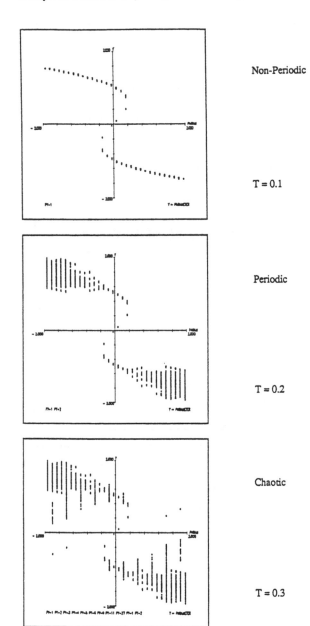

Non-Periodic

T = 0.1

Periodic

T = 0.2

Chaotic

T = 0.3

Figure 8 Cell diagram system.

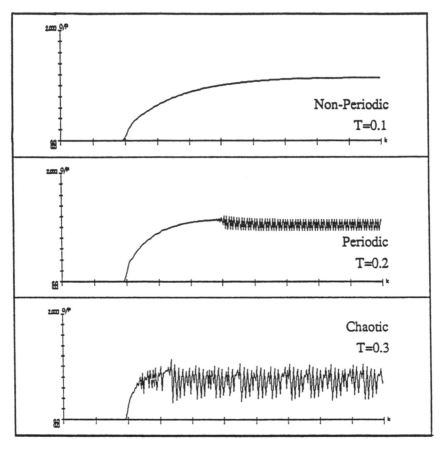

Figure 9 Step responses: $T = 0.1$, $T = 0.2$, and $T = 0.3$; $r = 0.4$.

nomena. However, it is usually a straightforward matter to decide whether periodic, P-K, behavior with K large, is actually periodic or indeed aperiodic/chaotic. Typically, either a phase plane, frequency domain or Poincaré map based analysis can be used to ascertain this. As it stands, the analysis of this simple nonlinear sampled data system has served again to emphasize the utility of the probing approach made possible by carrying out a parameterized cell map analysis within the framework of a CAD analysis/design tool.

7.8 CONCLUSIONS

The cell-to-cell mapping provides a convenient framework for the analysis of both continuous and discrete systems. Development of the suspended cell state-

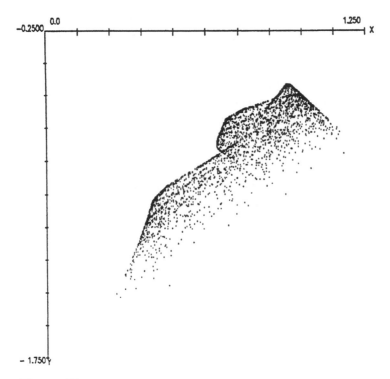

Figure 10 Chaos phase plane plot $T = 0.3$.

space system, enabling the analysis of systems of parameterized bifurcation problems, has proven successful. The preceding examples, although simple, succeed in illustrating the concepts behind this approach. The combination of qualitative and numerical techniques has provided a useful, and more importantly flexible, methodology for the analysis of nonlinear systems. Providing both local and global characterization of a wide range of systems models at a level of detail appropriate to the problem.

From the theoretical point of view, the methods are still in the early stages of development. The effectiveness of the algorithms shows itself in the concrete global results which are difficult to obtain by previous methods. This work has gone some way in achieving the aim of developing an integrated tool for the analysis and probing of nonlinear dynamics within different representations of a nonlinear system. At the moment, a bifurcation is detected, or inferred to exist, simply by evidence of change in the solution structure at some point within the cell state space. Further work is required to enable a more detailed, possibly analytical, classification at these points. The examples in this section have been chosen for their illustrative properties. Another productive area of application of the approach is in the analysis and qualitative validation of nonlinear models constructed using system identification techniques [28].

It has been shown how a qualitative approach based around using the parameterized cell map can provide useful information on periodic, and to some extent aperiodic, characteristics within a system. Furthermore, this study has served to illustrate just how important parameters such as sampling interval, set point, and input excitation can be in determining the qualitative characteristics of nonlinear systems. Both stability characteristics, periodic behavior, and, to some extent, aperiodic behavior can all be revealed in the analysis. The qualitative approach allows both a broad-based coarse probing of the nonlinear dynamics as well as more detailed focused analysis around any particular points of interest.

ACKNOWLEDGMENT

The authors gratefully acknowledge that this work has been supported in part by the SERC.

REFERENCES

1. Abed, E. A., Local feedback stabilisation and bifurcation control I: Hopf Bifurcation, *Syst. Control Lett.*, 7, 11–17 (1986).
2. Abed, E. A., Local feedback stabilisation and bifurcation control II: Stationary Bifurcation, *Syst. Control Lett.*, 8, 467–73 (1987).
3. Aeyels, D., Stabilisation of a class of nonlinear systems by a smooth feedback control, *Syst. Control. Lett.*, 5, 289–94 (1985).
4. Baillieul, J., Brockett, R. W., and Washburn, R. B., Chaotic motion in nonlinear feedback systems, *IEEE Trans. Circ. Syst.*, TCS-27, 990–997 (1980).
5. Bernussou, J., Point Mapping Stability, Pergamon Press, New York, 1977.
6. Brindley, J., Kass-Petersen, C., and Spence, A., Path following methods in bifurcation problems, *Physica D*, 34, 456–461 (1989).
7. Brockett, R. W., On the algebraic structure of bilinear systems, in *Theory and Application of Variable Structure Systems* (R. R. Mohler and A. Ruberti, eds.), Academic Press, New York, 1972).
8. Brockett, R. W., *Differential Geometric Control Theory*, Birkhauser, S. Boston, 1981.
9. Byrnes, C. I. and Lindquist, A., *Theory and Application of Nonlinear Control Systems*, North-Holland, Amsterdam, 1986.
10. Chang, H. C., and Chen, L. H., Bifurcation characteristics of nonlinear systems under conventional PID control, *Chem. Eng. Sci.*, 39, 1127–1142 (1984).
11. Chua, L. O., and Lin, G., Canonical realization of Chua's circuit theory, *IEEE Trans. Circ. Syst.*, TCS-37, 885–902 (1990).
12. Clancy, S. J., and Rugh, W. J., On the realisation problem for stationary homogeneous discrete time systems, *Automatica*, 14, 357–366 (1978).

13. Collet, P., and Eckmann, J. P., Universal properties of maps on the interval, *Commun. Math. Phys.*, *76*, 211–254 (1980).
14. Cook, P. A., Simple feedback systems with chaotic behaviour, *Syst. Control Lett.*, *6*, 223–227 (1985).
15. D'Alessandro, P., Isidori, A., and Ruberti, A., Realisations and structure theory of bilinear dynamical systems, *SIAM J. Control Optim.*, *12*, 517–535 (1974).
16. Desoer, C. A., and Lin, C., Tracking and disturbance rejection of MIMO nonlinear systems with PI controllers, *IEEE Trans. Auto. Control*, *TAC-30*, 860–867 (1985).
17. Doedel, E. J., AUTO—A program for the automatic bifurcation analysis of autonomous systems, *Congress. Numer.*, *30*, 265–284 (1981).
18. Doedel, E. J., AUTO—Software for continuation of bifurcation problems in ordinary differential equations, California Institute of Technology, 1986.
19. Fliess, M., Local realisation of linear and nonlinear time varying systems, in *Proc. IEEE Conference on Decision and Control, Orlando, Florida*, 1982, pp. 733–738.
20. Fliess, M., and Hazewinkel, M., (eds.), *Algebraic and Geometric Methods in Nonlinear Control Theory*, Reidel, Dordrecht, 1986, pp. 55–75.
21. Golubitsky, M., Stewart, I., and Schaeffer, P. G., *Singularities and Groups in Bifurcation Theory, Vol. II*, Springer-Verlag, New York, 1988.
22. Golubitsky, M., and Schaeffer, P. G., *Singularities and Groups in Bifurcation Theory, Vol. 1*, Springer-Verlag, New York, 1985.
23. Guckenheimer, J., Sensitive dependence on initial conditions for one dimensional maps, *Commun. Math. Phys.*, *70*, 133–160 (1979).
24. Guckenheimer, J., and Holmes, P. J., *Nonlinear Oscillations, Dynamical Systems and Bifurcation of Vector Fields*, Springer-Verlag, New York, 1983.
25. Gumowski, I., and Mira, C., *Recurrences and Discrete Dynamical Systems*, Springer-Verlag, Berlin, 1980.
26. Hahn, H., Computation of branching solutions for a nonlinear control system via dual input describing function and root locus techniques, *Int. J. Control*, *42*, 21–31 (1985).
27. Haynes, B. R., and Billings, S. A., *Global Analysis and Model Validation in Nonlinear System Identification* (submitted for publication).
28. Haynes, B. R., and Billings, S. A., A method for the global analysis of nonlinear systems, *Int. J. Control*, *55*, 457–482 (1992).
29. Hermann, R., and Krener, A. J., Nonlinear controllability and observability, *IEEE Trans. Auto. Control*, *TAC-22*, 728–740 (1977).
30. Holmes, P. J., Bifurcation and chaos in simple feedback systems, in *Proc. IEEE Conference on Decision and Control*, 1983, pp. 365–370.
31. Holmes, P. J., and Whitley, D. C., Bifurcations in one and two-dimensional maps, *Phil. Trans. R. Soc. London*, *A311*, 43–102 (1984). Erratum *A312*, 601–602 (1985).
32. Hsu, C. S., Cell-to-Cell Mapping: A Method for the global analysis of nonlinear systems, Springer-Verlag, New York, 1987.
33. Hsu, C. S., and Guttalu, R. S., An unravelling algorithm for the global analysis of dynamical systems. An application of cell-to-cell mappings, *Trans. ASME J. Appl. Mech.*, *47*, 940–948 (1980).
34. Iooss, G., *Bifurcation of Maps and Applications*, North-Holland, Amsterdam, 1979.

35. Isidori, A., *Nonlinear Control Systems: An Introduction*, Springer-Verlag: Berlin, 1985.
36. Isidori, A., *Nonelinear Control: An Introduction*, 2nd ed., Springer-Verlag, Heidelberg, 1989.
37. Jakubczyk, B., Feedback linearisation of discrete time systems, *Syst. Control Lett.*, 9, 411–416 (1987).
38. Keller, H. B., Numerical solution of bifurcation and nonlinear eigenvalue problems, in *Application of Bifurcation Theory, Proceedings Seminar, Math. Res. Center, University Wisconsin*, Academic Press, New York, 1977.
39. Li, T. Y., and Yorke, J. A., Period-3 implies chaos, *Amer. Math. Monthly*, 82, 985–992 (1975).
40. Mareels, I. M. Y., and Bitmead, R. R., Bifurcation effects in robust adaptive control, *IEEE Trans. Circ. Syst.*, TCS-35, 835–841 (1988).
41. Marsden, J., and Macracken, M., *The Hopf Bifurcation and its Applications*, Springer, New York, 1976.
42. Mayne, D. O., and Brockett, B. W., *Geometric Methods in System Theory*, Reidel, Boston, 1973.
43. Mees, A. I., *Dynamics of Feedback Systems*, Wiley, New York, 1981.
44. Mehra, R. K., Catastrophe theory, nonlinear system identification and bifurcation control, in *Proc. American Joint Automatic Control Conference*, 1976, pp. 823–831.
45. Mohler, R. R., Natural bilinear control processes, *IEEE Trans. SSC*, SCC-6, 192–277 (1970).
46. Mohler, R. R., *Bilinear Control Processes* Academic Press, New York, 1977.
47. Normand-Cyrot, D., and Monaco, S., On the realisation of nonlinear sampled data systems, *Syst. Control Lett.*, 5, 145–152 (1984).
48. Preston, C., *Iterates of Maps on a Interval*, Springer-Verlag, Berlin, 1983.
49. Rheinboldt, W. C., *Numerical Analysis of Parameterised Nonlinear Equations*, Wiley, New York, 1986.
50. Rheinboldt, W. C., On the computation of multidimensional solution manifolds of parameterised equations, *Numer. Math.*, 53, 165–181 (1988).
51. Rugh, W. J., *Nonlinear Systems Theory*, The John Hopkins University Press, Baltimore, 1981.
52. Salam, F. M. A., Feedback stabilisation, stability and chaotic dynamics, in *Proc. 24th IEEE Decision and Control Conference*, 1985, Vol. 1, pp. 467–472.
53. Salam, F. M. A., and Bai, S., Disturbance generated bifurcation in simple adaptive system: Simulation evidence, *Syst. Control Lett.*, 7, 269–280 (1986).
54. Salvadori, L. (ed.), Bifurcation of maps, in *Bifurcation Theory and Applications. CIME, Montecatine, Italy*, Springer-Verlag, Berlin, 1984.
55. Sandberg, I. W., A frequency-domain condition for the stability of feedback systems containing a single time-varying nonlinear element, *Bell Systems Tech. J.*, 43, 1601–1608 (1964).
56. Schwartz, C. A., and Dickinson, B. W., Some finite dimensional realisation theory for nonlinear systems, in *Theory and Application for Nonlinear Control Systems*, C. I. Byrnes and A. Lindquist (eds.), Reidel, Dordrecht, 1986.

57. Sontag, E. D., Realisation theory of discrete-time nonlinear systems: Part I: The bounded case, *IEEE Trans. Circ. Syst.*, TCS-26, 342–356 (1979).
58. Sparrow, C. T., Bifurcation and chaotic behaviour of simple feedback systems, *J. Theoret. Biol.*, *83*, 93–105 (1980).
59. Sparrow, C. T., Chaotic behaviour in a 3-D feedback system with piecewise linear feedback function, *J. Math. Anal. Appl.*, *83*, 275–291 (1981).
60. Sussmann, H. J., Existence and uniqueness of numerical realisations of nonlinear systems, *Math. Syst. Theory*, *10*, 263–284 (1977).
61. Ushio, T., and Hirai, K., Bifurcation and chaos in sampled data systems with nonlinear continuously differentiable elements, *Elect. Commun. Japan*, *66*, 1–9 (1983).
62. Vidyasagar, M., New Directions of research in nonlinear system theory, *Proc. IEEE*, *74*, 1060–1091 (1986).
63. Ydstie, B. E., and Golden, M. P. Bifurcation in adaptive control, *Syst. Control Lett.*, *11*, 413–430 (1988).
64. Zames, G., Functional analysis applied to nonlinear feedback systems, *IEEE Trans. Circ. Theory*, CT-10, 392–404 (1963).
65. Zames, G., Realisation conditions for nonlinear feedback systems, *IEEE Trans. Circ. Theory*, CT-12, 186–194 (1964).

8

PSI: An Established Block-Oriented Simulation Program

P. P. J. van den Bosch

Delft University of Technology
Delft, The Netherlands

8.1 INTRODUCTION

Simulation programs are accepted as flexible and powerful tools for analyzing dynamical systems. In general, they yield both close resemblance to the physical description of dynamical processes and a framework to express abstract mathematical models. Simulation programs allow almost any description of dynamical systems, being either linear or nonlinear, being continuous or discrete or any mixture of them. Consequently, their applicability is larger than linear computer-aided design (CAD) programs such as MATLAB and MATRIX$_x$. These CAD programs require linear models being either continuous or discrete. The price to be paid is that simulation is mainly used for analysis and not for design. Simulation can be considered as making several experiments with results that depend not only on the model but also on the simulation experiment itself. For example, the results are influenced by the selection of integration method, integration interval, selection of input signals, etc. Based on one simulation run, which yields an attractive step response, it is not allowed to conclude that the model is stable. There could be a very slow pole close to the origin, behavior of which becomes apparent only if the simulation run is extended to prolonged horizons. This drawback of simulation, namely, the limited validity of drawing conclusions from a limited number of experiments, has to be recognized. Then, a powerful and flexible analysis tool becomes available.

In this chapter, we will describe PSI [2], as it has been developed over the past 20 years. At the same time, a comparison with user requirements and other simulation programs is made. A simulation program is characterized by many entities, such as

- its ability to model or describe a dynamical system
- its toolkit of facilities to design and execute experiments with this model
- its calculation speed
- its user interface

After a short description of the evolution of PSI, these four topics will be discussed in more details in successive section of this chapter.

8.2 EVOLUTION OF PSI

The roots of PSI reach to times when digital computers, such as the PDP-9, were equipped with 8 or 16 kB and later even 64 kB (PDP-11) and when discussions went on for deciding whether an analog or a digital computer had to be purchased for simulation purposes. In these times, the beginning of the seventies, digital computers such as the PDP-9 were extremely expensive, were equipped with very little memory (16–32 kB), and small hard disks (1–2 MB). The quality of the software was poor and in no way comparable with present–day integrated language environments such as offered by Borland (Turbo C++) or MicroSoft (Programmer's WorkBench for C). Consequently, simulation programs yielded almost no user interface. The user had to adapt himself to the awkward and cryptic commands and model description of that time's simulation programs.

The first version of PSI, issued in 1973, was controlled by means of the hardware computer switches of a PDP-9. A model was described with only a few different block types such as integrator and summator, whereas variables or blocks were indicated by means of numbers. At that time, Fortran IV was the most convenient programming language, which had been adopted in creating PSI. When the PDP-11 became available with even 64 kB, a more sophisticated user interface could be created with, for example, names for variables. More block types could be supported and the graphics represented a higher standard. In that time, graphics were separated from the user terminal, yielding the comfortable situation of having access to two screens: one alphanumeric screen for communication between user and program and one graphic screen showing the responses as graphs. Although the PDP minicomputers of Digital Equipment Corporation were used at many places, the dedicated graphic device prohibited a widespread use of PSI.

A major change in simulation software, and so for PSI, has posed by the introduction of the IBM PC in the beginning of the eighties. For the first time, a well-defined open hardware and software environment became available. This open environment allowed developing software for a large market. Quickly, PSI was adapted, in Fortran 77, to cope with the use of one screen at a time, namely, both the alphanumeric information and the graphics had to be shown on the same single screen. At the same time, the memory limitation of 64 kB of the PDP-11 was extended to 640 kB of the IBM PC. This added memory was used for a better user interface, more block types, and more and better simulation facilities. These versions for the IBM PC found a widespread acceptance in the world. By gradually improving the software, for example, by introducing a command language, macro facilities, a PSI compiler that converts a simulation model into real Fortran code to be included in PSI for fast simulations and allowing the inclusion of Fortran code inside a model, PSI had reached a kind of maturity. At the end of the eighties, a new upgrade, called PSI/e became available, yielding an interactive model editor and the introduction of expressions or equations into the simulation model. These versions of PSI have turned out to be attractive for many users. About 5000 licenses have been sold all over the world. Together with unauthorized use, tens of thousands of users can be expected.

Now, in the beginning of the nineties a new major upgrade has been undertaken to cope with the high demands for a user-friendly interface and simulation capabilities. Although a command language yields attractive flexibility, the user of computers nowadays expects a clean, well-designed graphics user interface with windows, mouse support, menus, and some kind of hypertext help. This challenge has been accepted by selecting a new programming environment, namely, the language C and commercially available interface software. Consequently, PSI has been recoded completely in C, which relieves us of old Fortran restrictions such as a predefined maximum for all arrays and a predefined maximum of three inputs and three parameters per block. Functions are implemented, such as the MIN and MAX functions, that can have an infinite number of inputs.

In the succeeding sections of this chapter, this recent version of PSI, called PSI/c, will be discussed.

8.3 MODELING CAPABILITIES

8.3.1 Causality

Modeling physical, chemical, electrotechnical, or mechanical processes requires, in general, a modeling language that allows the use of acausal models. In acausal models, only relations among variables are given, no inputs or outputs. In causal models, relations have to be formulated as equations. At the left side of the equal sign of an equation is the output (effect) and at the right side the inputs (causes). For example, the behavior of a resistor is described by the law of Ohm. At any time, the relation between voltage V (V) across and current I (A) through the resistor equals the resistance R (Ω). This law assigns no input (cause) or output (effect). Both V and I can be either input or output. In an acausal model, this law could be represented by the relation

$$R = \frac{V}{I}$$

In a causal model, the user has to decide at the very beginning of modeling whether he selects the model

$$V = RI$$

or

$$I = \frac{V}{R}$$

Consequently, the use of acausal models reflects the nature of physical laws. Assigning causality in an acausal model is a tedious and not always recognized task which complicates modeling considerably. Still, present-day simulation programs require causal models, except simulation programs such as PHI [4] and OMOLA [5].

8.3.2 Differential Equations

Differential equations can be distinguished as ordinary differential equations (ODE), stiff ODEs, and differential algebraic equations (DAE). In the following simple example, consisting of two connected mass–spring systems, the differences can be explained. Suppose the two masses m_1 and m_2 are connected via a spring k with length L. Then the following equations can be derived; namely,

$$m_1 \ddot{x}_1 = k_1 x_1 - f_1 \dot{x}_1 + F_s$$
$$m_2 \ddot{x}_2 = k_2 x_2 - f_2 \dot{x}_2 - F_s$$
$$F_s = k(-x_1 + x_2 + L)$$

These equations form a fourth-order differential equation, and hence ODEs, which can be solved quite easily with any simulation program. When the stiffness of the spring k increases, these ODEs become stiff. The solution no longer can be calculated in an efficient way with explicit numerical integration methods. Variable step implicit methods then are to be preferred. When the spring k becomes very large or when the spring becomes a rigid connection, the following set of equations emerge:

$$m_1 \ddot{x}_1 = k_1 x_1 - f_1 \dot{x}_1$$
$$m_2 \ddot{x}_2 = k_2 x_2 - f_2 \dot{x}_2$$
$$x_1 = x_2 + L$$

These equations represent a dynamical model described via DAEs. The solution of DAEs also requires the use of implicit integration methods. Only a very few simulation programs support the solution of DAEs, for example, OMOLA and PHI.

8.3.3 State Events

Any numerical scheme for solving differential equations requires finite time integration steps. These finite time steps mean that the values of the continuous variables are only known at distinct points in time. The values of the variables in between are neither calculated nor known. This can introduce errors. For example, a diode conducts as long as the voltage across it is positive. As soon as the voltage crosses zero, the current has to become zero. Without accurate detection of the zero-crossing of the voltage, the current will become negative, which is physically not meaningful. The same situation arises if two masses collide with each other. In these types of systems, the state determines a time event at which the topology of the model is changed. In simulating power electronic circuits, these so-called state events occur frequently. Only a few simulation programs, such as ACSL and PSI, support the detection, location, and solution of state events.

8.3.4 Multirate Difference Equations

As soon as a digital computer becomes a part of a system, difference equations have to be supported to model the dynamic behavior of some kind of filter or control algorithm. Difference equations pose no additional difficulties, so they are generally supported. However, multirate difference equations require a centralized timing mechanism. Multirate models have different model parts that have to be executed at $k_i T$ (sec), so a different integer multiple k_i of some basic clock period T (sec) for each model part must be supported. Only a few simulation programs, among which is PSI, support an accurate and centralized timing mechanism for accurate multirate simulations.

8.3.5 Discrete Events

As soon as an operator becomes part of a simulation model, for example, in command and control systems, discrete events can be profitably utilized to model the human reaction. Also production systems, such as conveyor belts in assembly lines, can be described as discrete-event models. Discrete-event dynamic systems (DEDS) have become increasingly important for modeling event-driven, asynchronous dynamic systems [3]. Processes in discrete-event models can start at time t, other processes T_e sec in the future. Then, at $t+T_e$, the simulation will execute the intended action or process. Also for modeling timing

Table 1 Comparison among Simulation Programs

	ACSL	OMOLA	PSI	PHI	SIMNON	TUTSIM
ODE	*	*	*	*	*	*
Multirate	*	*	*	*	—	—
State event	*	—	*	*	—	—
Discrete event	*	—	*	*	—	—
DAE	—	*	—	*	—	—
Acausal model	—	*	—	*	—	—

sequences such as "after a trigger the switch should be closed for at least t_{min} sec and for a maximum time of t_{max} sec." This model can be solved easily, accurately, and efficiently with the aid of two discrete-event processes: one process that takes care of an event t_{min} sec after a trigger and another process that realizes an event t_{max} sec after a trigger. There are only a very few simulation programs, among which are ACSL, PSI, and PHI, that support models with both differential equations and discrete events.

8.3.6 Comparison

In Table 1, several modeling tools are shown for a number of well-known (ACSL, PSI, SIMNON [1], TUTSIM) or new (OMOLOA, PHI) simulation programs. Based on Table 1, PSI can model almost any dynamic system, except acausal models or models described with DAEs. By careful and proper action of the user, causality can be assigned such that the acausal model becomes causal. However, DAEs require quite elaborated numerical tools for their solution, as offered, for example, by DASSL. For models with DAEs, the new generation of simulation programs such as OMOLA or PHI has to be utilized.

8.4 BLOCK-ORIENTED VERSUS EQUATION-ORIENTED PROGRAMS

Up to now, a number of modeling tools have been described. However, the way an ODE has to be described differs considerably among the different simulation programs. There has been a long-lasting separation between block-oriented programs, such as PSI, and equation-oriented programs, such as ACSL. Block-oriented programs use blocks to shape a model. Block names are used both to define the topology (connections) of the model and the value of the output of a block. The inputs of these blocks arise from other blocks. Equation-oriented programs utilize variables and equations to express the model. However, nowadays, block-oriented programs also support equations or expressions, and equation-oriented programs support many dedicated functions or blocks to model a system. For example, a mass–spring system can be modeled both in PSI and ACSL as

$$x2dot = (k_1.x - f_1x1dot + f_1)/m_1$$
$$x1dot = INT(x2dot,0.)$$
$$x = INT(x1dot,0.)$$

The real difference between both types of programs is how the model is converted into numerical values giving the responses of the variables.

PSI simply is an interpreter of the model. Before a run is executed, the model is interpreted by PSI, which yields a calculation structure. This structure is interpreted during the simulation to obtain the required responses. The richness of the "language" of PSI, together with the power of the supported blocks and functions, pose the limits of the modeling capability. Because an interpreter is used, possible errors can be shown to the user in a simulation-oriented context. The user will receive meaningful error messages. ACSL is also an interpreter of the model. However, the result of the interpreter is a Fortran program. With the aid of a Fortran compiler and appropriate libraries, an executable program is created that calculates the required responses.

Consequently, programs such as PSI allow a fast cycle of modifying and calculating a model, whereas programs such as ACSL take more time to realize a modification. However, the calculation time needed for doing the calculations of a simulation run in ACSL can be less.

8.4.1 Expressions

All variables in PSI are real variables. An input variable x is assumed to be TRUE if $x > 0$ and it is assumed to be FALSE if $x \le 0$. If the output y is TRUE, $y = 1$, otherwise $y = 0$.

In PSI, expressions can be used consisting of about 20 different functions and about 15 different operators, as illustrated in Table 2.

Table 2 Functions and Operators in Expressions

Functions							
ABS	ACOS	ASIN	ATAN	ATAN2	COS	COSH	EXP
INT	MAX	MIN	MOD	LOG	LN	SQRT	TAN
TANH	XOR						
Operators							
$\wedge$	*	/	+	−	! (not)	I (or)	& (and)
>=	<=	>	<	<>	=		

8.4.2 Functions

In PSI, about 60 different functions (block types) can be used to realize simple or complex operations, among which are:

* Different types of integrators (e.g., limited, mode-controlled, resettable) and a differentiator.
* Discrete dynamic functions for multirate difference equations.
* Different controllers both continuous and discrete (PI, PD).
* Fixed and variable dead times [$y = \exp(-sT)$ and $y = exp(-su)$].
* State-event and discrete-event functions.
* Functions for solving nonlinear algebraic equations or algebraic loops.
* Bond graph functions, among which are a transformer and a gyrator.
* Electrical circuit functions (magnetic hysteresis, DC motor with stiction and friction, pulse-width modulation, and a thyristor).
* Input-output functions for AD and DA converters, for disk access, and for synchronizing the simulation with real time.
* One- and two-dimensional function generators. The data can be stored intables of any size. These function generators are ideally suited for parameter estimation if both input and output variables are stored in a function generator.
* Nonlinear functions such as dead space, limiter, hysteresis, quantizer.
* A user-defined function.

The user-defined function allows the inclusion in PSI of any user-defined function. The user can write his own procedure in the programming language C. After compilation and linking with an appropriate compiler, for example,MicroSoft C, the user will have an executable program. PSI will load this program in memory and supply this program with the required information and retrieve the calculated quantities. Of course, the user has to obey some rules concerning the interface of data between PSI and his own procedure. Although this approach requires some additional memory, it allows an easy way to extend the capabilities of PSI.

The facility relaxes some of the constraints imposed by an interpreterlanguage.

In PSI a model is inserted and edited as free-form text. The model can be expressed in a language, suited for the PSI interpreter. The following models are examples of the language of PSI/c.

% The famous Lorenz "butterfly"

```
X = INT(10.*(Y-X)    par: -9.9);    % x-variable
Y = INT(28*X-Y-X*Z   par: -7.4);    % y-variable
Z = INT(-8*Z/3+X*Y par: 31.5);      % z-variable
```

Design of a PI controller for a second-order model with time delay.
% Model description:

referenc	= 1.0;	% referencevalue
inf1	= INF(lim par: .0 4.0 1.0);	% first-order transfer function
inf2	= INF(inf1 par: .0 1.0 5.0);	% first-order transfer function
output	= TDE(inf2 par: .01 .3 1.0);	% time delay of 0.3 seconds
lim	= LIM(pic par: −1 1.0 1.0);	% limiter of u(t)

% PI controller

pic	= PIC(error par: .0 k tau);	% PI controller
error	= referenc − output;	% error signal
k	= PAR(1.);	% controller gain
tau	= PAR(3.);	% integrating constant

% Criterion

crit	= INT(abs(error) par: .0);	

8.5 CALCULATION SPEED

In executing simulations, the calculation speed is of major importance. Especially large models or models equipped with many state events and/or optimization may require large amounts of calculation time. Then, a fast program has to be preferred.

In PSI, several measures have been taken to reduce unnecessary operations.

a. An efficient calculation structure is used.
b. The interpreter uses a two-pass interpretation. In the first pass, the structure of the simulation model is derived. Then, the calculation sequence is determined and, if present, algebraic loops detected. If the user wishes, the algebraic loops are automatically solved by adding code to solve the associated nonlinear algebraic equations. In the second pass all expressions with multiple constants are reduced and simplified.

The interpreter searches for all variables needed to calculate a variable. So, all inputs and variables that are inputs for these inputs and so on, are known. This knowledge is used to distinguishes four types of variables, namely:

1. Variables that are never used to produce any useful information. These variables may result from old experiments or are temporarily not utilized. Only

variables that are visible for the user or that contribute to variables that are visible need to be considered in the calculation structure.

2. Variables that have a constant value during the whole simulation run need only be calculated for $t = 0$. Examples are constants, or variables calculated with only constants.

3. Discrete variables or variables that contribute as inputs for discrete variables only have to be considered at the end of an integration interval. At all intermediate points in time in an integration interval of Runge-Kutta methods, discrete variables are fixed. Consequently, they can be neglected and need not to be considered at these points in time.

4. Continuous variables, such as outputs of integrators and their inputs, have to be calculated at any point in time.

The resulting calculation sequences consists only of the variables of groups 3 and 4. At the intermediate points in time in an integration interval, only the variables of group 4 are considered.

This distinction can save a considerable amount of time, especially in mixed discrete–continuous models and in models with many constants used to initialize a model. In using optimization, measure 1 can be applied even more successfully. Only those variables have to be calculated that contribute to the criterion. All other variables can be disregarded.

This effect of these measures for reducing calculation time is quite impressive. PSI/c is the fastest of any previous version of PSI and faster than the majority of other simulation programs.

8.6 Additional Facilities

Two additional facilities will be described; namely, the possibility of executing optimization and solving nonlinear algebraic equations.

8.6.1 Optimization

In PSI, any variable can be assigned to be a criterion of an optimization problem. In fact, the value of that variable at the end of a simulation run is considered to represent the cost function.

Any parameter, defined via a PAR function, can be considered as an optimization parameter. These parameters form the parameter space in which the nonlinear optimization algorithm will search for a minimum value of the criterion. Any parameter p_i has a gain g_i or sensitivity and an upper p_i^{max} and a lower bound p_i^{min}. Suppose variable $c(t)$ is selected to be the criterion and the simulation is executed in the interval $[0, t_f]$. PSI then solves the following optimization problem with $p = (p_1, \ldots ,p_n)$:

$$p^{opt} = \arg \min_{p} c(t_f, p)$$

$$p_i^{\min} <= p_i <= p_i^{\max} \quad \text{for } i = 1, \ldots, n$$

During each iteration of the nonlinear optimization process, a simulation run is executed in the interval $[0, t_f]$. The optimization procedure determines the values of parameter vector p. The simulation "calculates" the value $c(t_f, p)$. Based on these values of $c(t_f, p)$, the optimization procedure is able to locate a local minimum p^{opt}. The sensitivity g_i is used to adjust the step sizes. For example, optimization can be used for identification of the two parameters K and τ of a first-order model. Suppose $K \approx 1300$ and $\tau = 0.015$. Without individual gain or sensitivity, the optimization will have additional difficulties in locating the optimum. In this situation, it is advisable to select 1000 for the gain of K and 0.01 for the time constant τ.

PSI supports three different optimization methods:

- Simplex, according to Nelder and Mead
- Pattern search, according to Hooke and Jeeves
- Gradient projection method, according to Rosen

The gradient method calculates the gradient via forward difference quotients. The line search utilizes quadratic interpolation. The linear constraints are taken into account via the projection of the gradient on all active equality constraints. The gradient projection method is, by far, the most efficient method for finding a local minimum if the cost function or criterion is smooth in the parameter space p.

Both Simplex and Pattern search are direct-search methods, without utilization of gradient information or line searches. Both are reliable and robust methods that can locate a local minimum of any cost function. The constraints $p_i^{\min}$ and $p_i^{\max}$ are taken into account via a penalty function $P(p)$:

$$P(p) = 10^{30}\{\min(\max(p_i^{\min}, p_i), p_i^{\max}) - p_i\}$$

The penalty function $P(p)$ has the value 10^{30} if p_i is not a feasible point; otherwise $P(p)$ has the value 0. Then the constrained optimization problem can be reformulated as the unconstrained optimization problem:

$$p^{opt} = \arg \min_{p}\{c(t_f, p) + P(p)\}$$

Both Simplex and Pattern search can be equipped with the facility of premature termination of a simulation run. If the criterion is a monotonous nondecreasing function of the simulation time t, a simulation run can be terminated prematurely if the value of the criterion $c(t,p)$ becomes larger than the best solution found up this iteration (Pattern search) or becomes larger than the maximum value of any corner point of the simplex (Simplex search). In such a situation, it can be decided that continuing the simulation run will yield no additional information. The parameter set p of this run will be disregarded anyway. Premature termination too yields a reduction of the time needed for solving the optimization problem.

The application of optimization in designing controllers or in parameter estimation is a powerful tool. Compared with linear design methods the following remarks can be made:

- Both the model and the controller can be nonlinear.
- The model can be continuous and the controller discrete or vice versa.
- Any criterion can be utilized, for example, the sum of the absolute values of some error variable.
- Parameters of a fixed controller structure can be found, instead of the parameters of some polynomial.
- Compared with, for example, LQG design methods, convergence cannot be guaranteed. It is possible that a local instead of a global minimum is located.
- Designing via simulation and optimization requires more calculation time compared with pole-placement techniques or LQG methods.

Compared with linear parameter estimation methods the following remarks can be made:

- The model can be nonlinear.
- The model can contain both continuous and discrete parts.
- Any criterion can be utilized, for example, the sum of the absolute values of some prediction error variable.
- A priori information concerning the structure and some parameters of the model can be exploited profitably.
- Convergence cannot be guaranteed.
- Parameter estimation via optimization and simulation requires (much) more calculation time compared with linear estimation methods.

In spite of some disadvantages, simulation and optimization are a profitable combination for solving estimation and design problems. Especially the possibility of using a priori information during estimation and the inclusion of non-linearities in the design makes this combination very attractive and powerful.

8.6.2 Solving Nonlinear Equations

PSI yields a facility for solving nonlinear algebraic equations via the Newton iterative method. The same facility is used for solving algebraic loops. The purpose is to make an input x of the nonlinear equation equal to the output ($f(x)$) of that equation; so solve

$$F(x) = f(x) - x = 0$$

via

$$x_{k+1} = x_k - \frac{F(x_k)}{F'(x_k)}$$

with the derivative $F'(x_k)$ being calculated via a difference quotient. If this Newton iteration does not converge, a more reliable but also slower optimization procedure starts that minimizes the absolute value of $F(x)$.

8.6.3 State Events

In this section, the influence of state events on both accuracy and calculation time is illustrated. Suppose $x(t)$ is a triangle-shaped variable with $x(0) = x(20) = 0$ and $x(10) = 1$. The variable $f(t)$ is defined via

$$f(t) = 1 \text{ if } x(t) > 0.359 \text{ and } x(t) < 0.391$$
$$= 0 \text{ otherwise}$$

By integrating $f(t)$, which yields $y(t)$, the value $y(20)$ is calculated, as elucidated in Figure 1. The analytical solution $y(20) = 0.64$. The value $y(20)$ has been calculated with and without state-event detection by using Euler (accurate enough) for several values of the integration interval T (sec), as elucidated in Table 3. The calculation time T_c (sec) is normalized to 1 for the simulation with state-event detection with $T = 3.0$ sec. All simulations with state-event detection yield the correct answer $y(20) = 0.64$. Only the simulation without state-event detection with $T = 0.01$ sec yields the correct answer with 63 times more calculation time than the use of state events with $T = 3.0$ sec. All other simulations have calculated a wrong answer.

This example illustrates that without state-event detection the integration interval has to be selected small enough to guarantee that all discontinuities coincide with the calculation points of the numerical integration process. With state-event detection, this is not required because these zero-crossings are detected and calculated. Solving state events can save large amounts of calculation time.

Table 3 Relative Calculation Time T_c and $y(20)$ as function of T

	T(sec)					
	0.01	0.05	0.1	0.5	1.0	3.0
$y(20)$	0.64	0.66	0.8	0.9	0.0	0.0
T_c	63	21	6.5	2.4	0.8	0.5

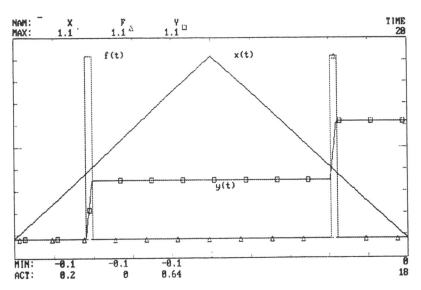

Figure 1 Variables $x(t)$, $f(t)$, and $y(t)$ testing state-event detection.

8.6.4 Numerical Integration Methods

PSI is equipped with six different explicit numerical integration methods, namely, the four fixed-step methods, Euler, Adams-Bashford 2, Runge-Kutta 2, Runge-Kutta 4, and two variable-step methods, Runge-Kutta-Fehlberg 2/3 and Runge-Kutta 4.

8.7 USER INTERFACE

8.7.1 Command Structure

All previous versions of PSI have been equipped with a command language to control the interaction between user and PSI. Although a command language exhibits an attractive performance for experienced users, the user gets accustomed to nice user interfaces such as those of Lotus 123, dBase, Windows, and programming environments offered by Borland and MicroSoft. Certainly, these new interface are attractive and self-explanatory. Even novices can find their way through these new programs.

Based on these observations, a new menu-oriented user interface for PSI has been designed, based on commercially available software. As this software is available for MS.DOS, UNIX, and VMS, it is easy to transport PSI from, for example, MS.DOS to UNIX (SUN) or VMS (VAX). This user interface is based on multiple windows. Via the mouse and/or cursor keys, a menu item can be selected and, subsequently, activated. Via pop-up and pull-down menus, addi-

```
                          File Model
% Design of a PI controller for ┌Optimization parameters... ┐me delay.
                                 │Blocks to optimize...      │
                                 │Start optimization         │
% Model description:             │Continue optimization      │
referenc = 1.0;                  │Transfer opt. parameters   │
infl     = INF(lim par: .0 4. └────────────────────────────┘      function
inf2     = IN┌──────────────Optimization parameters────────────┐tion
output   = TD│                                                  │
lim      = LI│           Criterion block:                       │
             │        Initial step size: 1                      │
% PI controll│          Stop accuracy: 0.03    ┌──────────────────┐
pic      = PI│ Maximum no. of iterations: 15   │Pattern Search I  │
k        = 1;│       Optimization method: Pattern Sea│Pattern Search II │
tau      = 3;│                                 │Simplex I         │
error    = re└Hit space bar to select a method─┤Simplex II        │
                                               │Gradient Projection│
% Criterion                                    └──────────────────┘
crit     = INT(abs(error) par: .0);
```

Figure 2 Screen layout with optimization method selection.

tional information can be requested. Full-screen editors are available to add or modify the model or any other data structure such as data tables or responses. The size of the windows can be customized by the user. All data entry is tested for its validity. For example, one specific window in PSI/c replaces about 15 commands in PSI/e. In Figure 2 the screen is shown after opening three windows, namely, Optimization, Optimization parameters, and Optimization methods. Figure 3 elucidates the screen after selection of Options and, subsequently, the Display options.

A hypertext-based help facility allows, at any time, to inquire for information concerning the actual command, facility, function, or operator. Each help screen again contains help-sensitive words that can be requested. This advanced help facility reduces, but not eliminates, the necessity of a manual. A facility to store and retrieve all key strokes is available as a macro facility, for example, to prepare demonstrations.

8.7.2 Graphics

Compared with other simulation programs, PSI always has offered an attractive graphics environment. Many simulation programs first calculate the responses and display, afterward, the responses on the screen. PSI offers the responses immediately during the simulation run. A user can observe the responses as they "grow" on the screen and stop and/or continue a run if he so wishes. PSI also allows the storage of, for example, 20, different variables in memories. After the run a selection of or all stored variables can be shown. Via a cursor, mouse-controlled or key-controlled, the user can inspect all responses by asking for the numerical values of each indicated point of the response. In Figure 4, the variables $y(t)$ and $z(t)$ of the Lorenz' butterfly are shown as they appear on the graphics screen.

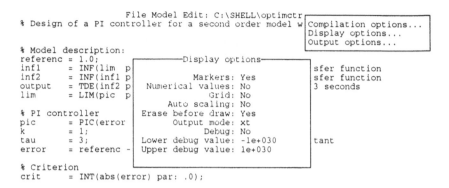

```
                        File Model Edit: C:\SHELL\optimctr
% Design of a PI controller for a second order model w|Compilation options...
                                                       |Display options...
                                                       |Output options...
% Model description:                                   └──────────────────────
referenc = 1.0;
inf1     = INF(lim  p┌──────────Display options──────────┐
inf2     = INF(inf1 p|                                   |sfer function
output   = TDE(inf2 p|          Markers: Yes             |sfer function
lim      = LIM(pic  p| Numerical values: No              |3 seconds
                     |      Auto scaling: No             |
% PI controller      | Erase before draw: Yes            |
pic      = PIC(error |       Output mode: xt             |
k        = 1;        |            Debug: No              |
tau      = 3;        | Lower debug value: -1e+030        |tant
error    = referenc -| Upper debug value: 1e+030         |
                     └───────────────────────────────────┘
% Criterion
crit     = INT(abs(error) par: .0);
```

Figure 3   Screen layout with Display options selection.

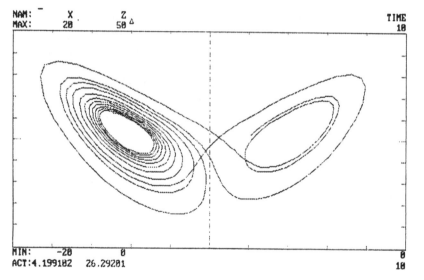

Figure 4   Screen layout of variables of Lorenz butterfly.

Old versions of PSI have supported "screen-dumps" on matrix or laser print-ers. Starting with PSI/c, publication quality graphics are supported also for a selected range of laser printers. These graphics can be printed or included in programs such as WordPerfect or DrawPerfect.

In spite of the appealing graphic user interface of SystemBuild (MATRIX$_X$) and Simulab (MATLAB) for "drawing" the graphical topology of a model, it has been decided to leave it out of PSI/c.

8.7.3. Default Values

Owing to the flexibility of C, a user can tailor PSI for his own requirements. For example, all default values at a start up of PSI, the number of memories, and the number of points of each memory can be installed by the user. In this way, a user can make his own tradeoff between memory requirements, calculation speed, and maximum model size. Via dynamic memory allocation, all available memory can be used by PSI.

8.7.4 Availability

PSI/c has been developed by Boza Automatisering in cooperation with the Delft University of Technology. Boza (P.O. Box 113, 2640 AC Pijnacker, The Netherlands) makes PSI available for customers. (For a PC, the price is about DM 1100, = . Educational institutes receive a discount of about 40%.)

8.8 CONCLUSIONS

PSI has emerged as an evolutionary development over the past 20 years. The program has been adapted to new developments, both in hardware and software. Although PSI is called a block-oriented simulation program, it exhibits nearly all the characteristics of an equation-oriented simulation language. The main difference that still remains is that PSI is an interpreter-based program instead of a language and compiler-oriented language. Apart from acausal models and DAEs, PSI yields almost all the required modeling tools. Its modeling language, determined by its syntax and the many functions and operators, together with user-definable procedures written in the programming language C, can create complex, yet surveyable, models.

The numerical integration methods, the nonlinear optimization methods, and the facility for solving nonlinear algebraic equations give PSI a sound numerical basis. The user interface makes it state-of-the-art, although some users will miss a graphics interface for drawing the model.

REFERENCES

1. Astrom, K. J., Computer aided tools for control system design, in M. Jamshidi and C. J. Herget (eds.), *Computer-Aided Control Systems Engineering*, North-Holland, Amsterdam, 1985, pp. 3–40.
2. Bosch, P. P. J. van den, Interactive computer-aided control system analysis and design, in M. Jamshidi and C. J. Herget (eds.), *Computer-Aided Control Systems Engineering*, North-Holland, Amsterdam, 1985, pp. 229–242.
3. Yu-Chi Ho (eds.), *Discrete-Event Dynamic Systems*, IEEE Press, New York, 1991.

4. Bujakiewicz, P., and van den Bosch, P. P. J., A structured language for modelling and simulation of mixed continuous and discrete-event systems, *Proceeding IFAC Symposium on Computer Aided Design in Control Systems*, *Swansea*, Pergamon Press, London, 1991.
5. Anderson, M., Omola, an object-oriented language for model representation. Lund Institute of Technology, Sweden, LUTFD2/(TFRT-3208).

9

Industrial Applications of SPEEDUP to Process Control

Mohammad Rahbar

AspenTech UK Ltd.
Cambridge, England

9.1 INTRODUCTION

There is increasing competition among process companies to strengthen their position in the global marketplace. Quality of operating performance, specification of products, and time to market availability are critical factors in achieving this. To stay competitive, they must maintain technical superiority, operate their plant efficiently, and keep project costs under control.

With increasing pressure from public and government, the process industry is also taking greater responsibility for the impact their products have on the environment. Safer production processes require greater understanding and better control and monitoring of the entire plant.

This means that process engineers and plant managers need more sophisticated tools to study the design and the behavior of a plant. They need technology which allows them to evaluate design options and operating strategies, perform optimization, evaluate control system performance, conduct in-depth HAZOP studies, train operators, and so on. They also need a tool which helps them to forecast and make decisions quickly based on the best available information.

Although pilot plants can be used to examine some of these issues, they are expensive to build and use, and are of limited capability. In many cases, a more complete solution is to be found in the use of mathematical modeling and computer simulation. This caters to the entire life cycle of a project from process development through plant management. Simulation can be applied to a wide range of engineering problems and has proved to be an indispensable decision support tool.

The process industries have used computational methods to model processes since the late 1950s. Process flowsheeting packages soon became a standard tool for steady-state simulation in the calculation of heat and material balances and for the study of the behavior of unit operations in the plant.

A typical characteristic of a continuous chemical process is that it is energy intensive with high material throughput. A small improvement to the plant design or its control systems can result in very large savings. Hence, process companies are always exploring ways to reduce the operating cost and improve the quality of their products. Well-known steady-state simulators such as ASPEN PLUS enjoy a widespread application to design analysis and optimization in the oil, gas, and petrochemical industries.

Batch and semicontinuous processes have different characteristics. They have lower energy consumption, but must produce high-quality products, allow fast product changeover, and operate safely. These processes are not only dynamic, but must also have a very flexible control system and operate optimally.

In recent years, the process industries have become more aware of the benefits of applying modern control theory to improve the plant performance. The dynamic simulators, multivariable control theory, and computer-aided control system design (CACSD) tools have matured to give industry and process engineers the confidence needed to invest in advanced control systems. The large increase in the number of companies using dynamic simulators and CACSD tools is indicative of their confidence and their commitment to improve the control and the operability of their plant.

The main challenge in the design and analysis of advanced control system is the understanding of the dynamic behavior of a plant because chemical plants are highly complex. Therefore, the dynamic simulator designed for use in process industry has to satisfy certain requirements.

This chapter explains some of the main attributes for a dynamic simulator. It describes SPEEDUP modeling and simulation environment which has been designed and implemented with these requirements in mind. It then describes the functionalities of the SPEEDUP dynamic simulator which supports the full cycle in a modeling and simulation project and has become the process industry standard. It describes the interfaces available and shows how they are used in industrial applications for control design and synthesis, operator training, and on-line simulation. This chapter briefly describes several industrial examples and considers the control design and analysis of a separation process in detail. It shows how a new advanced control system can improve the product quality in a distillation column separating a benzene-toluene-xylene feed into three product streams.

9.2 CHARACTERISTICS OF A DYNAMIC SIMULATOR

The dynamic simulator design for process industry must have the following main attributes:

First, it must allow a process to be defined by interconnecting units of operations together. That is, it must be a flowsheeting tool. Process flow diagrams (PFD) and pipe and instrumentation diagrams (PID) are used frequently by the process engineers at the design stage and during the operation of a plant. The flowsheeting package is, therefore, easily understood by a process engineer.

Second, it should be able to handle very large and complex systems. A chemical plant is an integrated system with a high degree of interactions among units of operations. A model of such a plant typically involves thousands of differential and algebraic equations.

Third, the simulator has to allow easy model definition and modularization. This requires acceptance of models given as differential and algebraic equations (DAEs), which makes the model readable and facilitates the development of a library of models of standard equipment. Mass and energy balances describing these equipments are easily defined as DAE equations and are recognized by process engineers.

Fourth, it should allow the same dynamic model to be used for steady-state simulation and optimization. Control engineers and chemical engineers have traditionally used different tools for a simulation study. A multifunction simulator would allow them to use the same system. This reduces the cost of simulation and also helps to bridge the communication gap between these two groups.

Fifth, it should contain a comprehensive library of models to allow faster problem description and to reduce the overall development time.

Sixth, the simulator should support the entire life cycle of a simulation project from the model development to on-line implementation of the models. Because the cost of developing models of a chemical plant is usually high, these models are not only used for control system design but also used for new or revamped process design, operator training, on-line optimization, and model-based control. A dynamic simulation system offers maximum benefits if it is used for all activities from basic design to plant operation.

A typical scenario of how models may be employed in a chemical plant follows:

- Develop dynamic model of a process.
- Run the simulation open loop and validate results. Modify the models and tune the simulation if needed.
- Linearize the nonlinear model at the operating point.
- Design the control systems, perform controllability studies, etc.
- Include controllers in the simulation and run it closed loop.
- Tune the controllers parameters.

- Produce standalone simulation programs for safety studies, operator training, plant commissioning, on-line optimization, and model-based control.

We, therefore, need interfaces to CACSD tools for design and analysis of control systems, distributed control systems (DCS) for on-line applications, and graphical interface tools for operator training.

9.3 SPEEDUP MODELING AND SIMULATION ENVIRONMENT

SPEEDUP is a DAE-based equation-oriented simulation language. It may be used for steady-state modeling, dynamic simulation, optimization, parameter estimation, and data reconciliation. Figure 1 shows the structure of the SPEEDUP environment. The executive maintains a database which contains the problem description defined by the user. The executive will, on command, translate the problem which is then passed to the run-time system to be solved by numerical methods which can be chosen by the user. The run-time system con-

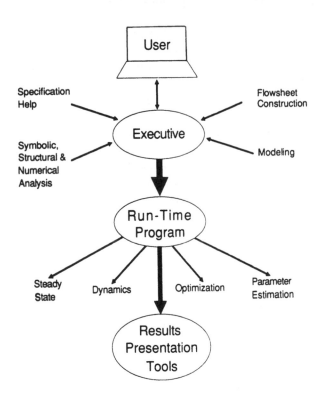

Figure 1 SPEEDUP environment.

sists of a collection of numerical methods for solving large systems of nonlinear DAEs, routines for physical property calculation, and a run-time database. The presentation tools provide facilities for tabulating and plotting results. It also allows the generation of user-defined reports for documentation purposes.

The executable program generated as a result of the translation may be used independently as a stand-alone program. In this case, the user cannot modify the problem structure, but can perform experiments on it. This is useful as the expert modeler develops the model and tunes the simulation and generates the executable program for use by the plant operator or a client. The stand-alone executable can be interfaced to a graphical frontend for operator training, an expert system for plant diagnostics, or distributed control systems for on-line simulation. The external data interface is used to interface these systems to the simulation program.

9.4 INTERFACES TO SPEEDUP

SPEEDUP has an open architecture, which includes a high-level language for defining models and generic interfaces for coupling the executable code to any external program or system. The generic interfaces available in SPEEDUP include:

- Generalized Physical Property Interface (GPPI) allows interfacing to commercial and in-house physical property packages. Standard interfaces already exist for packages such as PPDS and PROPERTIES PLUS.
- Graphical Management System (GMS) allows commercial graphical routines to be used with SPEEDUP.
- External Data Interface (EDI) is a versatile interface allowing SPEEDUP to communicate with a software package or a plant. For example, expert systems such as G2, control design packages such as CONNOISSEUR, distributed control systems such as Eckardt, IBM's Real Time Plant Management System (RTPMS), and Honeywell TDC3000.
- Control Design Interface (CDI) generates linearized state-space model of a nonlinear model. The state-space model can then be loaded into CACSD packages such as MATLAB, MATRIX$_x$, etc., for control design and synthesis.

Figure 2 shows the generic interfaces available in SPEEDUP. Here, we only consider the external data interface and the control design interface.

9.4.1 External Data Interface

The External Data Interface (EDI) is a mechanism for passing data to and from a simulation from other software tools. EDI may be used in steady state, optimization, and dynamic run modes. It is very flexible and has been used for a

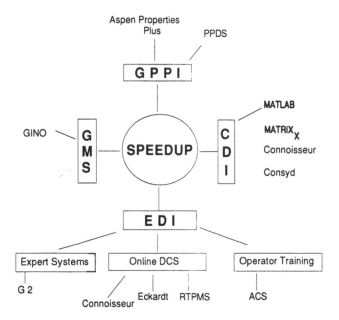

Figure 2 Interfaces to SPEEDUP.

wide range of applications. EDI consists of six main routines. The tasks that each routine performs may be divided into the following categories: initialization of the interface, data acquisition from external system/process, interrupt handling, data transmission to external system, synchronization, and termination. These routines are easily customized for a particular system. Three main uses of EDI in industry is to interface the simulation program to expert systems, DCS or DCS emulators for operator training, on-line optimization, and model-based control.

Expert Systems

An expert system allows easy description of qualitative knowledge which is difficult to express mathematically. Understanding the dynamics of chemical processes is a difficult task and a rule-of-thumb approach is used frequently by the experienced engineers and operators for optimal operation or in the case of emergency. Capturing these experience and expertise in a knowledge base and interfacing it to a simulator that handles and generates quantitative data can result in a very powerful system. Such systems are currently applied to on-line diagnostics and operator training.

In plant diagnostics, the results obtained from a simulator are compared with the data obtained from the plant. If and when necessary, the expert system can take corrective action or raise an alarm depending on the instructions stored in the knowledge base.

In the case of operator training, the expert system can explain how to achieve the objectives by performing certain tasks. It can also monitor the actions taken by the operators and provide advice.

SPEEDUP has an interface to the G2 (Gensym Corporation) expert system and is used for the above applications as well as for off-line testing of G2 knowledge bases where the simulator acts as replacement for real data source. The integration is achieved by producing a stand-alone simulation program of a process and customizing the EDI to transmit and receive the required information to/from the G2 expert system. The G2 knowledge base contains the relevant rules and facts for the analysis of the simulation results and giving advice. The user experiments with the process via a graphical frontend, which makes it easy for a nontechnical user to interact with the process. Figure 3 shows a typical diagrammatical representation of a process in a G2 integrated environment.

Operator Training

An operator-training simulator consists of two parts: a dynamic simulator and a distributed control system (or a DCS emulator) to diagrammatically mimic the process plant and its associated control instruments. The simulator running in the background acts like a real plant and the DCS provides the facilities that are similar to those that exist in a control room of a plant. The trainee interacts with the simulator as if it is a real plant. The display monitor graphically depicts the results and operation of DCS, allowing the trainee to understand operations and control effects as he gets the feel for the actual job conditions. The training simulator is generally used to educate operators to understand process control, to safely operate a plant equipment, and to become familiar with distributed control system.

SPEEDUP has been interfaced to the ACS (Advanced Control and Systems Inc.) and Eckardt DCS system. ACS emulates a variety of DCS systems from vendors such as Honeywell, Fisher Controls, and so on. Eckardt is a DCS system consisting of field instruments, process interfaces, control functions, and operator stations. Several process have been simulated and linked to DCS systems for operator training. Considering the complexity of the chemical processes, the operator training simulators allow a better understanding of the process and increase the operator's productivity.

On-line Simulation

Simulation is being seen increasingly not just as a design and development tool but as part of plant management. Linking the simulation model directly with other systems opens up a multitude of opportunities for wider applications. The trend towards the use of rigorous plant wide simulation as part of on-site

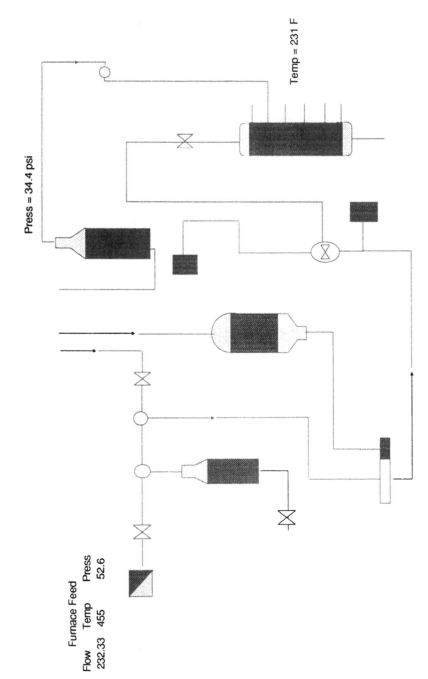

Press = 34.4 psi

Temp = 231 F

Furnace Feed

Flow   Temp   Press
232.33   455   52.6

Figure 3 Graphical environment for the nontechnical user.

systems is now achieving a significant momentum as its advantages for optimization and improved control become more widely recognized.

Major effort is currently underway in the process industry to implement and perform on-line optimization. To increase profit, optimizations are performed daily. Based on the price of the feed and the product in the market the optimum setpoints and the best operating policy is determined by the simulator and is conveyed to the operator for immediate planning and to the managers for long term operation strategy.

The heart of the simulation system is a detailed model of the process in question. The results from the simulation is communicated to the plant via the simulation interface and the real time database of a DCS. Figure 4 shows how communication between the plant and simulator can be established. EDI provides the channels to receive and send data. The best setpoints calculated are sent to the real time database and are ultimately define the setpoint for each local controller.

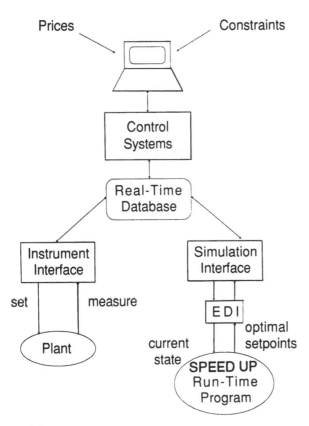

Figure 4   On-line simulation with SPEEDUP.

9.4.2 Control Design Interface

This facility enables a linearized version of a nonlinear model to be produced. The linearization is in the form

$$\frac{dx}{dt} = Ax(t) + Bu(t)$$

$$Y(t) = Cx(t) + Du(t)$$

where t is time, $x(t)$ is a vector of differential variables in the model, $u(t)$ is a vector of chosen control inputs, and $y(t)$ is a vector of chosen control outputs. A, B, C, and D are constant matrices of appropriate dimensions. They are, in fact, the products of CDI, written to a file for use by other packages. x, y, and u are the perturbation variables, representing the deviations of the original variables from the chosen point of linearization. The state-space matrices generated can be loaded to control design packages such as MATLAB, MATRIX$_x$, and so on, for controllability studies, controller design, and synthesis.

When used with a control design package such as MATLAB, CDI helps the user to select a controllable process design. It also helps to choose the best combination of controlled and manipulated variables for the control system. Both of these will result in better process control, leading to a better-quality product.

There is also available an interface to CONNOISSEUR (Predictive Control Ltd.) for designing advanced predictive controllers based on the plant model. CONNOISSEUR supports conventional PID as well as adaptive controllers for implementing optimal control. It collects the required data by random actuator adjustments and helps the user to design the controllers based on the statistical analysis of these model data. Different control schemes can be tried out without having to implement them on the plant. Once the best control scheme is evaluated on the rigorous plant model and satisfactory results are obtained, the CONNOISSEUR controller can then be put in the plant with confidence.

The connection with CONNOISSEUR can be done either via EDI facility or via files generated by CDI. EDI allows direct transmission of data between the software. The simulator acts as a plant, and CONNOISSEUR provides the controllers at run time. The state-space model of a process generated by CDI can also be entered to CONNOISSEUR for off-line controllability studies, control design, and analysis.

9.5 INDUSTRIAL APPLICATIONS OF SPEEDUP

SPEEDUP is designed to model systems which can be represented by a set of equations. These systems include chemical processes, food processing, biochemical processes, physical systems, financial systems, and any combination thereof. Yet the user interface has been designed for use by process engineers and, thus, employs a modular structure based on unit operations joined by streams. Users are not involved in how the numerical calculations are performed.

The following examples are representative of many projects in process industry.

9.5.1 Pipeline Depressurization

This project involved simulating the depressurization of a high-pressure (650-bar) compressible gas pipeline network. The requirement was to determine the appropriate sizing of the relief valves and vents. Of particular interest was the ability to accurately predict the likely skin temperature attained as a result of depressurization because this would determine the piping material required.

Thus, a highly rigorous dynamic simulation was needed that was capable of modeling, not only choking along with both compositional and enthalpy balances but also the difficult problems associated with reversible flows. A rigorous SPEEDUP model of the system was developed and was simulated dynamically.

As a result of this project, the design was modified to ensure that lower-cost materials could be used. This led to cost savings of several million dollars.

9.5.2 Adsorption Systems

As adsorption technology comes of age it will inevitably find its way onto off-shore platforms. Currently, the industrial scale systems are found onshore and are used to separate methane from natural gas. Such processes which by their very nature are cyclic can only be modeled rigorously using a dynamic simulator. ADSIM/SU is an extension to the SPEEDUP system which uses its equation-solving and dynamic simulation technology to enable the rigorous simulation of adsorption processes.

Developed in collaboration with leading cryogenics companies, ADSIM/SU has been successfully applied to a wide range of gas-separation systems both for process design and control system design. Figures 5a and 5b show a rapid-pressure-swing adsorption system and the changes in the pressure at each cycle.

9.5.3 Slugging

Slugging poses major design and control challenges for offshore installations. In this study, it was necessary to evaluate whether the multistage separation system on the platform could cope with anticipated slugging or would it be necessary to install a slug catcher (Figure 6). By very rigorous analysis of the behavior of the installation under various conditions using SPEEDUP, it was established that the separation system was capable of handling likely upsets, and major additional capital expenditure was avoided.

Another application concerned an existing platform where compressor surge control was required to deal with the results of severe slugging. This included

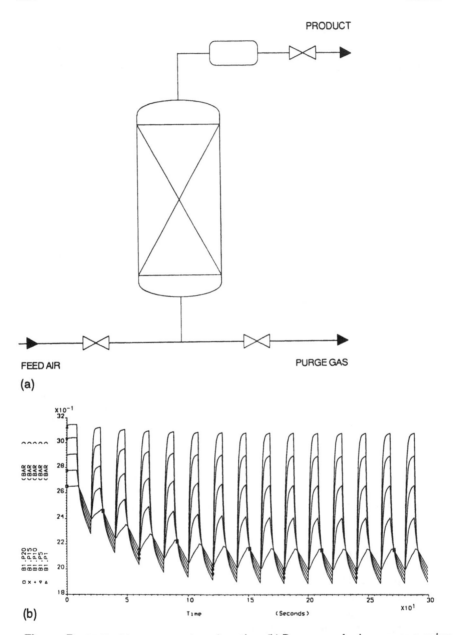

Figure 5 (a) Rapid-pressure-swing adsorption. (b) Pressure cycles in a pressure-swing adsorption system.

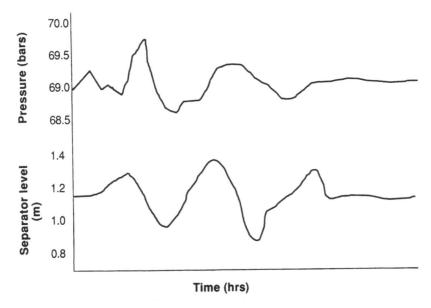

Figure 6. Transient response of gas/liquid separator to arrival of liquid slugs.

simulation around an individual compressor and the overall modeling of a compressor train (four machines, five stages). SPEEDUP has been used at the design stage of a new platform to determine the control strategy required to balance well production rates to a gas processing platform and the effect of transients on the processing trains and final gas specification.

9.5.4 Onshore Treatment Plants

The impact of shutting and restarting wells offshore can create major operating problems for the receiving onshore treatment plant, especially if compositions differ widely from well to well. For normal operations the co-mingling of well fluids is strictly controlled in an attempt to present a uniform fluid composition onshore. However, this is not always possible when wells are shut down for maintenance or restarted.

A dynamic model of the gas treatment plant has been developed. The plant consists of three trains each with compressors, separators, and exchangers. Dynamic simulation enables the effect of changing feed compositions on the plant to be predicted and the most appropriate course of action identified. Similarly, the same model draws on powerful optimization facilities to enable the operators to determine setpoints required to prepare the plant plant for a given average composition before it arrives onshore. This ensures that the sales gase specification is maintained within contract limits and penalties are not incurred.

9.5.5 Parallel Pumping

SPEEDUP has been used in a control study to evaluate the dynamic character-istics of a potable system with changing end-user demand. The potable water systems consisted of two storage tanks, each tank having three sump pumps. Normally, three pumps were duty and three standby. One pump is in continuous operation; a minimum flow valve allows water to be circulated back to the stor-age tanks. The recycle opening and closing is controlled by a pressure transmit-ter located downstream of the pumps. The number of pumps in operation depends on the potable water consumption; the start and stop of the pumps is controlled by the same pressure transmitter.

The pressure buildup and drop due to the start and stop of pumps was inves-tigated (Figure 7). This included the effect of running various combinations of pumps, that is, just one pump, just two, etc., and the effect of a blockage in the pump header while the pumps are running and the dynamic effect on pressure and flowrate. Using this simulation, various control scenarios were investigated, such as the choice between fixed speed pumps versus variable speed pumps.

9.5.6 Benzene–Toluene–Xylene Separation Process

In simulation of chemical engineering processes, the distillation has always been recognized as one of the most difficult unit operation to model accurately. In-deed, simulation programs are often judged on how well they can model distil-lation columns.

The dynamic column models in SPEEDUP offer the engineer a powerful tool for studying the transient behavior of distillation columns. The calculation methods are fully rigorous and the model includes the detail required for spe-cific tray designs. Multiple column distillation trains may be modeled as easily as single columns.

Figure 8 shows the column configuration for the separation of benzene, tol-uene, and xylene mixture. The benzene and xylene are products from the main column and the toluene is the product from the side stream stripper. The required purities of the products are: benzene—95%; toluene—86%; xylene—95%.

The aim of the study is to evaluate the controllability of the control scheme shown in Figure 8. The study is carried out in three stages:

1. Simulate the dynamic column in steady state.
2. Run the dynamic simulation of the column configuration open loop.
3. Study the control system.

Dynamic Column Models

The column is made up of trays, feed tray, reboiler, and so on. MACROs are used to simplify the input. A MACRO allows a number of units and their connec-tions to be combined and treated as a single unit operation. All of these models

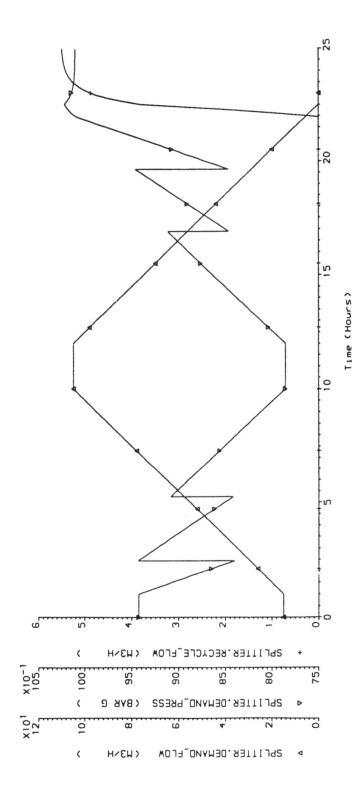

Figure 7 Pump switching in response to consumer demand.

227

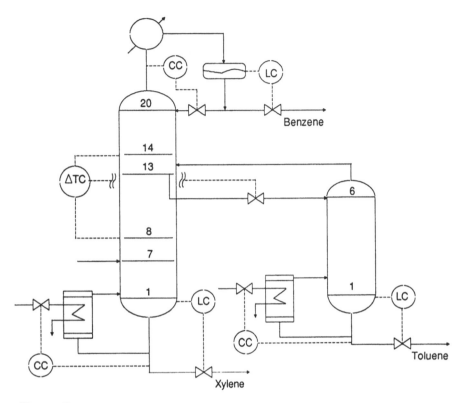

Figure 8 The BTX separation configuration.

are available in the library. The dynamic tray model includes the tray hydraulic effects such as tray holdup, weir dimensions, hydrostatic and plate pressure drop, and the liquid level on the plate. The dynamic model in this example assumes perfect mixing on the tray, instantaneous equilibrium is attained, and no significant effect from the vapor phase dynamics. The tray hydraulic and pressure drop equations are clearly dependent on the particular tray design. If a known type of tray is being used, the user may modify the library model or add a totally new model to represent it accurately.

Steady-State Solution of Dynamic Column Model

The dynamic model was first solved in the steady state to match the known data. When a steady-state run is requested, the d/dt terms are set to zero. The results were compared with the previous steady-state solutions and the agreement was good. The temperatures showed some differences as the tray pressure drop had previously been ignored.

Dynamic Studies

The process change was an increase of 20% in the toluene concentration in the feed after a period of 12 min. The benzene mole fraction was kept at the previous value and so the xylene fraction was reduced.

Mole fractions in feed:

Benzene 0.45	12 min	0.45
Toluene 0.10	$\longrightarrow$	0.12
Xylene 0.45	$\longrightarrow$	0.43

The column response to the change was followed over a 5-h period. The reboiler duties, reflux ratio, and side draw were fixed at the steady-state values.

Figure 9 shows the variation of the three product compositions over the simulated period. After 5 h, they reached steady state.

Controller Study

The basic control scheme for the column configuration is shown in Figure 10. The measured and manipulated variables are as shown below:

Measured variables	Manipulated variables
Condenser holdup	Benzene product rate
Main reboiler holdup	Xylene product rate
Side stream reboiler holdup	Toluene product rate
Benzene composition	Reflux ratio
Toluene composition	Reboiler duty
Side draw rate	Temperature difference

The holdups are controlled by level controllers. These respond very quickly and so have little effect on the changes observed in the products. The other controllers maintain the three product compositions and the side draw rate at the specified values.

The control response was followed after the same feed composition changes in the dynamic (open loop) study. That is, a 20% increase in the toluene concentration with a corresponding decrease in the xylene concentration. Figure 11 shows the changes in the three product compositions which have almost returned to the set point values after the 5-h period.

The changes in the corresponding manipulated variables are shown in Figure 12. The reflux rate and the column and side stream stripper duties all increase to meet the controller specifications.

Several weaknesses were identified in the control scheme used in Figure 10. First, the flow controllers are cascaded from the composition control. The composition analyzers give discrete measurements, typically every 5 min, whereas

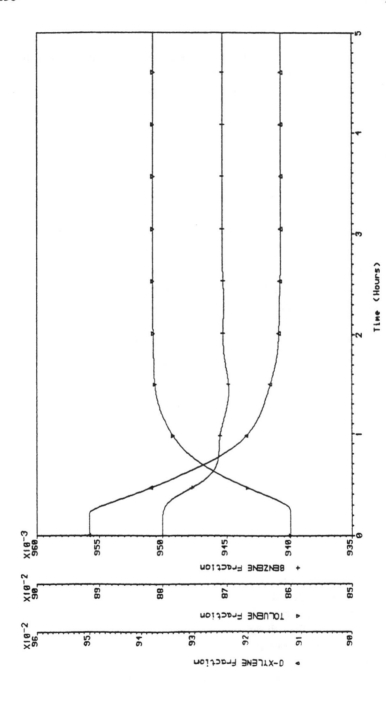

Figure 9 Open-loop response.

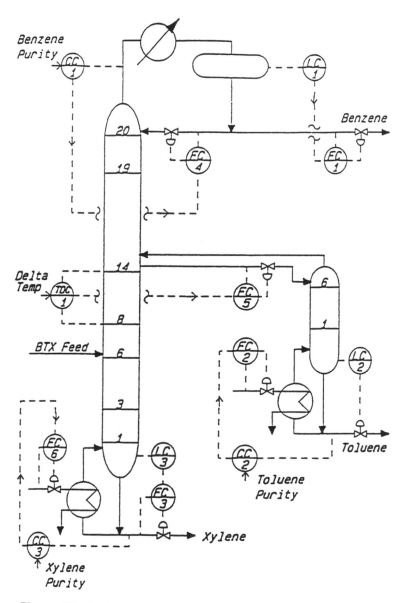

Figure 10 Basic control scheme.

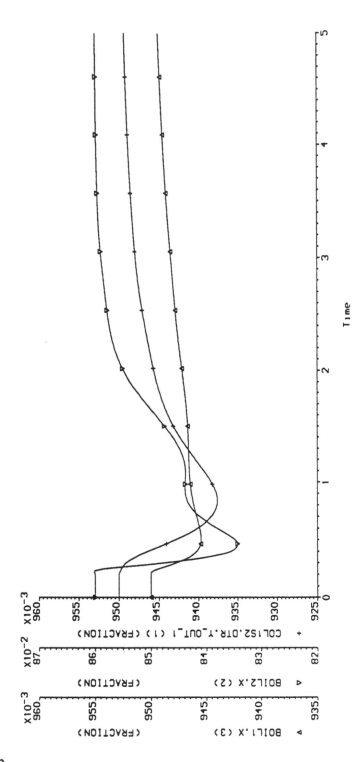

Figure 11 Change in product consumption.

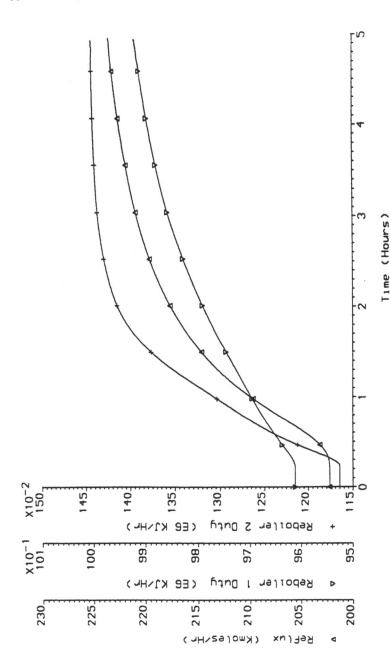

Figure 12 Change in manipulated variables.

the flow is continuously measured. The flow controllers have to be detuned to-make them compatible with the composition control. As a result, the controllers' response to a change in the process is slow.

Second, the composition of the toluene in the side stream stripper is difficult to control because the composition of neither impurities is known. Without knowing the composition of either of the impurities, it is impossible to determine whether to increase or to decrease the steam flowrate to the reboiler.

Third, the liquid level in the reflux drum is controlled by the rate of benzene produced. The rate of reflux back down the column is cascaded from the benzene composition out of the top of the column. This can cause the reflux drum to empty. In fact, it was discovered that the control system is unstable with a reflux ratio between 1 and 4. Also, manipulating the reflux flow does not control the composition very well.

An alternative control scheme was then designed and implemented. Figure 13 shows the new control system. Several modifications were made. The top product composition is now controlled by manipulating the reflux ratio and not the reflux flow. The control on the reboiler of the side-stripper now has two composition controllers as opposed to one in the basic design. The reason for this is that measuring the two components gives more information on all three. If the impurity in the stripper is the lighter component, then more steam is used. If it is the heavier component, then the liquid take off from the side draw is reduced.

The results of the new system were compared with the basic control system. Figure 14 shows how the benzene composition is controlled by the two systems when the feed flow to the column is reduced. The figure shows the second control system to be superior. To give a better comparison between the two control scheme, the Integrated Total Square Error (ITSE) index was introduced to the simulation. Figure 15 shows a plot of the ITSE index for the two systems. It shows that the advanced control scheme reduces the standard deviation in the error in the product composition. This tighter control allows the customer demand to be met exactly when a high-purity product is required and removes the need for a mixing tank downstream of the column.

9.6 CONCLUSION

The efficient operation of a plant and the quality of the product are dictated by good control strategy. It is, therefore, very important that the control system is designed and implemented correctly.

An increasing number of companies are now aware that dynamic simulation is a powerful tool for the total control system design and implementation cycle. The flowsheeting capabilities of SPEEDUP provide the process engineer with the flexibility to investigate systems to any level of complexity and he or she can

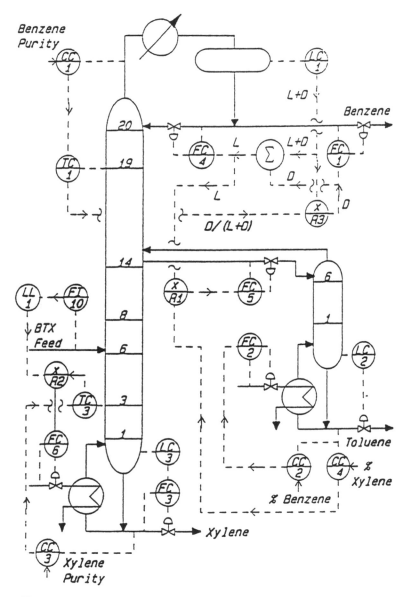

Figure 13 Advanced control scheme.

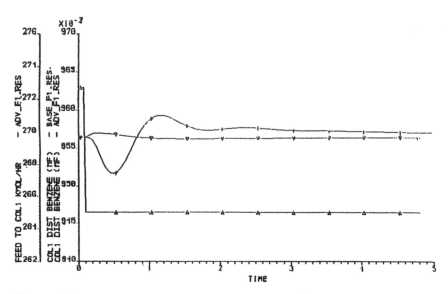

Figure 14 Response in composition change for basic and advanced control systems.

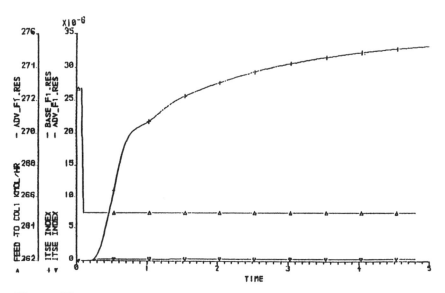

Figure 15 ITSE index for basic and advanced control systems.

quickly create, adapt, or expand simulations of subsystems into full-system simulations and ultimately complete plants.

Generic interfaces make it easy to integrate with existing control system design tools as well allowing the user to satisfy individual requirements and applications.

SPEEDUP simulation environment offers the flexibility to perform dynamic and optimization studies in addition to steady-state design. It provides a consistent framework within which these tasks can be performed enabling the various disciplines to work together more effectively in the provision of higher-quality engineering solutions.

TRADEMARKS

SPEEDUP   is a trademark of Aspen Technology Inc.
ADSIM/SU   is a trademark of Aspen Technology Inc.
ASPEN PLUS   is a trademark of Aspen Technology Inc.
PROPERTIES PLUS   is a trademark of Aspen Technology Inc.
MATLAB   is a trademark of MathWorks Inc.
MATRIX$_x$ is a trademark of Integrated Systems Inc.
Connoisseur   is a trademark of Predictive Control Ltd.
RTPMS   is a registered trademark of International Business Machines Corp.
ACS   is a trademark of Advanced Control and Systems Inc.
Eckardt   is a trademark of Eckardt AG.
G2   is a trademark of Gensym Corporation.
Consyd   is from Department of Chemical Engineering, University of
     Wisconsin.
Gino   is a registered trademark of Bradly Associates Limited.
PPDS   is from National Engineering Laboratory.
TDC3000   is a trademark of Honeywell Inc.

10

Industrial Plant Models for Research in Control System Design

Dean K. Frederick

Rensselaer Polytechnic Institute
Troy, New York

Percival H. Hammond

University of Wales at Swansea
Swansea, Wales

10.1 INTRODUCTION

In the control system design process, and especially where academic research for a new control system technique is concerned, a model representing the controlled plant or process is normally employed. Typically, such models are linear and of low order, a choice which can normally be justified in the absence of any generally accepted nonlinear or more complex model structure.

Unfortunately, a new design method based on such simplified approximations to engineering practice may not be accepted as realistic by industrial control engineers who may doubt the validity of the method when applied to a real process.

One way of increasing the validity and potential applicability of research results is to provide researchers with a set of realistically complex models with pedigrees traceable to actual industrial plant and processes, and for their design ideas to be demonstrated on these models. Such a course of action has other benefits because it provides the opportunity for researchers to compare different design methods on the same plant model and, conversely, to assess the robustness of a given control system proposal over a range of different types of plant.

It should be noted that the provision of a set of models as proposed here differs in purpose and intent from that of the benchmark problems initiated, in 1985, by the Benchmark Working Group of the IEEE Committee on CACSD

[10]. The purpose of these benchmark problems is to provide a quantitative means of comparing the capabilities and limitations of CACSD computer software packages for simulation and analysis. Nevertheless, it was foreseen by Rimer, Frederick, and Huang [23] that a future development might be the formulation of benchmark routines which allowed comparison of design and synthesis techniques.

In early 1989, work began on a UK-based portfolio of well-tested and thoroughly documented plant and process models. The work, supported by the UK Science and Engineering Research Council, was undertaken at University College, Swansea, as described in Barker et al. [4] and Frederick et al. [12]. The model portfolio includes both linear and nonlinear models.

The seven linear models are written in a combination of PRO-MATLAB [28] and ECSTASY [18]. They have been carefully selected from published sources and are based on nonlinear models or real process data. They are available in computer readable form. Four of the seven nonlinear models currently form the basis of a directly accessible Model Library. They are written in two nonlinear simulation languages in common use in the UK; these are TSIM 2 [5] and ACSL [16]. The nonlinear models are provided with files giving full descriptions of each model and examples of dynamic responses, which enable a user to verify that correct results are produced on his or her computer system.

A nonlinear model derived from the Library may be linearized at any desired operating point to derive a linear form suitable for the application of a control system design algorithm. Model-order reduction may be carried out on the linearized form if necessary. When the control system design is complete, it may be implemented on the original nonlinear model.

The Library models are available by electronic mail over the UK Joint Academic Network (JANET), thus eliminating the need for keyboard entry of data or code. The JANET network can also be accessed by other network users in overseas academic communities.

In the following sections the development of both linear and nonlinear models will be presented, followed by a description of the Model Library directory structure and protocols.

10.2 Model Development

10.2.2 Linear Models

A comprehensive literature survey of published reports on dynamic models of industrial plant was carried out for a master's degree thesis at UMIST by Singh [26]. Seven of the models identified by Singh have been used in the portfolio. In each case, the equations have been translated into a computer language, generally into MATLAB, though some equations were written in ECSTASY form.

Table 1 Linear Models

Plant	Equation type	Inputs	Outputs	States	Delays
Distillation column	Transfer function matrix	3	2	—	Yes
Ore pelletization	do.	2	2	—	Yes
Two-bed catalytic reactor	State variable	4	2	15	No
Ships boiler	do.	8	5	12	No
Automotive gas turbine	do.	2	2	12	No
F100 turbo fan engine	do.	5	5	16	No
Lateral aircraft dynamics	do.	2	—	4	No

Conversion to other languages should not be difficult. Each of the models will be described with enough details to indicate to a potential user the relevance of the model to his or her work.

Table 1 gives an overview of the linear models.

Distillation Column

The physical plant is an eight-tray pilot scale distillation column with condenser and reflux arrangements at column top and reboiler at bottom. The model equations [17] are derived from extensive tests carried out on the actual column, the layout of which is illustrated in Figure 1.

The model is in transfer-function form and has three inputs and two outputs. Responses from the actual column were used to derive model transfer functions that are approximated by a first-order lag and a dead time. The model is implemented with a MATLAB file.

The control inputs are the reflux flowrate and the reboiler steam flowrate. The feed rate can be regarded as a disturbance. The outputs are the compositions of the top and bottom products.

Ore Pelletization Plant

A plant designed to convert crushed ore into pellets for subsequent processing is described by Wellstead and Munro [30]. A rotating drum is fed with crushed ore at the rate of 90 tons/h and water spray at 6.4 tons/h. Pellets discharged from the end of the drum are screened and the rejected product is recycled back to the ore feed stream. The control objectives are a stable and fast response to changes in the input variables. A schematic of the plant is shown in Figure 2.

The model type is a 2 × 2 transfer function matrix with time delays. It was derived by linearizing a nonlinear model about a specific operating condition.

The inputs are the ore feed rate and the water spray rate. The outputs are the pellet production rate and the moisture content of the pellets.

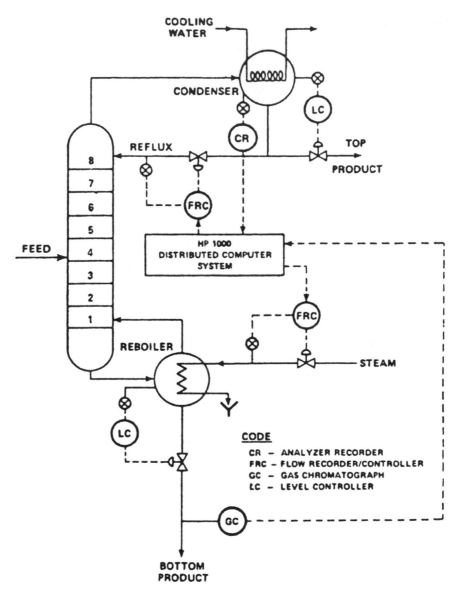

Figure 1 Column arrangement from Ref. 17, used with permission.

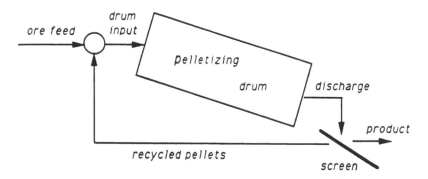

Figure 2 Ore pelletization plant from Ref. 30, used with permission.

Two-Bed Catalytic Exothermic Reactor

As described by Foss, Edmunds, and Kouvaritakis [9], reactant gases are fed into the first catalyst bed through the main feed heater. On emerging from the first bed, the reactor products are mixed with a quench stream and then pass to the second catalyst bed where further reaction takes place. The emerging product passes into an analyzer where product concentration is measured.

The control objectives are to meet defined product concentration and temperature specifications.

The reactor layout is illustrated in Figure 3.

State-variable equations describe the four system components. These are as follows:

1. The main feed stream heat exchanger (one input, one output, and one state)
2. The upper reactor bed (two inputs, two outputs, and seven states)
3. The mixing chamber (four inputs, three outputs, and zero states)
4. The lower reactor bed (three inputs, two outputs, and seven states).

Control inputs are the feed temperature (controlled by the feed heat exchanger), the quench flowrate, and the quench temperature (controlled by the quench heat exchanger). The feed concentration is a disturbance input. The outputs are the temperature and concentration of the product. In addition, several intermediate temperatures are measured and can be used for control purposes.

Ship Boiler

This system is a water-tube boiler producing steam for a marine propulsion turbine [29]. The assembly comprises a steam drum with downcomer and riser tubes. Steam from the drum enters a heat exchange system from whence it is supplied to the turbine. Control objectives are described in Ref. 29. Figure 4 illustrates the boiler and its associated units.

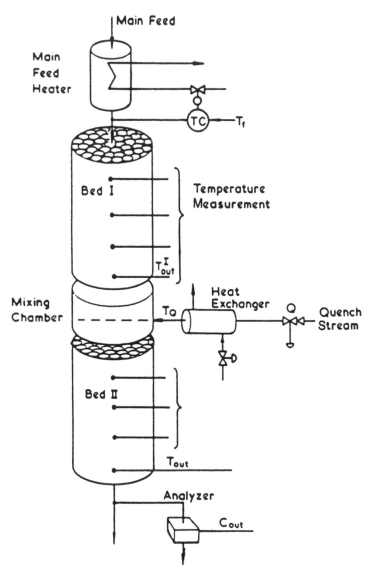

Figure 3 Two-bed reactor from Ref. 9, used with permission.

The model consists of state-variable equations with 12 states. The A matrix has one eigenvalue at the origin and one just inside the right half-plane. The linear model was derived from a 20th-order nonlinear model which was verified by comparison with trial results. The control inputs are the fuel flowrate, the feedwater flowrate, and the position of the bypass valve. Disturbance inputs include the excess air, the air temperature, the feedwater temperature, the fuel

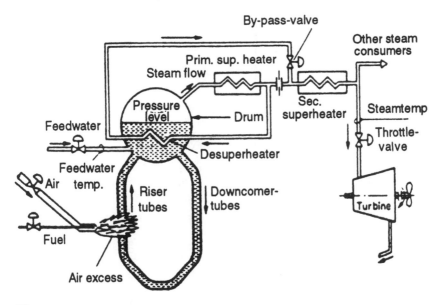

Figure 4 Ship boiler from Ref. 29, used with permission.

temperature, and the steam flowrate from the secondary superheater. The outputs of the model are the drum steam pressure, the steam temperature after the secondary superheater, and the water level in the drum.

Automotive Gas Turbine

This model describes a gas turbine engine for road transport applications. The basic modeling work was done by Nuske [20] and the equations are given by Singh [26], where they are attributed to Hung [14]. A 2 × 2 transfer function is also given by Singh [26] that is attributed to Patel and Munro [22]. The model is in state-space form and its inputs are the excitation signals to the fuel pump and the nozzle actuator; the outputs are the gas generator speed and the inter-turbine temperature.

F100 Turbofan Engine

Sain [24] and Hackney, Miller, and Small [13] describe the Pratt and Whitney F100-PW-100 after-burning, low bypass ratio, twin spool, axial flow turbofan engine. The model is of the state-space type with 16 states.

Transfer functions for the actuators and sensors are provided in the reference source but are not included in the state-variable equations.

The linear model was derived from a nonlinear model, with operating point close to maximum power at zero altitude, without afterburning. Constraints and control objectives are given in the references. A simpler linear model with five states is also given by Sain [24]. A sectioned drawing of the engine with an indication of control inputs is shown in Figure 5.

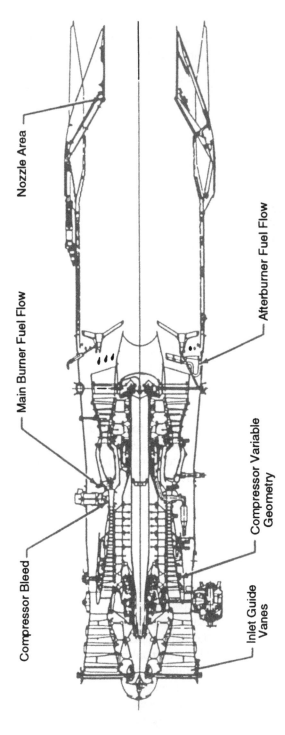

Figure 5 Turbofan engine from Ref. 24, used with permission.

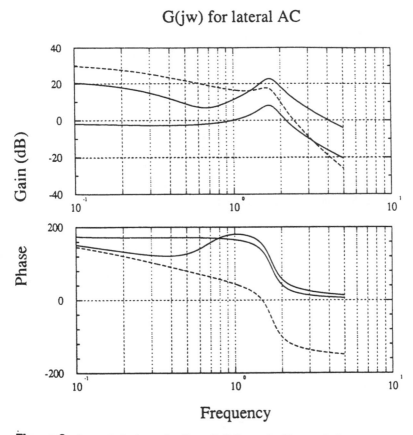

Figure 6 Lateral a/c dynamics from Ref. 7, used with permission.

The inputs are the main burner fuel flow, the nozzle jet area, the inlet guide vane position, the stator position, and the compressor bleed air flowrate. The outputs are the net engine thrust level, the total engine airflow rate, the turbine inlet temperature, the fan stall margin, and the compressor stall margin.

Lateral Aircraft Dynamics

The lateral dynamics of a slender delta wing aircraft are described by Crossley, Munro, and Henthorn [7]. The model is in state-space form, having four state variables, namely, the sideslip angle, the roll rate, the yaw rate, and the bank angle.

The inputs are the aileron and rudder angles. The outputs are the sideslip and bank angles. Examples of responses for three of the four input/output pairs are shown in Figure 6.

10.2.2 Nonlinear Models

To obtain a representative set of nonlinear models, visits were made by the authors to several industrial organizations where it was known that modeling of plants and processes had taken place. As a result of these visits, agreement was reached to incorporate selected models in the portfolio. These were all well-tried models which had been verified against operating data from the plant or process concerned. The industrial models were in a variety of formats ranging from sets of equations to computer programs. Each model was rewritten in two commonly used simulation languages—TSIM 2 and ACSL.

In this section a brief account of seven nonlinear models will be given. More detailed information is available in the report by Frederick [11]. Table 2 presents an overview of the models to be described.

Anydym

This represents a 660-MW coal-fired electricity generation plant operated by the Central Electricity Generating Board (CEGB) [25]. The components of the model, which has 73 state variables, include (1) two coal-grinding mills with air fans, (2) the boiler assembly with its risers, downcomers, superheaters, reheaters, economizer, attemporators, and spray valves, (3) the steam distribution

Table 2 Nonlinear Models

Model Title	Simulation Language(s)	No. of States	Look-up Tables	Details of Plant
Anydym	ACSL	73	—	Coal fired power station with coal mills, boiler, and turbines
Point reactor	ACSL, TSIM2	13	—	Advanced gas cooled reactor
Vaporizer	ACSL	9	—	Unit of chemical plant showing limit cycles at high feed rates
Condenser	ACSL	6	—	Chemical process unit with unconventional configuration
Extruder	ACSL	5	—	Plastics wire-coating extrusion process with variable time delay and high measurement noise
Propfan	ACSL, TSIM2	22	40	Jet engine driving contrarotating propellors
Turbofan	ACSL, TSIM2	9	19	Pegasus vectored thrust turbofan engine with fuel system and digital control system

main with its safety valves, and (4) the high- and low-pressure turbines with their governors.

The model was originally written in double precision, in the CEGB Plant Modelling System Program [19], in the form of macros, each representing a plant unit. The translated version, in ACSL, uses a combination of macros and in-line code. The simulation represents an open-loop system except for a drum-level controller which was included to hold the boiler water level within necessary tight constraints.

The control inputs available to the user are (1) the coal and air feed valve positions, (2) the high- and intermediate-pressure turbine governor valve positions, and (3) the primary and secondary attemporator spray valves.

Several outputs can be measured; examples are (1) the power generated by the high- and low-pressure turbines, (2) the drum level and pressure, (3) the burner combustion temperature, and (4) the high-pressure-main steam pressure and flowrate.

When the model is run, it is initialized at a power output level of 600 MW. It can be initialized alternatively at 300 MW by loading a data file containing the values of 82 parameters.

The model can be linearized by the use of the ACSL linearization command to create the matrices of a linear state-space model at either the 300-MW or 600-MW levels. These matrices have been used with MATLAB to simulate the time response of the linearized model and to calculate the eigenvalues of the A matrix. For sufficiently small changes in the inputs, the linearized model gives excellent agreement with the nonlinear one. Also, all 73 of the eigenvalues of the computed A matrix were in the left-half of the complex plane.

Tests on the model comprise runs at each of the two power levels and with two sets of conditions. The first of these simulates a stored-energy test. For this test, the coal-feed and air rates are maintained at their equilibrium values and the high-pressure turbine governor valve is ramped, over a 30-sec interval, from its nominal position to fully open, and held there. The resulting transient is driven by the energy stored in the steam and in the kinetic energy of the turbine. Figures 7a and 7b show the responses of several key variables when the equilibrium power level is 600 MW. After an initial sharp rise as the stored energy is released, the power decays toward a lower equilibrium value. The drum level exhibits the characteristic "shrink and swell" behavior; the initial rise is due to the reduction in the boiler pressure and the subsequent fall to the action of the controller in reducing the feedwater inlet rate.

The second test simulates the behavior when a step reduction of 5% is applied to the coal-feed and primary-air flowrates. Figures 8a and 8b show examples of response transients over a 1000-sec (16.67-min) interval. The decreased energy flow to the boiler shows up in the rapid drop in combustion temperature. As expected, both the pressure in the steam main and the power generated fall off.

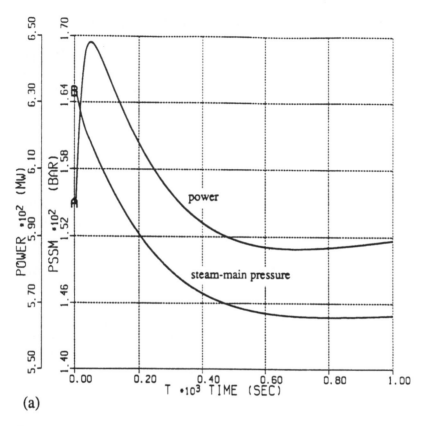

(a)

Figure 7   Stored energy test on ANYDYM model at 600 MW over 1000 sec.

Point Reactor

This is a model of an advanced gas-cooled nuclear reactor (AGR) having a rated
power output of 1550 MW. The key variables and gas flow details of the reactor
are indicated in Figure 9. The model was derived within the United Kingdom
Central Electricity Generating Board (CEGB) by Dulson and Merriman [8]. It
assumes radial symmetry and adopts lumped approximations to the distributed
dynamic relationships within the reactor. The model comprises 25 nonlinear differential and algebraic equations and contains 32 variables and 43 parameters.

The data for the model were received in the form of equations from
the CEGB and has been implemented in both the ACSL and TSIM2 simulation
languages.

There are 13 state variables in the AGR model, namely, the neutron flux, five
gas temperatures at different points in the gas path, the fuel can temperature, the
uranium fuel temperature, three fission product heating power variables related

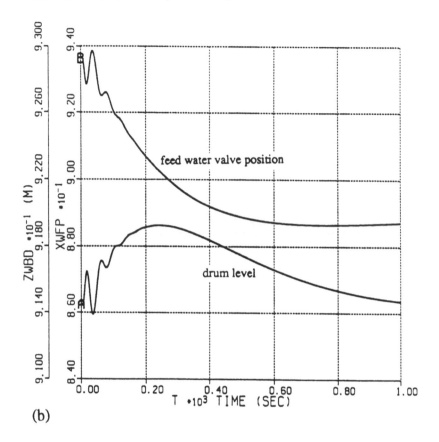

(b)

to the control rod positions, and two delayed neutron precursor variables related to control rod positions.

The inputs available for control are the coolant gas flowrate and the positions of a selected number of the control rods. A disturbance input is the temperature of coolant gas at the reactor inlet. Outputs that can be measured and used for control are the coolant-gas inlet temperature and the delayed coolant-gas outlet temperature.

Variables that cannot be directly measured but which are vital to safe operation are the reactor power output and the temperatures of the coolant gas outlet, the gas in the annulus, the uranium fuel, the fuel can, and the moderator. There are also several constraints on the variables and these must be observed during transient changes.

The ACSL implementation has 27 files. These include a number of ACSL command files for simulating step changes in the three input variables, for determining equilibrium values at four power levels, and for linearizing the model to derive a state-space form.

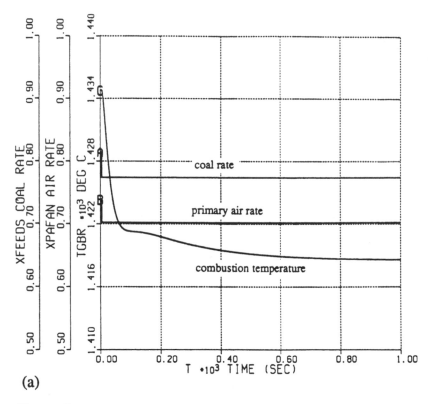

(a)

Figure 8 Step input test on ANYDYM model at 600 MW over 1000 sec.

Dulson and Merriman [8] give numerical values of all variables for an equilibrium power level of 85.19% rated power. These values have been used in both the ACSL and TSIM2 simulations.

In the case of the ACSL simulation, other equilibrium conditions can be computed corresponding to power levels of 100%, 70%, and 50% of rated power. Linearized models in the form of MATLAB files have been derived at each of these power levels and used to calculate the eigenvalues of the *A* matrices. A variety of ACSL command files have been generated to assist the user in defining simulation parameters, running step responses for each of the three inputs, trimming the model at any of the four equilibrium positions, and linearizing the model.

Figure 10 illustrates a simulated transient following a step reduction in the coolant-gas flowrate. It shows that the temperature of the gas in the main channel rises rapidly to a new level, and the temperature of the uranium becomes unstable.

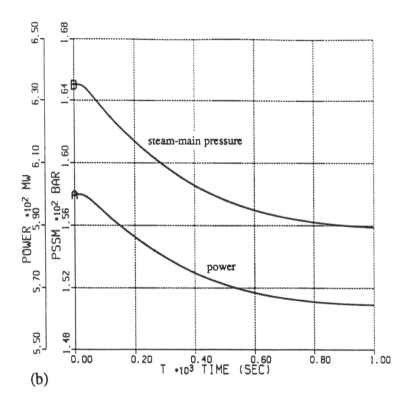

(b)

Vaporizer

Anderson [1] describes an industrial chemical process unit that comprises a vaporizer and superheater feeding a chemical reactor. The key variables in the unit are defined in Figure 11. When the feed rate exceeds a certain level, the response of the system enters a limit cycle. The process, as simulated in ACSL, has five state variables; an additional four states are associated with the three valve positioners and the feed controller. The feed rate to the preheater can be controlled as can the vapor rate to the reactor. The reactor product flow is split, some passing through the vaporizer heat exchanger and the rest bypassing the vaporizer. The latter, bypassed flow, is controllable. The use of heat from the reactor product to raise the temperature of the reactor inlet stream clearly saves energy. However, it can cause dynamic instability in the operation of the plant under certain load conditions.

There are nine state variables in the model, namely, liquid volume hold-up in the vaporizer, the liquid temperature in the vaporizer, the vapor temperature in the vaporizer, the vaporizer outlet temperature, the mass hold-up of vapor in the reactor, the integrator constant of the PI controller of vaporizer feed valve, and

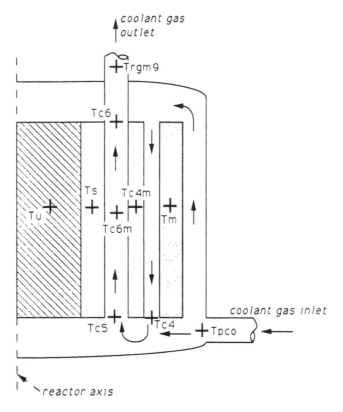

Figure 9   Advanced gas-cooled reactor diagram. (+), Points at which temperatures are defined; (→), direction of coolant gas flow.

the positions of the valves controlling the vaporizer feed, the reactor feed, and the vaporizer bypass flow.

The inputs are the demanded input feed flowrate, the demanded vapor flow rate to the reactor, and the demanded bypass valve position. The outputs are the temperature of the feed to the distillation column and the vapor flowrate from the vaporizer to the reactor. An important constraint is that the liquid level in the vaporizer must remain positive at all times.

The control objective is to provide good regulation of the feed stream temperature to the distillation column. This must hold over a wide range of feed rates and in the presence of ambient changes in feed temperature and reactor outlet temperature.

In Figures 12a and 12b are shown responses following a 5% increase in feed rate over a nominal rate of 20 kg/sec. It can be seen that limit-cycle oscillations occur in all variables. At feed rates lower than 18 kg/sec oscillations do not occur.

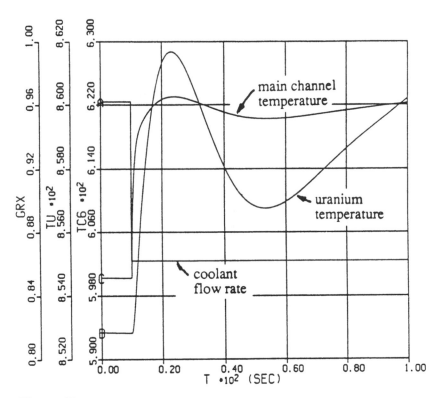

Figure 10 Advanced gas-cooled reactor model response to step in coolant gas flow.

Condenser

This nonlinear model represents the overheads condenser and reflux drum of an industrial distillation process. A schematic view of the process is shown in Figure 13, together with a list of the variables. Vapor from the column top is condensed and collected in a reflux drum. A proportion of the condensate is drawn off as product and the remainder is fed back to the column top as reflux. In this particular plant, the configuration is unusual in that the condenser is mounted at ground level and the reflux drum is near the column top. In a normal configuration, the condenser is mounted above the reflux drum near the column top. The problems of controlling this plant are discussed by Oglesby [21].

For the purposes of simulation using the ACSL language, the process is represented by three interacting nonlinear models as shown in Figure 14. The distillation column section represents the dynamic relationships in the top of the column. There are six state variables, namely, the column pressure, the liquid exit temperature and liquid mass hold-up of the condenser, the cooling water outlet temperature, and the temperature and liquid mass hold-up of the reflux drum.

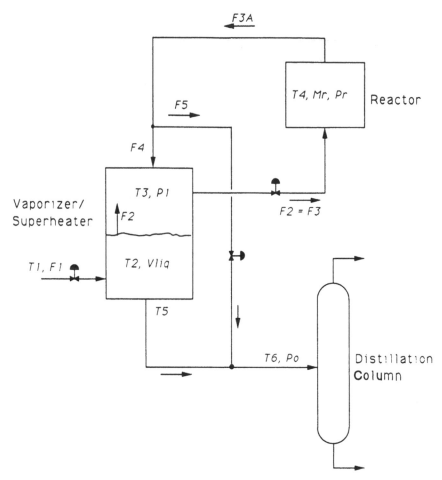

Figure 11   Vaporizer layout.

The control inputs are the top-product flow valve position, the vapor bypass valve position, and the column reboil rate. The measurable outputs are the column pressure, the reflux drum pressure, the reflux drum level, the reflux mass flowrate, and the top-product mass flowrate. Tests on the real plant showed that when the top-product and reflux mass flowrates were held constant, lightly damped oscillatory behavior followed the application of a short pulse in feed rate. The ACSL simulation of this phenomenon is shown in Figure 15.

Extruder
This model relates to a plastic extrusion process [15, 27]. It is used for continuously coating wire with plastic. The control objective is to achieve a given

overall coated-wire diameter. A schematic view of the process is shown in Figure 16.

Polymer pellets from a hopper are plasticized and mixed to form a homogeneous melt by means of a screw feeder inside a heated barrel. The screw action forces the molten plastic through a die onto a copper wire which is fed through the die. Coated wire then passes through a cooling trough after which the overall wire diameter is measured.

For modeling purposes, the extruder process has been represented by three submodels which generate three measurable outputs from two control inputs. The line speed and screw speed submodels are linear and the extrusion submodel is nonlinear.

The line speed submodel is of second order and generates the measurable output (line speed) from a control input (demanded line speed). There are two state variables: line speed itself and the intermediate, unmeasurable state variable. The second linear submodel is also of second order and generates screw speed from demanded screw speed. Again there are two state variables: screw speed and the unmeasurable variable. The nonlinear submodel generates coated-wire diameter from line speed and screw speed. This submodel exhibits a substantial time delay which is proportional to line speed and is subject to disturbances arising from fluctuations in coating thickness due to plastic melt temperature variations and to wire diameter sensor noise. Control of the process presents interesting problems due to the speed-dependent transport time delay and the high level of wire diameter measurement noise. A typical set of responses to changes in line speed are given in Figure 17.

The ACSL simulation uses macros to simulate the model and submodels and to provide test inputs; the TSIM2 implementation uses subroutines for the same purposes.

Propfan Engine

A Rolls Royce RB509 propfan engine is illustrated in Figure 18 and the model structure, comprising submodels of engine components, is shown in Figure 19. The model has 22 state variables which include the integrators in PI controllers. A novel feature of the model is the use of 40 look-up tables implementing functions of one and two independent variables. The engine has been simulated in TSIM2 code.

Turbofan Engine

This model relates to a Rolls Royce Pegasus vectored thrust turbofan engine, together with its fuel system and digital controller. The engine layout is shown in Figure 20. The air entering the engine is initially compressed by the low-pressure (LP) fan. Two-thirds of the LP delivery is discharged to atmosphere to generate the front nozzle thrust, XGF; the remainder enters the high-pressure

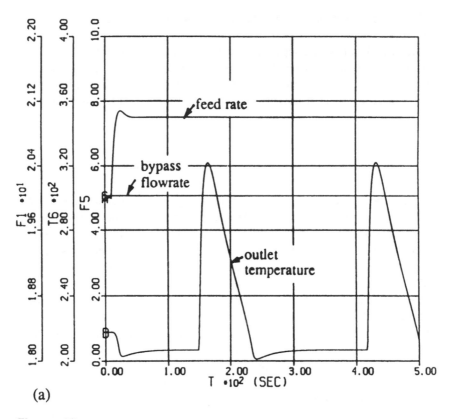

Figure 12 Vaporizer model responses to 5% feed rate change.

(HP) compressor. The HP delivery is discharged to the combustion chamber where a fuel–air mixture is burned to produce the exhaust gases. These gases pass through the HP turbine which is coupled to the HP compressor and then through the LP turbine which is coupled to the LP compressor. The gases leaving the LP compressor produce the rear nozzle thrust.

The engine model comprises eight nonlinear submodels, shown in Figure 21, generating eight measurable outputs from the single control input represented by the delayed fuel flow (F). The model uses 19 look-up tables of nonlinear functions which take into account the environmental factors influencing engine performance. Fuel is delivered to the engine through a control valve driven by a stepper motor. The stepper motor position is modulated by a signal from a sensor of the HP compressor delivery pressure as a protection against compressor surge. The stepper motor is controlled by a digital controller which uses the sampled difference between low-pressure turbine spool speed, NL, and the demanded speed to generate a signal to drive the fuel valve motor.

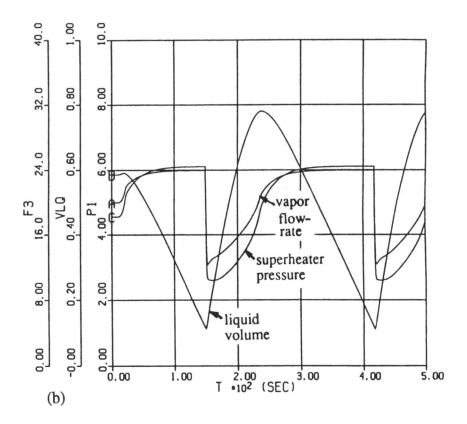

(b)

The overall model, whose structure is indicated in Figure 21, has nine state variables. Two of these are in the engine itself (the speeds of HP and LP turbine spools), two are in the fuel system, and five are in the digital controller.

10.3 THE MODEL LIBRARY

A Control Information Database has been established at the University of Glasgow. This is described by Ballance [2,3]. This database will hold a wide range of information of use to the academic control community and is accessible over the Joint Academic Network (JANET). Included in the database is the Model Library which now contains the Anydym, Condenser, Extruder, and Turbofan engine nonlinear models as described in the previous section.

The structure of the Model Library is designed to provide a remote user with all the necessary information relating to a chosen model. Code can be downloaded into the remote user's machine and can be run, provided that the appropriate version of the ACSL or TSIM2 simulation language is present.

Variable	Name
Column reboil rate	FCOLRB
Vapor bypass valve position	XBPV
Cooling water inlet temperature	TCOOLI
Cooling water mass flow	FCOOLW
Top product mass flow	FTP
Reflux mass flow	FRX
Column pressure	PCOL
Column temperature	TCOL
Cooling water outlet temperature	TCOOLE
Condenser liquid exit temperature	TCONLE
Reflux drum pressure	PRDM
Reflux drum temperature	TRDM
Reflux drum level	LRDM
Vapor bypass mass flow	FVBP
Condensation rate	FCOND
Condenser to reflux drum heat flow	QCONRD
Condenser to reflux drum mass flow	FCONRD
Reflux drum mass hold up	MRDM

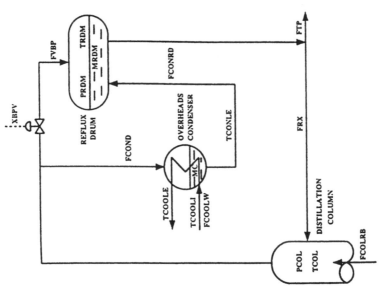

Figure 13 Condenser and reflux drum layout with list of variables.

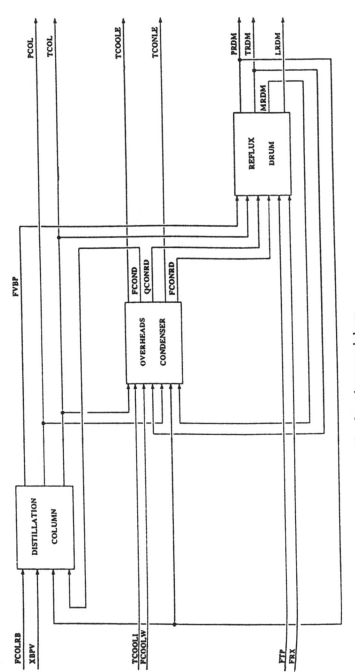

Figure 14   Interacting nonlinear submodels of condenser and drum.

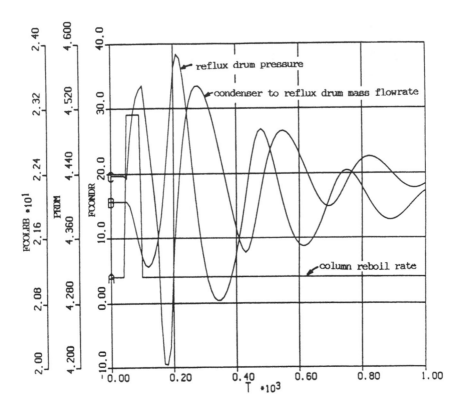

Figure 15 Condenser model transient response.

The Model Library files are organized into a Top directory and Model directories. The Top directory comprises (1) a file showing the structure of the Library and how to access it and (2) a Model Index file which lists all the models currently implemented. Each Model directory comprises a Contents file which lists all files in the directory, a Postscript file containing the model documentation, and directories containing implementation codes, including Make files, command files, and example runs.

The Model Documentation file contains a title page, including version and date, a summary and introduction, the model structure, and model implementations, including user guide and graphical results, acknowledgments, and references.

The Model Implementation Directory comprises the model source code (in ACSL, TSIM2, or MATLAB), a Make file that can be used to produce executable code for running the model, command or script files to assist in running the code, a contents file listing and briefly describing contents of command or script files, and diary files of sample runs.

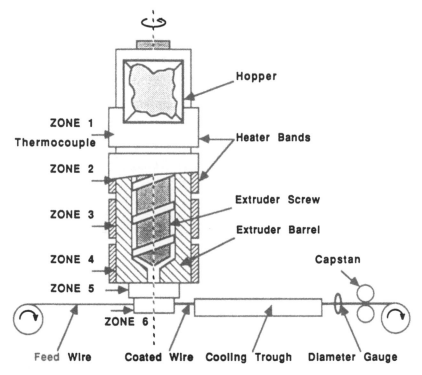

Figure 16 Plastics extruder system.

The model source code files contain a model header, including model name and version, a list of model variables, and the source code itself. The command files contain a command header, description, and list of commands.

The Control Information Database, on which the Model Library is held, can be accessed in a number of ways as follows:

1. By electronic mail-server:
    archive-server@uk.ac.glasgow.eng.control (via JANET in the UK)
    or
    archive-server@control.eng.glasgow.uk (from outside the UK)
    The archive server will respond to a message body containing the word HELP.
2. By JANET NIFTP interactive access:
    Host: uk.ac.glasgow.eng.control   (DTE 000071105002)
    Username: guest   Password: your e-mail address
3. By Internet ftp interactive access:
    Host: dodo.ctrl.eng.gla.ac.uk   (130.209.160.14)
    Username: anonymous   Password: your e-mail address

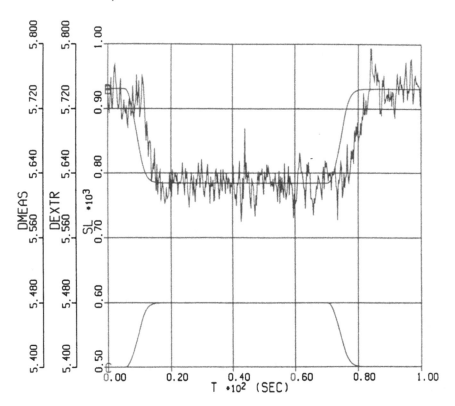

Figure 17 Extruder model responses.

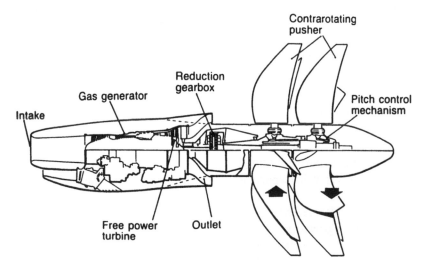

Figure 18. Propfan engine outline (with acknowledgement to Rolls Royce PLC).

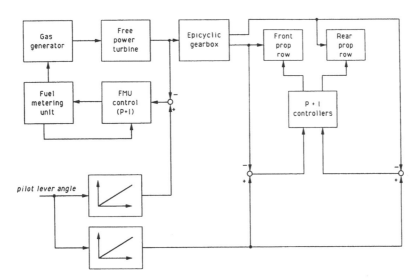

Figure 19. Propfan engine block diagram.

10.4 CONCLUSION

The need for an accessible portfolio of well-tested and nontrivial dynamic models of industrial plant has been identified. The work described attempts to satisfy this need by providing seven linear and seven nonlinear models. The linear models can be obtained directly from the references included here. Four of the nonlinear models are now available by electronic mail access to the Control Information Database now established at the University of Glasgow.

It is not intended that the Library shall be exclusive to the models described here. For this reason, the format of the Library models and the associated descriptive material has been standardized so that users may contribute their own models to the Database. In this way, the Library may be extended to include further models. Suggested subjects for further models are helicopter dynamics, robotics, space structures, chemical reactors, and steel-making plant.

The simulation languages ACSL and TSIM2 were chosen for the nonlinear models in the Library. Models may be entered into Library written in other simulation languages, provided that the basic protocols are adhered to.

ACKNOWLEDGMENTS

The authors gratefully acknowledge the assistance of those staff members of Imperial Chemical Industries plc, Nuclear Electric plc, Rolls Royce plc, and Eurotherm International Ltd plc whose enthusiastic cooperation made this work possible.

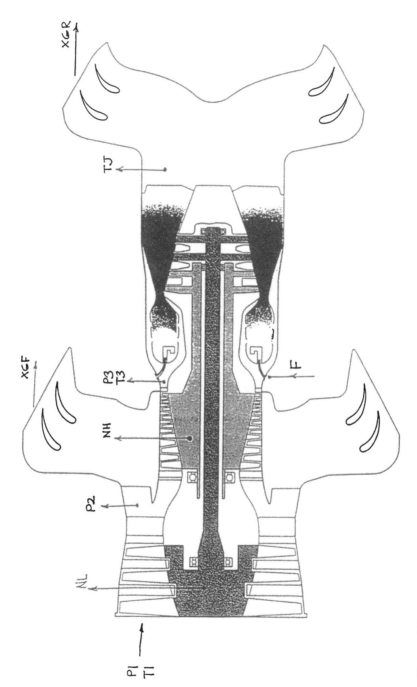

Figure 20 Turbofan engine outline (with acknowledgment to Rolls Royce PLC.)

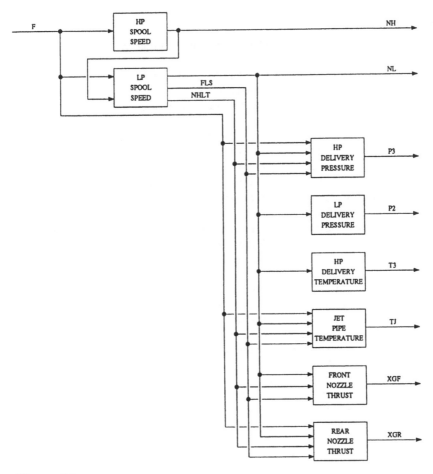

Figure 21 Turbofan model layout.

The work described was a team effort and the extensive contributions of colleagues of the authors at University College, Swansea, UK and the Rensselaer Institute, Troy, NY, USA are gratefully acknowledged. These colleagues include Professor H. A. Barker, Dr. C. P. Jobling, Dr. T. Huynh-Quoc, F. M. Seiler, and I. Harvey. The contribution of Dr. D. J. Ballance at the University of Glasgow is also acknowledged.

Acknowledgment is also made to the UK Science and Engineering Research Council for their financial support of this work through a grant (ref. SO/8/90) and a visiting research fellowship for one of the authors (ref. GR/F/28960).

REFERENCES

1. Anderson, J. S., A practical problem in dynamic heat transfer. *The Chemical Engineer* (May 1966).
2. Ballance D. J., Control Information Database, University of Glasgow, Dept. of Mechanical Engineering, Research Report A-91, 1991.
3. Ballance D. J., Control information database, *American Control Conference*, *Chicago*, 24–26 June 1992.
4. Barker, H. A., Hammond, P. H., Huynh-Quoc, T., Jobling, C. P., and Frederick, D. K., A model library for the control engineering community. *Proc. IEE Conference, Control 91, Edinburgh, UK*, 1991.
5. *TSIM User's Guide*, Cambridge Control, Ltd., Cambridge, UK, 1990.
6. *The CEGB Plant Modelling System Program. An Introductory Guide*, Central Electricity Generating Board.
7. Crossley, T. R., Munro, N., and Henthorn, K. S., Design of aircraft autostabilisation systems using the inverse Nyquist array method, *Proc. of IFAC Symposium on Multivariable Technological Systems, Fredericton, New Brunswick, Canada*, Pergamon Press, Oxford, 1977.
8. Dulson, P., and Merriman, G. P., Specification of equations, data, and nomenclature for the A. G. R. design and simulator reference models—Hinkley Point B, Central Electricity Generating Board (CEGB) Report PKR/SE/155, 1978.
9. Foss, A. S., Edmunds, J. M., and Kouvaritakis, B., Multivariable control system for two-bed reactors by the characteristic locus method, *Ind. Eng. Chem. Fund.*, *19*, pp. 109–117 (1980).
10. Frederick, D. K., and Rimer, M., Control system design benchmark problems, *IEEE Contr. Syst. Mag.*, 7 (No. 5), 19 (1987).
11. Frederick, D. K., A Library of plant models for use with computer-aided control system analysis and design packages, Control and Computer Aided Engineering Group, Dept. of Electrical and Electronic Engineering and Dept. of Mathematics and Computer Science, University of Wales, Swansea, UK. Report TR-1989 No. 14, 1989.
12. Frederick, D. K., Barker, H. A., Hammond, P. H., Jobling, C. P., and Seiler, F. M., Plant models for use with computer-aided control system design packages, *Proc. 5th IFAC/IMACS Symposium on Computer Aided Design in Control Systems, Swansea, UK*, July 1991.
13. Hackney, R. D., Miller, R. J., and Small, L. L., Engine criteria and models for multivariable control system design, in *Alternatives for Linear Multivariable Control*, (M. K. Sain, J. L. Peczkowski, and J. L. Melsa, eds.), National Engineering Consortium, Inc., Chicago, 1977.
14. Hung, Y. S., *Multivariable Feedback: A Quasi-classical Approach*, Springer-Verlag, Berlin, 1982, p. 27.
15. Merki, H. A., Control of diameter and capacitance of products with cellular insulation, *Wire Ind. Mag.*, Jan. 1983.
16. *ACSL Reference Manual*, Mitchell and Gauthier Associates, Concord, MA, 1986.
17. Morris, A. J., Nazer, Y., and Wood, R. K., Multivariable self tuning control—theory and experimental evaluation, *Control and its Applications*, IEE, London, 1981, pp. 106–112.

18. Munro, N., ECSTASY—A control CAD environment, *Proceedings of the International Conference, Control 88*, IEE, London, 1988, pp. 76–80.
19. *Dynamic Modelling—PMSP*, Nuclear Electric PLC. 1990.
20. Nuske, D. J., Multivariable Control Studies on an Automotive Gas Turbine, Ph.D. thesis, UMIST, Manchester, UK, 1976.
21. Oglesby, M. J., Use of dynamic simulation to solve a distillation column pressure control problem. in *Mathematical Modelling of Industrial Processes*, (P. C. Hudson, and M. J. O'Carroll eds.), EMJOC Press, Northallerton, UK., 1983, pp. 199–217.
22. Patel, R. V., and Munro, N., *Multivariable System Theory*, Pergamon Press, Oxford, 1982.
23. Rimer, M., Frederick, D. K., and Huang, C. Y., Solutions of the second IEEE benchmark problem, *IEEE Control Syst. Mag.*, August, pp. 33–39 (1990).
24. Sain, M. K., The theme problem, in *Alternatives for Linear Multivariable Control*, (M. K. Sain, J. L. Peczkowski, and J. L. Melsa, eds.), National Engineering Consortium, Inc., Chicago, 1977.
25. Sidders, J. A., ANYDYM—A Fossil Fired Total Plant Mathematical Model, Central Electricity Generating Board Report PKR/SE/255, 1989.
26. Singh, K., Examples of multivariable Control Systems, S.M. thesis, UMIST, Manchester, UK, 1987.
27. Smith, L., Applications of bond graphs to modelling industrial processes and manufacturing systems, *IEE Colloquium on Bond Graphs in Control*, 12 April 1990.
28. *PRO-MATLAB User's Guide*, The MathWorks, Inc., South Natick, MA, 1987.
29. Tysso, A., Brembo, J. C., and Lind, K., The design of a multivariable control system for a ship boiler, *Automatica*, *12*, pp. 211–224 (1976).
30. Wellstead, P. E., and Munro, N., Multivariable control of a cold iron ore agglomeration plant, *Proc. of IFAC Symposium on Multivariable Technological Systems, Fredericton, New Brunswick, Canada*, Pergamon Press, Oxford, 1977, pp. 447–450.

11

MATLAB: Its Toolboxes and Open Structure

P. J. Fleming and C. M. Fonseca

University of Sheffield
Sheffield, England

T. P. Crummey

University of Wales
Bangor, Wales

11.1 INTRODUCTION

MATLAB [1] is a high-performance interactive software environment for numeric computation. The package enables the user to express problems and solutions as if they are written mathematically, without the use of conventional programming. From the command line prompt the user can easily address problems of numerical analysis, matrix computation, and signal processing, and obtain results in both numeric and graphic formats. These are some of the features which make MATLAB so attractive for scientific and engineering applications in general and Control Systems Engineering in particular.

MATLAB was originally written in Fortran by Cleve Moler, so that matrix software developed by the LINPACK [2] and EISPACK [3,4] projects could be easily accessed. The new MATLAB program has been written in C by *The MathWorks* and has evolved over the years with input from many users, from both academia and industry.

Being an open environment, users are encouraged to write their own application tools, thus becoming contributing authors. This is certainly one of the most attractive features of MATLAB. Its user extensibility has led to the production of several MATLAB Toolboxes, covering fields such as control systems analysis and design, identification, optimization, and signal processing. The ability to interface directly with Fortran and C subroutines bridges the gap between MATLAB and the physical computing system, enabling the MATLAB environment to communicate effectively with, for example, data acquisition and signal processing platforms, parallel platforms, and graphical devices.

This chapter has two parts. In the first part, we provide an overview of MATLAB, its facilities, and Toolboxes. In the second part, we demonstrate how its open structure allows users to customize the software for their own purposes. This can be achieved through the development of special-purpose Toolboxes. Here we extend its adaptability by building a customized interface and linking it to parallel processing software as well as providing new tools to support our work with multiobjective optimization in control system design.

11.2 HARDWARE ENVIRONMENT

There are four versions of MATLAB for IBM and other MS-DOS compatible Personal Computers. These are PC-, AT-, 386-, and 386/Weitek-MATLAB, and they differ only in terms of performance and memory management. All of them require the presence of a numeric coprocessor and support all commonly used IBM display adaptors.

On Vax computers and Sun Workstations the software is called Pro-MATLAB and can drive several graphic terminals. It can also take advantage of both SunView and X-Windows graphical environments. Finally, there are versions of MATLAB for Apple Macintosh, Alliant, Apollo, Convex, Cray, DEC-stations and DECsystems, HP Workstations, IBM RS/6000 and Silicon Graphics. All versions will drive several hardcopy devices, including laser and dot matrix printers, pen plotters, and PostScript devices.

Another good feature of the package is that all of the data files it generates and/or uses are machine interchangeable. Some of these are ASCII files and therefore, are machine independent. The remainder contain a machine signature in the file header which enables MATLAB to perform the appropriate format conversion when necessary.

11.3 FEATURES

11.3.1 Data Structures

MATLAB provides essentially only one data type, a rectangular matrix of real or complex numeric elements, of variable size. Scalars are treated as 1×1 and vectors as $1 \times n$ or $n \times n$ matrices. Scalars and vectors may be assigned a special meaning according to the situation. Figure 1 shows examples of a scalar, a 5×1 vector, and a 4×3 matrix as they would be stored in MATLAB.

String variables can also be used, these being stored as row vectors containing the ASCII codes of the characters in the string. A special flag within the variable description caters for it to be displayed as text. Although being ineffi-

K = 0.5000 V = 2 M = 5 6 0
4 2 3 2
1.5000 1 −3 −1
3.1000 6 9 7
1

Scalar   Column vector   Matrix

Figure 1 Data structures in MATLAB.

cient in terms of storage space, this is acceptable given the reduced role strings play in MATLAB. Other entities, such as polynomials, are represented by storing the appropriate coefficients in matrix format.

11.3.2 Inputting and Outputting Data

MATLAB provides an interactive environment similar to that of an interpreted language such as BASIC, with a command prompt. Commands are typed in and results are displayed immediately, as shown in Figure 2. All variables are held in what is known as the workspace.

The workspace can be saved to disk either in binary (in MAT format) or ASCII, totally or partially, through the SAVE command. The converse action is performed by the LOAD command, which reads variables from disk into the workspace.

Variables can be displayed numerically on screen, or graphically. MATLAB's graphic abilities include linear, logarithmic, polar, mesh, and contour plots. Several line and marker types can be used in a single plot, and the graphics screen

```
>> M = [ 5  6  0
         2  3  2
         1 -3 -1
         6  9  7 ];
>> M * [ 1 3 2 ]'

ans =

    23
    15
   -10
    47
```

Figure 2 Data input example.

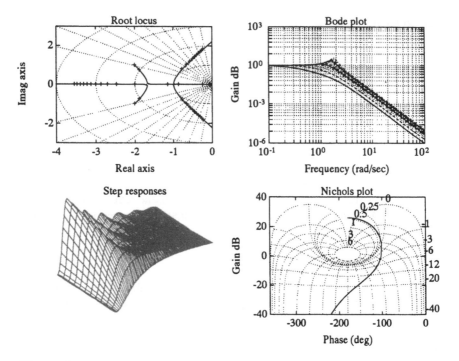

Figure 3 Plot examples.

can be divided to display up to four subplots simultaneously (see Figure 3). Data can also be input by driving a cross-hair pointer displayed over the graphics screen with the mouse and clicking at the relevant points.

11.3.3 Built-in Functions

The basic primitives for matrix manipulation and graphical output in MATLAB are implemented as built-in functions. Functions can be called with any number of input and output arguments, of any size. It is up to the function to display an error message if the information passed to it does not conform to the function's specifications.

Functions calls are of the following form:

$$[op1, op2, \ldots, opm] = \text{function name}(ip1, ip2, \ldots, ipn)$$

where op$i$ are the output and ip$j$ the input arguments. The dimensions of the output arguments are generally a function of those of the input, in a way which is transparent to the user. The user can then concentrate on what to do instead of having to be concerned with memory allocation and the like. In some cases, arguments are optional, which means that, if unspecified, they take default values. This allows a function to be easily tried out, whereas the default arguments can be adjusted at a later stage.

Also built-in are flow control structures such as if, for and while, constants such as π and $i = \sqrt{-1}$, and a whole set of other primitives.

11.3.4 M-files

M-files are text files containing MATLAB commands. They provide the majority of the more complex functions and allow the user to extend the library of functions available. The name M-file comes from the fact that they are stored on disk with an extension .m. There are two types of M-files:

Script files consist simply of a list of commands to be executed as though they were entered one at a time at the command prompt. Script files work on the main workspace and will be able to read and modify any current variables. The main workspace will reflect what the script has done on its completion.

Function files implement new MATLAB commands, which can take input and output arguments as in the case of built-in functions. Input arguments are passed by value, so no modification of their values within the function is visible from the environment. Also, all function variables are, in principle, local to the function.

MATLAB does allow the user to specify variables as global in scope. However, the use of global variables is discouraged.

When M-files are used, they are compiled into memory for faster execution. Thus, an M-file which is repeatedly used will run faster the second and subsequent times it is invoked.

11.3.5 MEX-files: Calling Fortran and C Subroutines

It is possible to call C and Fortran subroutines from within MATLAB, just as if they were built-in or M-file functions. These have to be compiled into MEX-files (files with .mex extension) and will be linked dynamically with MATLAB the first time they are invoked.

MEX-files enable large preexisting C and Fortran code to be called from MATLAB without having to be rewritten as MATLAB code. They can also be used to speed up bottleneck computations which would take too long to run in MATLAB, especially when for loops are involved. However, care should be taken not to use the MEX facility unless it is indeed necessary. M-files will be appropriate in most of the situations.

One of the most attractive ways in which MEX-files can be used is to directly access hardware such as A/D and D/A cards, parallel platforms, and graphical devices.

11.4 TOOLBOXES

MATLAB offers the possibility of using extra sets of commands, known as tool-boxes, which can be purchased separately. A toolbox generally consists of a collection of M-files which implement application-specific functions. MATLAB toolboxes currently available which are relevant to control systems design and analysis include:

Control Systems Toolbox [5]. Implements common control system design, analysis, and modeling techniques. It computes time and frequency responses, and root-locus measure of both continuous- and discrete-time model systems, represented in either state-space or transfer function format. Format conversion tools are also provided.

Robust-Control Toolbox. To be used in conjunction with the Control Systems Toolbox just described, this set of routines implements algorithms for LQG-based optimal control synthesis using loop transfer recovery and frequency-weighted methods, $\mathcal{H}_2$ and $\mathcal{H}_\infty$ optimal control synthesis, singular-value-based model reduction, and spectral factorization and model building. Richard I. Chiang and Michael G. Safonov co-authored this toolbox.

System Identification Toolbox [6]. Written by Lennart Ljung, this toolbox matches directly his textbook [7]. It incorporates the most common and useful parametric and nonparametric techniques available for estimation and identification. Models estimated using this toolbox can be handled directly by the Control Systems Toolbox.

Multivariable Frequency Domain Toolbox. This toolbox is a set of functions for the frequency analysis and design of MIMO systems. It supports both continuous- and discrete-time systems, including those containing transport delays, and will also handle nonsquare plants. This toolbox was co-authored by J.-M. Boyle, M. P. Ford, and J. M. Maciejowski.

MMLE3 State-Space Identification Toolbox. This toolbox is an enhanced MATLAB implementation of the classic MMLE3 computer code developed by Richard Maine and Kenneth Iliff, written by Garth Milne. It enables the user to identify just the unknown parameters of a partially known system while preserving its essential physics, as opposed to the black-box model implemented by the Identification Toolbox.

Spline Toolbox. Written by Carl de Boor, the toolbox implements the essential programs of the B-spline package described in his textbook [8].

Optimization Toolbox. [9] This toolbox is a set of linear, quadratic and nonlinear optimizers. It includes constrained and unconstrained optimization, least squares, minimax, and the goal attainment method for multiobjective optimization problems.

Signal Processing Toolbox [10]. This toolbox implements a large variety of modern signal processing techniques. Features include filter design, analysis and implementation, parametric modeling, FFT processing, power spectrum estimation, cepstral analysis, and two-dimensional signal processing.

Data Acquisition Toolbox. A collaborative development between Signal Processing Technology and Cambridge Control, this toolbox is for use with the PC range of MATLAB software products and comprises all the necessary software and hardware in a single package. MATLAB can be used to generate a test signal which is directed to one of the outputs of the analog interface. At the same time, real-time data can be acquired and readily made available as a variable in the MATLAB workspace.

μ-Analysis and Synthesis Toolbox [11]. Also known as μ-Tools, this toolbox was developed by MuSyn and is oriented toward the analysis and synthesis of robust control systems. It implements $\mathcal{H}_\infty$ optimal control and μ analysis and synthesis techniques over a consistent set of data structures which allow the unified treatment of systems in either a time or frequency domain, or in a state-space representation.

Neural Network Toolbox. This toolbox provides functions for designing and simulating neural networks. The toolbox, by Howard Demuth and Mark Beale, implements several learning rules, training, and transfer functions.

SIMULAB [12]. A program for simulating dynamic systems. It adds a new set of windows to MATLAB, known as block diagram windows, as illustrated in Figure 4. These enable the user to build up models mainly by using the mouse. After a model has been defined, it can be analyzed and simulated. Simulation results can be viewed while running and made available in the MATLAB workspace after completion. The standard block library provides the means of easily exciting a system with different inputs, for example, by using the Signal Generator and monitoring the output by using the Scope. The user can also define his/her own blocks by writing MATLAB functions. A block diagram can be grouped together in a single block to form a *subsystem*. Subsystems can be used in conjunction with standard blocks to build more complex systems, in a hierarchical approach.

11.5 EXAMPLE OF USE OF MATLAB—MULTIOBJECTIVE OPTIMIZATION IN CACSD

The design of system controllers to satisfy many, often competing, criteria is a demanding task, usually involving a trial-and-error approach to achieve a compromise over the initial design criteria. Multiobjective optimization (MO) is an attractive approach to such problems, affording the designer the opportunity to select the level of tradeoff between the different design objectives [13]. This approach has two practical drawbacks. The control engineer may be unfamiliar with numerical optimization and MO, in particular, and may need assistance in casting the problem in this form. Also, the numerical complexity of such an

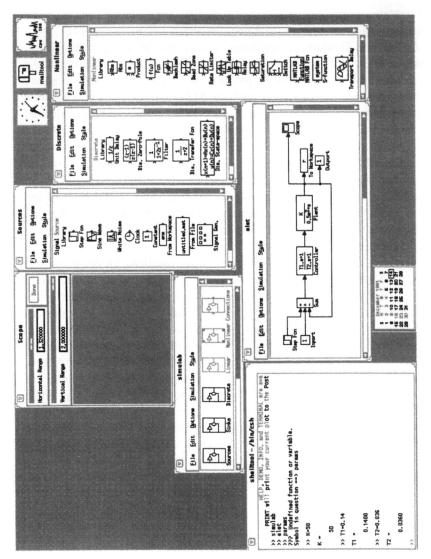

Figure 4   SIMULAB block diagram windows.

optimization approach may not be suitable for truly interactive computer-aided control system design (CACSD). In this second half of the chapter we describe how we have built upon MATLAB software to address these shortcomings.

The use of MO in control systems design recognizes that, in general, practical problems require a variety of often conflicting design objectives to be satisfied. In our approach using MATLAB, an X-Windows-based interface permits easy access to the MO design process and problem formulation. M-files and MEX-files are introduced to support the approach and we also describe how we can address the attendant computational burden of the MO approach by using a parallel processing platform attached to the workstation hosting MATLAB.

11.6 INTRODUCTION TO MULTIOBJECTIVE OPTIMIZATION

The application of traditional optimization techniques to control system design poses several difficulties for the design engineer. Usually (s)he is trying to satisfy several competing design goals and the traditional approach involves the aggregation of these different objectives into a single cost function (for example, the LQG approach). This obscures the impact of each design criterion on the optimization process and the solution. It is not apparent how well particular objectives have been met or how the solutions are constrained by the criteria. Furthermore, methods such as LQG often have the drawback of requiring linear system descriptions and quadratic integral measures of performance.

Multiobjective optimization, however, permits the separate design objectives to be simultaneously optimized. This allows the designer to see how each objective affects the final solution and also enables the prioritization of the different goals so that more important performance goals are met at the expense of less important ones. Mathematically, the process is described as

$$\min_{x \in \Omega} \mathbf{f}(x) \tag{1}$$

where x is the design parameter vector, Ω the feasible parameter space, and $\mathbf{f}$ the set of objective functions.

There is no unique solution to this form of problem; rather a set of "noninferior" [13] solutions is generated from which the designer can choose the compromise most suited to the system (s)he is designing.

MO can be implemented by several algorithms, but the goal attainment method due to Gembicki [14] has been found to be suitable for CACSD and is available in the MATLAB Optimization Toolbox [15,16]. Using this method to obtain a noninferior solution for the MO problem, Eq. (1), the following nonlinear programming problem is solved.

$$\min_{\lambda, x \in \Omega} \lambda$$

such that

$$f_i - w_i\lambda \leq f_i^*$$

where f_i^* are prespecified goals for the design objectives f_i, and $w_i \geq 0$ are designer-selected weights. Minimization of the scalar, λ, tends to force the objectives (or design specifications) to meet their goals. It is possible for the designer to set unrealizable goals and still obtain a solution; in this case, the goals will be underattained. The quantity $w_i\lambda$ represents the degree of underattainment or overattainment of the corresponding goal, f_i^*

11.7   CACSD AND OPTIMIZATION

Optimization may be attractive to the control system designer in theory, but when it comes to taking a system description and reformulating it in a form suitable for the application of standard numerical routines, the effort involved may make the process less appealing. To aid the designer, we have developed a user interface shell to integrate with MATLAB/SIMULAB which permits the automatic use of SIMULAB models and the easy expression of most common system design goals.

To make best use of this optimization approach, especially for nonlinear systems, it is necessary to be able to easily integrate the optimization software with a simulation package since the system may be simulated many times during each optimization process. The combination of SIMULAB and MATLAB gives the designer the ability to model and analyze nonlinear systems in one environment.

The user interface runs on a UNIX workstation using X-windows (as does SIMULAB) and is directly linked to SIMULAB via the MEX-file utilities of MATLAB. The interface fits into an optimization environment as shown in Figure 5. Here it can be seen that the interface communicates only with MATLAB/ SIMULAB and calls MATLAB commands to perform all its functions. This architecture means that the implementation of the optimization commands is encapsulated and thus could be written as MATLAB M-files, C, or FORTRAN MEX-files, or as parallel routines on a parallel processing platform.

11.8   USER INTERFACE FOR MO

This interface is designed to allow the user to perform the following tasks:

- Formulate the MO problem, where the control system (plant and controllers) may be described in a SIMULAB block diagram.
- Specify system design criteria in the time and frequency domain.
- Manage the optimization process allowing starting, pausing, and restarting of the procedure.

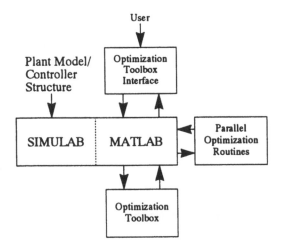

Figure 5 The control system design environment.

- View the results of the optimization in the same domains as the original criteria (numerical or graphical).
- Provide context-sensitive help to the user when required.

These tasks are implemented in a tool that interfaces closely with SIMULAB via the MEX-file facility provided in MATLAB and SIMULAB. The interface code is run as a separate process to SIMULAB, with communication between the two being achieved via Berkeley UNIX-style "sockets."

The SIMULAB end of the communication channel is handled by a MEX-file (see Figure 6). The interface process sends commands and data to the communication MEX-file which calls the appropriate MATLAB/SIMULAB function. If results are expected, the interface process will request them to be read back through the pipe.

The use of a MEX-file to handle communication offers advantages over the more usual method of interfacing with MATLAB/SIMULAB. That involves the invocation of MATLAB/SIMULAB as an inferior process communicating via standard input/output. The advantages are that a proper communication protocol can be established between the two processes such that synchronization can be maintained.

In the past, environments such as ECSTASY [17] and PROTOBLOCK [18] have exhibited problems with errors being generated by MATLAB and subsequent loss of synchronization between MATLAB and the environment. Also, MATLAB does not always execute the first command after an error and these conditions have led to this form of communication being unreliable. The MEX-file route is reliable and offers the possibility of an error protocol for automatic exception handling.

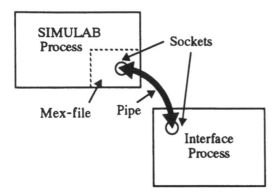

Figure 6 Communication between SIMULAB and the interface.

The interface is implemented using the C language and the XView toolkit on a SUN Workstation. The resultant program follows the rules of the Open Look standard [19]. There are several windows belonging to the optimization interface (see Figure 7). The main control window contains five buttons in start-up mode. These allow the user to read and save matrix workspaces which are compatible with the SIMULAB format, use a window-based matrix editor, start and control an optimization run, select an appropriate optimization routine, and quit the tool.

Figure 7 shows the MO routine window. Also visible is the time domain window where time response plots are displayed. This window allows the user to draw time-related criteria on the initial system time response so that (s)he can easily specify objectives graphically and monitor how the current response relates to these desired targets.

All the criteria for the optimization are displayed in equation form in the Optimization Control window and can be modified by the user if required. Control of the optimization process is achieved through a series of pop-up windows, one of which is shown in the diagram. These allow the termination criteria, the number of iterations, and various other control parameters specified in the Optimization Toolbox to be set. Context-sensitive help is available. An example is shown in Figure 7, where the Help help was pressed when the mouse was placed over the Optimizer Routine button. The Help window is at the bottom of the screen.

The tool also has a matrix editor which allows the user to view and edit matrices. It is invoked from a Workspace manager window and both are shown in Figure 7. Workspace 1 is shown with two matrices contained in it. One is the time response shown in the Time Domain window and the other is shown in the edit window which is below the Workspace manager. Individual elements of

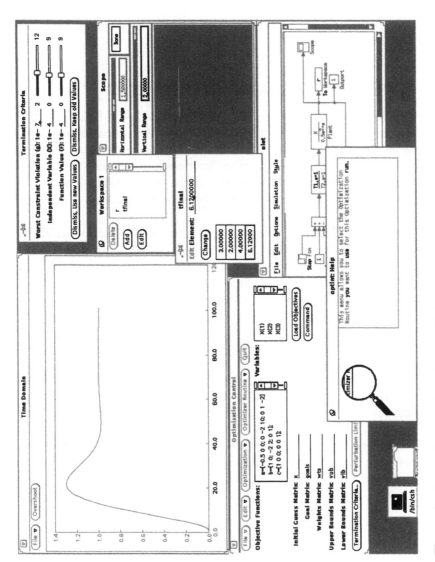

Figure 7 Screen dump obtained during MO design process.

the matrix (tfinal) can be chosen using the mouse and edited in the Edit item at the top of the window.

11.9 STRATEGIES FOR PARALLEL PROCESSING

Since the computation routines are all called via MATLAB/SIMULAB, it is possible using the MEX-file interface to implement the various optimization routines on a parallel processing platform. One implementation strategy involves mapping the individual objective functions onto separate processors. These objectives comprise the major part of the computational load of the optimization process since each one must be evaluated at every iteration of the algorithm.

Distributing the functions over several processors can lead to significant speed increases. However, because the objective functions need not be of similar complexity, such a simplistic arrangement is unlikely to lead to a well-balanced system. A method of identifying and further parallelizing the more complex objective functions is also being investigated.

At present, the parallel processing of the optimization algorithm is achieved using MEX-files. When the parameters and functions for the optimization are defined, say, using the user interface, a gateway routine, which is a MEX-file, constructs a message which is passed to the root parallel processing node. This then farms out each objective to the available processors. The gateway program passes the new estimates of the solution to the farmer node which passes these on to the individual processors and collects the results for return to the optimization program (see Figure 8).

The hardware currently used in the environment comprises a Sun Workstation coupled to a transputer-based platform consisting of 20 parallel processors. The arrangement is shown in Figure 9.

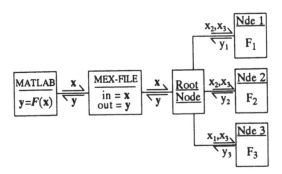

Figure 8 Mapping objective functions onto parallel processing nodes.

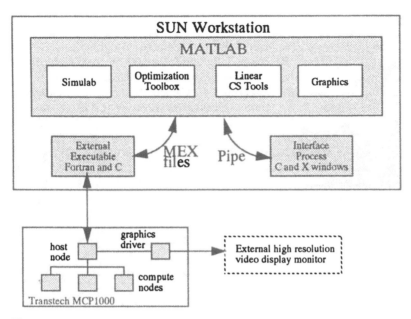

Figure 9 MATLAB/transputer platform interface.

11.10 CONCLUDING REMARKS

MATLAB is widely accepted by control engineers as an attractive environment for analysis, design, and simulation of control systems. Its "open" structure can be effectively used to construct customized "solution domains." Its utility has recently been significantly enhanced by the addition of the simulation package, SIMULAB, to the MATLAB Toolbox portfolio. To illustrate the extent to which the MATLAB environment can be adapted to meet individual user's needs, we have described our work in providing tools to support the use of multiobjective optimization in control system design. A powerful and flexible interface buffers the user from the intricacies of the optimization process and facilitates the formulation of the problem and monitoring of the design process. Using this basic framework, it has been possible to alleviate the computational burden by accessing a parallel processing platform.

ACKNOWLEDGMENTS

The authors wish to acknowledge the support of this research by UK SERC grants, GR/F23866 and GR/F94064, the EC ESPRIT Parallel Computing Action, and a NATO Collaborative Research award with Professor J. L. Junkins (Texas A&M University). Carlos Fonseca gratefully acknowledges support by the Programa CIENCIA, Junta Nacional de Investigação Científica e Tecnológica, Portugal.

REFERENCES

1. *MATLAB User's Guide*, The MathWorks, Inc., Concord, MA, 1991.
2. Dongarra, J. J., Moler, C. B., Bunch, J. B., and Stewart, G. W., *LINPACK User's Guide*, Society for Industrial and Applied Mathematics, Philadelphia, 1979.
3. Smith, B. T., Boyle, J. M., Dongarra, J. J., Garbow, B. S., Ikebe, Y., Klema, V. C., and Moler, C. B., *Matrix Eigensystem Routines—EISPACK Guide*, Springer-Verlag, Berlin, 1976.
4. Garbow, B. S., Boyle, J. M., Dongarra, J. J., and Moler, C. B., *Matrix Eigensystem Routines—EISPACK Guide Extension*, Springer-Verlag, Berlin, 1977.
5. Grace, A., Laub, A. J., Little, J. N., and Thompson, C., *Control System Toolbox User's Guide*, The Math Works, Inc., South Natick, MA, 1990.
6. Ljung, L., *System Identification Toolbox User's Guide*, The MathWorks, Inc., South Natick, MA, 1991.
7. Ljung, L., *System Identification—Theory for the User*, Prentice-Hall, Englewood Cliffs, NJ, 1987.
8. de Boor, C., *A Practical Guide to Splines*, Springer-Verlag, New York, 1978.
9. Grace, A., *Optimization Toolbox User's Guide*, The MathWorks, Inc., South Natick, MA, 1990.
10. Little, J. N., and Shure, L., *Signal Processing Toolbox User's Guide*, The Math-Works, Inc., South Natick, MA, 1988.
11. Balas, G. J., Doyle, J. C., Glover, K., Packard, A., and Smith, R., μ-*Analysis and Synthesis Toolbox*, MuSyn Inc., and The MathWorks, Inc., South Natick, MA, 1991.
12. *SIMULAB User's Guide*, The MathWorks, Inc., South Natick, MA, 1991.
13. Fleming, P. J., and Pashkevich, A. P., Computer aided control system design using a multiobjective optimisation approach, in *Proceedings of the Control 85 Conference (Cambridge, UK)*, 1985, pp. 174–179.
14. Gembicki, F. W., *Vector Optimization for Control with Performance and Parameter Sensitivity Indices*, Ph.D. thesis, Case Western Reserve University, Cleveland, Ohio, 1974.
15. Grace, A. C. W., *CACSD using Optimization Methods*, Ph.D. thesis, University of Wales, Bangor, 1989.
16. Crummey, T. P., Farshadnia, R., Fleming, P. J., Grace, A. C. W., Hancock, S. D., An optimization toolbox for MATLAB, in *Proceedings IEE Control '91* (Edinburgh, UK), 1991, Vol. 2, pp. 744–749.
17. Munro, N., An Overview of the ECSTASY environment, in *Colloquium on Computer Aided Control System Design—Environments and Methods*, *(London, UK)*, 1987, Digest 1987/97.
18. *Protoblock*, Grumman Aerospace Corporation Bethpage, NY, 1987.
19. *Open Look Graphical User Interface Functional Specification*, Sun Microsystems, Inc., Reading, MA, 1990.
20. Chipperfield, A. J., Fleming, P. J., and Jones, D. I., Optimization and computer aided control system design: Parallelisation strategies, in *Proceedings of the European Workshop on Parallel Computing (EWPC 92)*, IOS Press, 1992.

12

Programming Language MaTX for Scientific and Engineering Computation

Masanobu Koga and Katsuhisa Furuta

Tokyo Institute of Technology
Tokyo, Japan

12.1 INTRODUCTION

The interactive matrix manipulation package MATLAB [4,5] has become a standard among users and developers of numerical software. It has found widespread acceptance in application areas in which linear algebra plays an important role, such as signal processing, process identification, and control system design. For these applications, special toolboxes are available, which are collections of small procedures (m-files) written in the MATLAB language. Since programming m-files is very easy and also because the MATLAB environment is very well-suited for testing new ideas interactively, many developers of numerical software today prefer programming in MATLAB over traditional languages such as Fortran, Pascal, and C.

However, there are at least two points in MATLAB which should be improved. One is concerned with limited data structure. MATLAB has only one data type, that is, the Matrix; thus the other kinds of data are supposed to be coded using matrices. Such limited data structure does not allow a natural symbolic representation of fundamental mathematical objects such as polynomials and rational functions. This makes problem descriptions hard to understand. The other is concerned with speed of computation. It takes much time to calculate complicated algorithms because m-files are interpreted rather than executed after compilation.

We have developed a high-performance programming language MaTX for scientific and engineering computation [2,3]. Many of the important ideas of matrix operations of MaTX stem from the matrix manipulation language MATLAB, developed by Cleve Moler [4,5]. Although it shares several characteristic features with MATLAB, MaTX is in no sense a dialect of it. MATLAB is "typeless" languages: The only data type is the Matrix, and operations of other kinds of objects is by special operators or function calls. MaTX is equipped to recognize several primitive data types such as Integer, Real-Number, Complex-Number, String, Polynomial, Rational-Function, Matrix, Array, and List. Each class object is stored in an appropriate data structure and treated suitably. It deals with symbolic formulas as well as numbers. It can solve polynomial equations and evaluate derivatives and integrals symbolically. MaTX has inherited the idea of C language such as declaration of variables, control flow, style of function, and structure of a program. It is written all in C language and the parser has been developed by using the well-known parser generator "yacc."

Most implementations of MATLAB have a special interface which makes it possible to call separately compiled and linked Fortran (or C) routines (MEX-files) from the MATLAB environment, just as if they were ordinary m-files. This feature is useful if the m-file implementation of an algorithm is not efficient enough. Creating MEX-files, however, is no simple task. One needs to write a special "gateway function" for passing the function parameters from MATLAB to Fortran and back. The MATLAB gateway compiler has been developed [6] that automatically creates a gateway function for almost Fortran routines to be used in MATLAB.

MaTX provides not only command-line *interpreter* **matx** whose interfaces are similar to the use of MATLAB but also *compiler* **matc** which accepts the same syntax as that of matx and outputs portable C language code. The users can extend the functionality of programs by realizing algorithms as functions in "mm-files" which correspond to "m-files" of MATLAB. It is very easy to call your own C functions from MaTX functions by linking C language routines to the output of matc. This allows us to utilize huge preexist C routines and speed up the rate of computation and promote the efficiency of memory usage.

The translator which translates m-files to mm-files is under development. When the translator comes out, the users of MATLAB benefit from the MaTX interpreter and compiler facilities. But it is tough work to make the translator because the syntax of MATLAB is designed for interpreter rather than compiler. There is no class declaration statements in the MATLAB language because the interpreter knows everything about all variables at run time. Compilers, however, cannot get class information of variables and functions at compile time, unless there are class declaration statements.

The disposition of the chapter is as follows: The architecture of MaTX is outlined in Section 12.2, and the program developing process in MaTX is presented in Section 12.3. The matrix manipulation and the symbolic computation are described in Sections 12.4 and 12.5, respectively. Section 12.6 deals with the List class and the multiple output function. Section 12.7 explains what programming facilities are available in MaTX. Section 12.8 make a comparison of the performance between the interpreter (matx) and the compiled program which is produced by the compiler (matc). In Section 12.9, we touch on the graphical drawing environment XPLOT. Section 12.10 concludes with some perspective remarks.

12.2 ARCHITECTURE OF MaTX

The architecture of MaTX is shown in Figure 1. At the moment, MaTX provides three useful user-interface tools: "interpreter "matx," compiler "matc," and matrix-editor "mated." They are built on MaTX-Lib which is a collection of useful class libraries such as **Complex-Number, Polynomial, Rational-Function, Matrix,** and **List** [2,3]. The library set MaTX-Lib can be utilized to make programs in C language by itself. The matrix class library is composed of a higher-level library and a lower-level library. The lower-level library consists of nearly 300 functions, which provide fast access to the objects but do little to make programming easier. It handles the operation of the objects and includes some optimizations that encourage efficient memory usage. The higher-level library is built on the lower-level library and consists of about 200 functions. The purpose of the higher-level library is to provide the supervisory layer that supports the user-interface facilities. It handles the memory allocation of object and calls suitable lower-level functions according to the attribute of the object.

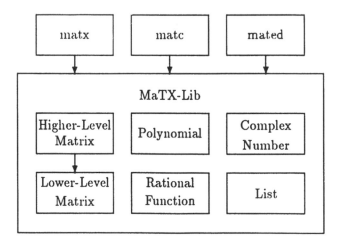

Figure 1 The architecture of MaTX.

Figure 2 indicates the internal process of MaTX. A source file (test.mm) written in the MaTX language is given to the lexical analyzer which recognizes the expressions in the input stream and partitions it into strings matching the MaTX expressions. The Parser calls the lexical analyzer to pick up the partitioned basic items (tokens). These tokens are organized according to the MaTX input structure rules; when one of these rules is recognized, the corresponding action for the rule is invoked. In each action, both the interpreter and the compiler defect type mismatches, inconsistent argument usage, unused or apparently uninitialized variables, potential portability difficulties, and the like. In case of operations between distinct class objects, "Class Conversion" is done according to the "Automatic Class Conversion Rule" shown in Figure 3. After the strict check of many aspects of a program, **matc** calls the "C Language Code Generator" while **matx** calls the "Internal Code Generator" and puts the generated code into execution.

12.3 PROGRAM DEVELOPING PROCESS

Figures 4 and 5 show the program developing process in MATLAB and MaTX, respectively. In the program developing process in MATLAB, the users usually realize algorithms in m-files (test.m) according to the requirements (specs). The m-files are tested several times until they satisfy the requirements. It is possible to call the user's C or Fortran subroutines (ctest.mtex), which are produced from source code (ctest.c) by C or Fortran compiler from the MATLAB

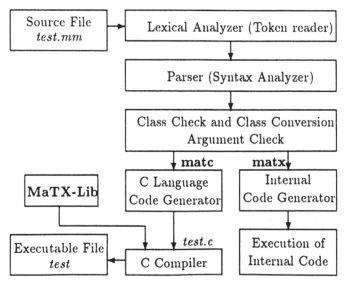

Figure 2 The internal process of MaTX1.

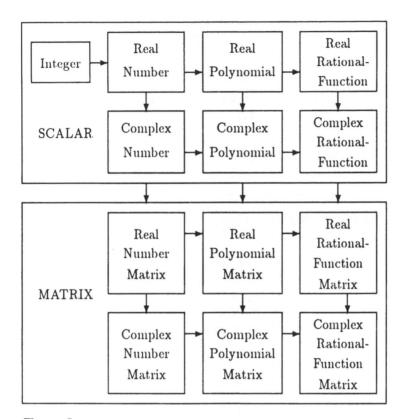

Figure 3 The "Automatic Class Conversion Rule."

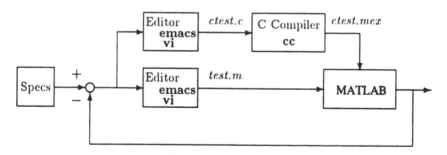

Figure 4 Program developing process in MATLAB.

function. In the program developing process in MaTX, the users ordinarily make several procedures (mm-files) using a favorite editor (**vi**, **emacs**, **or so**) according to the requirements (**specs**). The interpreter (**matx**) enables us to test out the mm-files interactively file-by-file or line-by-line. The examined mm-files are compiled to portable C language files by the compiler (**matc**). It is possible

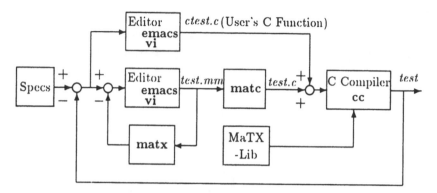

Figure 5   Program developing process in MaTX.

to call your own C functions from MaTX functions by linking C language files (ctest.c) to the output of matc (test.c). When all C language files are completed, they are compiled and linked up with the run-time library (MaTX-Lib) by C compiler (cc) to produce the desired executable program (test). Finally, the executable file is executed to check whether all requirements are satisfied.

12.4   MATRIX AND ARRAY

MaTX distinguishes Array class objects from Matrix class objects. Array class objects have the same data structure as that of Matrix class objects, but arithmetic operations of arrays are element-by-element instead of the usual algebraic matrix operations. For array operations of matrices, Matrix class has array operators denoted by the symbols .*, ./, .\, .^. The syntax of matrix manipulation in MaTX is very similar to that used in MATLAB. The subscripting abilities of MaTX allow manipulation of rows, columns, individual elements, and submatrices of matrices.

12.4.1   Entering Simple Matrices

The easiest method of entering a small matrix is to use an explicit list whose elements are separated by blanks, tabs, commas, or carriage returns, and surround it by square brackets, "[" and "]." Brackets are also used for indicating the beginning and end of the rows. For example, entering a statement

   A = [[1 2 3][4 5 6][7 8 9]];

or in more natural fashion:

   A = [[1 2 3]
        [4 5 6]
        [7 8 9]];

results in the output

```
=== [A]: ( 3, 3) ===
              ( 1)                ( 2)                ( 3)
( 1) 1.00000000E+00   2.00000000E+00   3.00000000E+00
( 2) 4.00000000E+00   5.00000000E+00   6.00000000E+00
( 3) 7.00000000E+00   8.00000000E+00   9.00000000E+00
```

12.4.2 Matrix Elements

Matrix elements can be any MaTX expressions except List class objects; for example,

$$x = [-1.3 , sqrt(3), (1+2+3)*4/5];$$

results in

```
=== [x]: ( 1, 3) ===
              ( 1)                ( 2)                ( 3)
( 1) -1.30000000E+00   1.73205081E+00   4.80000000E+00
```

Individual matrix elements can be specified with indices separated by a comma "," inside parentheses, "(" and ")", that is, $x(1,1)$ is equal to -1.3. Large matrices can be constructed by using small matrices as elements. For example, we could attach another row vector to our matrix **A** with

$$B = [10\ 11\ 12]; C = [[A][B]];$$

which result in

```
===[c]: ( 4, 3) ===
              ( 1)                ( 2)                ( 3)
( 1) 1.00000000E+00   2.00000000E+00   3.00000000E+00
( 2) 4.00000000E+00   5.00000000E+00   6.00000000E+00
( 3) 7.00000000E+00   8.00000000E+00   9.00000000E+00
( 4) 1.00000000E+01   1.10000000E+01   1.20000000E+01
```

Any submatrix can be extracted from the larger matrix, specifying the subportions with the left-top and the right-bottom locations, or the intervals of rows and columns. For example,

$$A = C(1,1,3,3);$$

or equivalently

$$A = C(1:3,1:3);$$

takes the 3×3 matrix to give us back the original matrix **A**. Furthermore, the dimensions of matrices can be changed. For example,

$$D(3:4) = C;$$

results in

```
=== [D]:( 3, 4) ===
         ( 1 )           ( 2 )           ( 3 )           ( 4 )
( 1) 1.00000E+00  2.00000E+00  3.00000E+00  4.00000E+00
( 2) 5.00000E+00  6.00000E+00  7.00000E+00  8.00000E+00
( 3) 9.00000E+00  1.00000E+01  1.10000E+01  1.20000E+01
```

The element ordering in the matrix C is left to right and top to bottom, which explains the ordering of the rearranged matrix in D. The *i*th row vector and the *j*th column vector can be extracted by A(i,*) and A(*,j) [or A(i,:) and A(:,j)], respectively.

12.4.3 Matrix Operations

Matrix operations are fundamental to MaTX; they are indicated the way they would be in a textbook or on paper, subject only to the character-set limitations of the computer.

The special character tilde "~" and the prime (apostrophe) "'" denote the inverse of a matrix and the transpose of a matrix, respectively. The special character sharp "#" denotes the complex conjugate transpose of a complex matrix; if A is a real-valued matrix, that is, real-number matrix, real-polynomial matrix, or real-rational-function matrix, the A# is its transpose.

Addition and subtraction of matrices are denoted by " + " and " − ". The operations are defined whenever the matrices have the same dimensions. Multiplication of matrices is denoted by "*." The operation is defined whenever the inner dimensions of the two operands are equal.

There are two matrix division symbols in MaTX, "/" and "\" as in MATLAB. If A is a nonsingular square matrix, then "A\B" and "B/A" correspond formally to left and right multiplication of B by the inverse of A. The expression "A^p" raises A to the pth power and is defined if A is a square matrix and p is an integer greater than zero.

12.4.4 Array Operations

We use the term **array operations** to refer to element-by-element arithmetic operations, instead of the usual linear algebraic matrix operations denoted by the symbols " +, /, \, ^, ', #, ~." Preceding an operator with a period " . " indicates an array or element-by-element operation.

For addition and subtraction, the array operations and the matrix operations are the same, so " + " and " − " can be regarded as either matrix or array operations. Array, or element-by-element, multiplication is denoted by ".*." If A and B have the same dimensions, then A .* B denotes the array whose elements are simply the products of the individual elements of A and B. The expressions A ./ B and A .\ B give the quotients of the individual elements.

12.4.5 Arithmetic Sequence

Arithmetic sequence can be created by using the special character, the colon ":" in side square brackets, "[" and "]." The statement

A = [1:5];

generates a row array containing the number from 1 to 5 with unit increment:

```
=== [A]:( 1, 5) ===
       ( 1)            ( 2)          ( 3)           ( 4)
( 1) 1.0000E+00    2.0000E+00    3.0000E+00    4.0000E+00
       ( 5)
     5.0000E+00
```

Increments other than 1 can be used. The statement

B = [0 : PI/4 : PI];

results in

```
=== [B]:( 1, 5) ===
       ( 1)            ( 2)          ( 3)           ( 4)
( 1) 0.0000E+00    0.7854E+00    1.5708E+00    2.2356E+00
       ( 5)
     3.1416E+00
```

It should be noted that the vectors created by the colon notation are not matrices but arrays. The colon notation allows the easy generation of tables. To get a vertical tabular form, transpose the row array obtained from the colon notation, compute a column of function values, then form a matrix from the two columns. For example,

x = [0.0 : 0.2 : 1.0]'; y = exp(x) * sin(x);
z = [x y];

produce

```
===[z] :( 6, 2) ===
       ( 1)                ( 2)
( 1) 0.00000000E+00    0.00000000E+00
( 2) 2.00000000E-01    2.42655269E-01
( 3) 4.00000000E-01    5.80943901E-01
( 4) 6.00000000E-01    1.02884567E+00
( 5) 8.00000000E-01    1.59650534E+00
( 6) 1.00000000E+00    2.28735529E+00
```

12.4.6 Piece-Matrix

Matrices may be considered as one-dimensional or two-dimensional arrays or a certain object such as real numbers, complex numbers, polynomials, and rational functions, whereas arrays of a matrix can be represented by introducing the piece-matrix notion. The piece-matrix is used to specify the unit size of arrays of matrix. For example,

A = [[1 2][3 4]]; AA = Z(2, 3, A);

lead to a 4 × 6 zero matrix AA.

AA(0, 0, A) = [[1 2][3 4]];
AA(1, 2, A) = [[5 6][7 8]];

result in

$$
AA = \begin{bmatrix} 1 & 2 & & & & \\ & & Z(2) & Z(2) & & \\ 3 & 4 & & & & \\ & & & & 5 & 6 \\ Z(2) & Z(2) & & & & \\ & & & & 7 & 8 \end{bmatrix}
$$

where matrix A is called piece-matrix of matrix AA, and the index of arrays of a matrix starts from 0. Using another piece-matrix B (= [1]) instead of A and

AA(0, 0, B) = [[1 2] [3 4]];
AA(1, 2, B) = [[5 6][7 8]];

yield another result:

$$
AA = \begin{bmatrix} 1 & 2 & 0 & 0 & 0 & 0 \\ 3 & 4 & 5 & 6 & 0 & 0 \\ 0 & 0 & 7 & 8 & 0 & 0 \\ 0 & 0 & 0 & 0 & 0 & 0 \end{bmatrix}
$$

12.4.7 Matrix-Editor (mated)

Figure 6 shows a screen-dump of the matrix-editor mated. The larger the matrix, the more complicated it is to enter a matrix. If you make a slight mistake in entering a matrix, you must start from the beginning. The matrix-editor "mated" provides us with a very convenient user interface to enter and revise matrices. Even if the users enter wrong characters such as alphabetic characters

```
Name :  [ A ]

Size :  [  50, 10 ]    Filename : A


              ( 3)           ( 4)           ( 5)           ( 6)

( 15)  -1.96996845E-01  8.96403785E-02 -1.96996845E-01  8.96403785E-02

( 16)  -1.96593060E-01  9.16088328E-02 -1.96593060E-01  9.16088328E-02

( 17)  -1.96189274E-01  9.35772871E-02 -1.96189274E-01  9.35772871E-02

( 18)  -1.95785489E-01  9.55457413E-02 -1.95785489E-01  9.55457413E-02

( 19)  -1.95381703E-01  9.75141956E-02 -1.95381703E-01  9.75141956E-02

( 20)  -1.94977918E-01  9.94826498E-02 -1.94977918E-01  9.94826498E-02

( 21)  -1.94574132E-01  1.01451104E-01 -1.94574132E-01  1.01451104E-01

( 22)  -2.08454259E-01  8.37854890E-02 -2.08454259E-01  8.37854890E-02

( 23)  -2.08050473E-01  8.57539432E-02 -2.08050473E-01  8.57539432E-02

( 24)  -2.07646688E-01  8.77223975E-02 -2.07646688E-01  8.77223975E-02

( 25)  -2.07242902E-01  8.96908517E-02 -2.07242902E-01  8.96908517E-02
```

Figure 6 Matrix-Editor (mated).

and special characters, it will cause no problem because mated takes in only suitable characters. The matrix-editor mated has the following features:

- To handle any size of matrices and change their size arbitrarily.
- Automatically scrolls if necessary.
- To cut and paste any subportion of the matrix.
- The key binding is similar to that of famous editors, **vi** and **emacs.**
- To read and write matrices from and to files.
- Do not require special window-system facilities.
- To do a few matrix operations.

12.4.8 Example 1

The MaTX has been used to illustrate the capabilities and features described above. The demonstration involves obtaining the solutions of the Riccati equations of continuous-time and discrete-time linear systems.

Discrete-Time Riccati Equation

List 1 is the program which obtains the solution of the Riccati equation of a discrete-time linear system by iteration. The following statements are explanations of List 1:

- On the 1st and 16th line, beginning of functions
- On the 3rd, 19th, and 20th line, declaration of local variables
- On the 6th to 8th, 10th, and 23rd line, entering constant matrices
- On the 9th, 27th, and 28th line, calculation of matrices
- On the 12th line, calling the function DRiccati( )
- On the 13th line, displaying the solution matrix P
- On the 17th line, declaration of arguments of the function
- On the 25th line, examination of frobenius norm of P − PP.

The solution matrix P and a temporary matrix PP are initialized with the identity matrix which has the same size as matrix A. To reduce the calculation error, matrix P is symmetrized. The calculation of the discrete-time Riccati equation is iterated until the Frobenius norm of the difference between matrix P and the previous solution matrix PP is less than eps.

```
     Func void main()
     {
       Matrix A, B, C, Q, R, P, PP;
       Matrix DRiccati();
05:
       A = [[0 1][-3, 4]];
       B = [0 1]';
       C = [-2, 1];
       Q = C'*C;
10:    R = I(1);

       P = DRiccati(A, B, Q, R);
       print P;
     }
15:
     Func Matrix DRiccati(A, B, Q, R)
       Matrix A, B, Q, R;
     {
       Matrix P, PP;
20:    Real eps;

       eps = 1.0E-7;
       P = PP = I(A);

25:    while (frobnorm(P - PP) > eps) {
         PP = P;
         P = Q + A'*P*A - A'*P*B*(R + B'*P*B)~ * B'*P*A;
         P = (P' + P)/2.0;
       }
30:
       return P;
     }
```

List 1 Solution of discrete-time Riccati equation.

Continuous-Time Riccati Equation

List 2 is the program which obtains the solution of the Riccati equation of a continuous-time linear system by Potter's method [1]. The following statements are explanations of List 2.

- On the 1st and 11th line, beginning of functions
- On the 3rd and 14th, declaration of local variables
- On the 6th line, reading matrices from standard input
- On the 7th line, call the function CRiccati ( )
- On the 8th line, displaying the solution matrix P
- On the 12th line, declaration of arguments of the function
- On the 16th line, making a Hamilton matrix [1]
- On the 19th line, calculation of eigenvectors
- On the 20th and 21st line, partitioning matrix vec using a piece-matrix A

Because the eigenvectors are ordered according to the magnitude of the real part of the eigenvalues, the statements on the 20th and 21st line extract the eigenvectors corresponding to the stable eigenvalues form the matrix vec.

12.5 SYMBOLIC COMPUTATION IN MaTX

Symbolic expressions such as polynomials, rational functions, polynomial matrices, and rational-function matrices are essential to the study of system theory and signal processing. MaTX permits the definition and manipulation of these

```
    Func void main()
    {
      Matrix A, B, Q, R, P;
      Matrix CRiccati();
05:
      read A, B, Q, R;
      P = CRiccati(A, B, Q, R);
      print P
    }
10:
    Func Matrix CRiccati(A, B, Q, R)
      Matrix A, B, Q, R;
    {
      Matrix H, U, V, vec, P;
15:
      H = [[ A   , -B*R~*B']
           [-Q   , -A'     ]];

      vec = eigvec(H);
20:   U = vec(0, 1, A);
      V = vec(1, 1, A);
      P = V*U~;
      return P;
    }
```

List 2. Solution of continuous-time Riccati equation.

objects in a natural symbolic fashion. It can solve polynomial equations and evaluate derivatives and integrals symbolically.

12.5.1 Polynomials and Rational Functions

MaTX handles polynomials and rational functions of one variable with real and complex coefficients. Polynomials and rational functions can be added, subtracted, multiplied, and divided as usual. At the moment, the coefficients of a polynomial are stored as a row vector containing the coefficients ordered by ascending powers, and a rational function is stored as a pair of polynomials, that is, a numerator polynomial and a denominator polynomial. Polynomials and rational functions are easily created and manipulated in MaTX. At least one polynomial variable must be defined in advance to describe symbolic expressions. For example, the statements

> x = $;
> A = 4 * x^2 + 5 * x + 6;

results in the output

> A = 4 s^2 + 5 s + 6

where x = $ is the definition of the polynomial variable and the character s is always used for the output polynomial variable. The class conversion function **Polynomial ( )** and **Matrix ( )** can be used to convert a row vector to a polynomial and vise versa. The coefficients of the polynomial are identical to the elements of the row vector in ascending order. For example, the statement

> B = Polynomial ([6 5 4]);

produces the same polynomial as the previous one. Individual polynomial coefficients can be specified with the index inside parentheses, "(" and ")." For example, the statements

> A(2) = 1.0; A(1) = 2.0; A(0) = 3.0;

results in the output

> A = s^2 + 2 s + 3

A rational function is described as a fraction of a pair of polynomials. For example, the statements

> x = $;
> B = (x + 1)/(4*x^2 + 5*x + 6);

result in the output

$$B = \frac{s + 1}{4s^2 + 5s + 6}$$

The numerator and the denominator of a rational function can be extracted by using the functions, Nu( ) and De( ), respectively. Nu(B) extracts a one-degree polynomial s + 1 and De(B) takes a two-degree polynomial 4s^2 + 5s + 6. Polynomials and rational functions, like real numbers and complex numbers, can be used as elements in matrices. This is a very useful feature of MaTX for system theory and is presented in the next section.

12.5.2 Symbolic Matrices

Essentially, the manipulation of symbolic matrices, that is, polynomial matrices and rational-function matrices, is identical with that for numerical matrices, that is, real matrices and complex matrices. There are at least two convenient ways to describe a polynomial matrix. They are illustrated by the following statements:

```
x = $;
A = [[1 2][3 4]]*x^2 + [[5 6][7 8]]*x
    + [[9 10][11 12]];
```

and

```
x = $;
A = [[ x^2+5*x+9, 2*x^2+6*x+10]
     [3*x^2+7*x+11, 4*x^2+8*x+12]];
```

which produce the same result. The former statement uses three matrices as coefficients of a polynomial. The latter one explicitly uses four polynomials as elements of a matrix.

A method of describing rational-function matrices is to use explicitly rational functions as matrix elements. For example, the statements

```
x = $;
A = [[(x+1)/(x+2) , (x+3)/(x+4)]
     [(x+5)/(x+6) , (x+7)/(x+8)]];
```

result in

$$A = \begin{bmatrix} \dfrac{s+1}{s+2} & \dfrac{s+3}{s+4} \\ \dfrac{s+5}{s+6} & \dfrac{s+7}{s+8} \end{bmatrix}$$

Another convenient way to describe rational-function matrices is to utilize the array operation of matrices. For example, the statements

```
x = $;
A = [[1 3][5 7]]; B = [[2 4][6 8]];
C = (ONE(2)*x + A) ./ (ONE(2)*x + B);
```

result in the same rational-function matrix as the previous one. The symbol "./" denotes the array operator for element-by-element division of matrices. The functions Nu( ) and De( ) extract the numerator part and the denominator part from rational-function arrays, respectively. For example, the statements

```
N = Nu(C); D = De(C);
```

result in

```
===[N]:( 2, 2) ===
   [  ( 1)  ]  [  ( 2)  ]
( 1)  s + 1        s + 3
( 2)  s + 5        s + 7

=== [D]:( 2, 2) ===
   [  ( 1)  ]  [  ( 2)  ]
( 1)   s + 2       s + 4
( 2)   s + 6       s + 8
```

12.5.3   Evaluation of Symbolic Expressions

Evaluation of symbolic expressions at a specified value can be done by using the function eval ( ). The evaluation of symbolic expressions at an array results in the array whose elements are simply the results of the evaluation at an individual elements. For example, the statements

```
x = $;                    A = 3*x^2 + 4*x + 5;
B = eval(A, 2.0);         C = eval(A, x+1);
D = eval(A, (x+1)/(x+2)); F = eval(A, [1:4]);
```

result in

```
B = 25
C = 3 s^2 + 10 s + 12
```

$$D = \frac{12\ s^3 + 62\ s^2 + 107\ s + 62}{s^3 + 6\ s^2 + 12\ s + 8}$$

```
F = [12 25 44 69]
```

respectively.

12.5.4 Derivative and Indefinite Integral

Derivatives and indefinite integrals of symbolic expressions can be obtained by using the functions derivative ( ) and integral ( ), respectively. The order of the derivative and the indefinite integral can be specified as the argument of the functions. For example, the statements

```
x = $;
A = 12*x^2 + 18*x + 10; M = [A A];
DM = derivative(M);   DA2 = derivative(A,2);
IM = integral(M);     IA2 = integral(A, 2);
```

result in

```
DM = [24 s + 18, 24s + 18]
IM = [4 s^3 + 9 s^2 + 10 s, 4 s^3 + 9 s^2 + 10 s]
```

and

```
DA2 = 24
IA2 = s^4 + 3 s^3 + 5 s^2
```

respectively.

12.5.5 Shift of Coefficients

The coefficients of the symbolic expressions can be shifted toward the lower and higher order by using the functions lower ( ) and higher ( ), respectively. The number of shifts can be specified as the argument of the functions. For example, the statements

```
x = $;
A = 12*x^2 + 18*x + 10; M = [A A];
LM = lower(M);     LA2 = lower(A,2);
HM = higher(M);    HA2 = higher(A,2);
```

result in

```
LM = [12 s + 18, 12s + 18]
IM = [12 s^3 + 18 s^2 + 10 s, 12 s^3 + 18 s^2 + 10 s]
```

and

```
LA2 = 12
HA2 = 12 s^4 + 18 s^3 + 10 s^2
```

respectively.

12.6 LIST CLASS

In the syntax of the MATLAB language, functions have the general form [a, b, c, ...] = fun(d, e, f, ...)

where the . . . denotes ellipsis. The a, b, c, . . . are left-hand-side arguments
and the d, e, f, . . . are right-hand-side arguments. In other words, functions take
multiple input arguments and return multiple output arguments. In the syntax of
traditional languages such as Fortran, Pascal, and C, functions use multiple input arguments and return only one value. To implement the multiple output
functions, we introduced the "List class" which is a collection of objects not
necessarily of the same type. Lists can contain any of the already discussed data
type as well as other lists. Functions can return multiple output values as a list
which contains them.

12.6.1 Entering Lists

The method of describing a List class object is to use an explicit list of objects
which are separated by the commas "," and surrounded by a couple of braces,
"{" and "}." For example, entering the statement

A = {4, (3,4), [1 2]};

results in a list which contains an integer, a complex number, and a real matrix.

12.6.2 List Elements

Any MaTX expressions can be used as elements of lists; for example, the
statement

A = {sqrt(3.0), ((1+2)*3, 4), [1 2] + [3 4]};

results in a list which contains a real number, a complex number, and a real matrix. Individual List elements can be extracted with the index and the class name
of the element separated by a comma inside parentheses, "(" and ")." For example, the expression A(1,Real) is equal to 1.732 and the statements
A(2) = (3,4); x = A(2,Complex); result in x = (3,4). Class name is required for compilation. Long lists can be produced from several short lists. For
example, we can append another list B to our list A by

B = {"Hello", 2.0};
C = {A, B};

which result in a five-element list

C = {1.732, (9,4), [4 6], "Hello", 2.0};

Any part of a list can be extracted by specifying the interval with two indices.
For example,

A = C(1,3);

extracts a three-element list to give us back the original list **A**.

12.6.3 Multiple Output Functions

In the syntax of the MaTX language, multiple output functions have the form

{a, b, c, ... } = fun(d, e, f, ...)

where the . . . denotes ellipsis. The a, b, c, ... are left-hand-side arguments and the d, e, f, ... are right-hand-side arguments. The function fun( ) returns multiple output values as a list which contains them. The number of left-hand-side arguments must be less than or equal to the number of the elements of the returned list. If the number of left-hand-side arguments is less than that of the elements of the list, then only the top elements of the list which correspond to the left-hand-side arguments are used and the rest are discarded.

12.6.4 Example 2

The demonstration, which is shown in List 3, involves calculation of Bode frequency response and Nyquist frequency response. The following statements are explanations of List 3.

- On the 1st, 19th, and 33rd line, beginning of functions
- On the 3rd to 5th, 23rd, 24th, 37th, and 38th line, declaration of local variables
- On the 6th line, declaration of functions
- On the 8th line, definition of the polynomial variable
- On the 9th line, generation of a logarithmically spaced vector
- On the 10th line, entering a three-degree real rational function
- On the 12th and 13th line, calling the multiple output functions Bode ( ) and Nyquist( )
- On the 20th, 21st, 34th, and 35th line, declaration of arguments of functions
- On the 26th and 40th line, definition of the imaginary unit
- On the 27th and 41st line, evaluation of rational-function arrays
- On the 28th and 29th line, array operations
- On the 30th and the 42nd line, returning the list which contains two elements

12.7 PROGRAMMING IN MaTX

In this section, we explain several programming facilities in MaTX.

12.7.1 Complex Valued Expressions

Complex valued expressions are allowed in all operations in MaTX. They are entered as a couple of real-valued expressions separated by a comma "," inside parentheses, "(" and ")". For example, entering the statement

```
    Func void main()
    {
      Polynomial s;
      Rational G;
05:   Array w, ga, ph, re, im;
      List Bode(), Nyquist();

      s = $;
      w = logspace(0.01, 100.0);
10:   G = (s^2 + 3*s + 4)/(s^3 + 5*s^2 + 6*s + 7);

      {ga, ph} = Bode(G, w);
      {re, im} = Nyquist(G, w);

15:   print w, ga, ph -> "bode-resp.mx";
      print re, im    -> "nyquist-resp.mx";
    }

    Func List Bode(G, w)
20:   Rational G;
      Array w;
    {
      Complex j;
      Array resp, ga, ph;
25:
      j = (0, 1);
      resp = eval(G, j*w);
      ga = 20.0 * log10(abs(resp));
      ph = arg(resp) * 180.0 / PI;
30:   return {ga, ph};
    }

    Func List Nyquist(G, w)
      Rational G;
35:   Array w;
    {
      Complex j;
      Array resp;

40:   j = (0, 1);
      resp = eval(G, j*w);
      return {Re(resp), Im(resp)};
    }
```

List 3 Bode and Nyquist frequency response data.

```
a = (3,4);
```

results in the output

```
a = (3.00000000E+00, 4.00000000E+00)
```

There are at least two convenient ways to enter complex matrices. They are illustrated by the statements

```
A = ([[1 2][3 4]], [[5 6][7 8]]);
```

and

```
A = [[(1,5) (2,6)][(3,7) (4,8)]];
```

which produce the same result. The former statement uses a pair of real matrices as the real part and the complex part of the matrix. The latter one uses four complex numbers as elements of the matrix. Entering the statements

```
x = $;
A = 3*x^2 + 4*x + 5; B = 6*x^2 + 7*x + 8;
C = (A, B);
```

results in the output

```
C = (3,6) x^2 + (4,7) s + (5,8)
```

The real part and imaginary part of complex valued expressions can be extracted using the functions Re( ) and Im( ), respectively.

12.7.2 Class Conversion Functions

Although the operations between distinct class objects invoke the automatic class conversions, some operations require explicit class conversions. The class conversion functions can be used for this. Their name is identical with the name of the destination class. For example, the statements

```
A = [[1 2][3 4]]; B = [[5 6][7 8]];
C = A .* B;
D = Array(A) * Array(B);
```

make matrix C equal to matrix D. The following statement uses the function Real ( ) to convert a 1 × 1 real matrix to a real number:

```
A = Real([1 2 3] * [4 5 6]') * [[1 2][3 4]];
```

The function String ( ) in the statements

```
A = [[1 2][3 4]];
for (i = 1; i <= 10; i++) {
    print A * i >> "A-" + String(i) + ".mat";
}
```

is used to convert an integer to a string.

12.7.3 Assignment Expressions

The statement

    A = B = C = Z(2,3);

sets all three variables to 2 × 3 zero matrix. This is not a special case but a consequence of the fact that an assignment has a value and assignments associate right to left.

12.7.4 Reuse of Expressions

Once an expression is described with some variables, this expression is reusable with other parameters like a macro since the compiler matc holds its definition as well as the value in the effective scope.

    x = 1.0;
    A = [x 2.0];

result in

    === [A] : (1, 2) ===
                (1)   (2)
      (1) 1.00000000E+00   2.00000000E+00

and

    x = 2.0;
    B = @A;

result in

    === [B] : (1, 2) ===
                (1)   (2)
      (1) 2.00000000E+00   2.00000000E+00

where the special symbol @ denotes the use of the definition of the variable rather than the value.

12.7.5 Input and Output

All class objects can be inputted from standard input and outputted to standard output by the **read** and **print** command, respectively. Multiple objects can be specified by separating with commas. If **read** command is used with matrices, the matrix-editor (mated) is invoked to read them:

    print a, A 1, (2,3), "Hello world.\n";
    read a, A, b, B;

To input objects from files and output objects to files, the **read** and **print** commands are used with the **redirect** symbols $<-$ and $->$ like the UNIX shell:

print a, A, 1, (2,3) $->$ "filename.mx";
read a, A, b, B $<-$ "filename.mx";

12.7.6 Control Flow

MaTX has control flow statements which are almost same as that of C language. It provides the fundamental flow-control constructions required for well-structured programs: statement grouping, decision making (if); looping with the termination test (while,for) such as **for** statement, **while** statement, and **if-then-else** statement. The **break** statement provides an early exit from for, while. They cause the innermost enclosing loop to be exited immediately. The continue statement causes the next iteration of the enclosing loop (for, while) to begin. In the while, this means that the test part is executed immediately; in the for, control passes to the reinitialization step.

12.7.7 Reference to Functions

In MaTX, a function is not a variable, but it is possible to pass its reference to functions. The reference to a function behaves like its name. We will illustrate this facility by showing the program of the continuous-time and hybrid-time simulation of a linear system

$$\dot{x} = \begin{bmatrix} 0 & 1 & 0 \\ 0 & 0 & 1 \\ -2 & -3 & -4 \end{bmatrix} x + \begin{bmatrix} 0 \\ 0 \\ 1 \end{bmatrix} u$$

to which the state feedback control

$$u = [\; -1 \;\; -1 \;\; -1 \;]x$$

is applied. List 4 is the program of the simulation. The functions odesolve( ) and odediscsolve( ) in MaTX are capable of integrating a broad range of nonlinear system functions of both explicit and implicit nature. The function odesolve( ) does the continuous-time simulation and the function odediscsolve( ) does the hybrid simulation, that is, the simulation of continuous-time plants with the digital controller. The former function takes five arguments: the initial time (0.0), the terminal time (5.0), a vector initial condition (x0), the reference (plant) to the function which calculates the value of the derivative (dx) from the time t and the values of x and u at this time, and the reference (controller) to the function which calculates the value of the input (u) from the time t and the value of x at this time. The last argument (controller) can be omitted in case of zero input. In addition to the five arguments, the function

```
    Func void main()
    {
      Real dt;
      List LC, LH;
05:   Matrix x0, TC, XC, UC, TH, XH, UH;
      void plant(), controller();

      dt = 0.01;      /* Sampling interval */
      x0 = [1 1 1]';  /* Initial state     */
10:
      LC = odesolve(0.0, 5.0, x0, plant, controller);
      LH = odediscsolve(0.0, 5.0, dt, x0, plant, controller);
      {TC, XC, UC} = LC;
      {TH, XH, UH} = LH;
15:
      print TC, XC, UC -> "cont-simu.mx";
      print TH, XH, UH -> "hybr-simu.mx";
    }

20:Func void plant(dx, t, x, u)
        Real t;
        Matrix x, dx, u;
    {
      Matrix A, b;
25:
      A = [[ 0    1    0]     /* System matrix   */
           [ 0    0    1]
           [-2, -3,  -4]];
      b = [0 0 1]';          /* Input matrix    */
30:   dx = A * x + b * u;    /* System equation */
    }

    Func void controller(u, t, x)
        Real t;
35: Matrix u, x;
    {
      Matrix f;

      f = [-1, -1, -1];     /* State feedback control-law */
40: u = f * x;
    }
```

List 4   Simulation of continuous-time systems.

odediscsolve( ) requires the sampling interval (dt) between which the constant input (u) is applied to the system.

The functions (in this example plant ( ) and controller ( ) which calculate the derivative and the input must have certain formats. The format of the former function specifies that the arguments list contains four variables and the latter one specifies that the arguments list contains three variables. In the derivative function, the first variable dx is the calculated derivative and has the same dimension as x, the second variable t is a real number which represents time, the third variable x is a vector (or matrix) of the state values, and the fourth variable u is the input to the system. In the input function, the first variable u is the calculated input, the second variable t is a real number which represents time, and the third variable x is a vector (or matrix) of the state values.

12.8 INTERPRETER VERSUS COMPILER

Interpreters usually do not require the users to declare class of variables because they know everything about all variables at run time. Compilers, however, cannot get the information about variables and functions at compile time unless there are declaration of variables. It seems reasonable to expect that the more information, in addition to algorithms themselves, given to the interpreters or the compilers, the more the rate of computation may be sped up and the efficiency of memory usage can be promoted. It will eliminate the abuse of variables. Of course, the more information required, the more time it takes the users to make programs. The relation between the speed of computation and the amount of information is depicted in Figure 7.

12.8.1 matx Versus matc (Comparison 1)

Here we compare the performance of the interpreter **matx** and the compiled program which was produced from the same source code by using the compiler

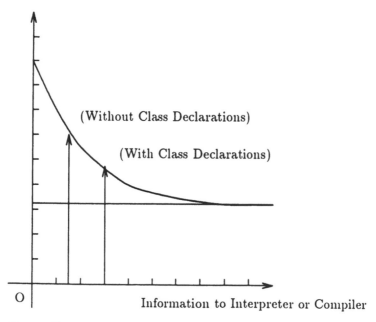

Figure 7 Computation time and amount of information.

matc. We choose for this the pole-assignment function PoleAssign.mm which is similar to the control toolbox algorithm acker.m [4] for single-input pole placement via linear state feedback. We compare the memory requirements and the execution times by taking random (A, b) pairs of state dimensions $n = 1, 2, 3,$ ..., 16 and asking for placement of the eigenvalues of $A + b*f'$ on prescribed locations, where it is a vector of random (but complex conjugate) eigenvalues. Figures 8 and 9 show the memory requirements and the computation times of the interpreter and the compiled program, respectively.

The memory requirement of the interpreter is about twice as much as that of the compiled program. In the case of the small-order system, the compiled program runs more than twice as fast as the interpreter, but the difference vanishes as the order of the system becomes larger. The reason is that the computation time for matrix calculation dominates, as the order of the system becomes larger.

The following statements may help account for the results.

1. The interpreter (matx) translates functions to the internal code and stores it in the memory. In other words, the sequence of program is predetermined in advance of the execution, though the interpretation of the objects in all statements are done for every execution. This preprocessing mechanism saves matx a lot of time.
2. The algorithm of PoleAssign.mm is written by using a powerful MaTX matrix notation, and there are few loops with scalar operations. If there are

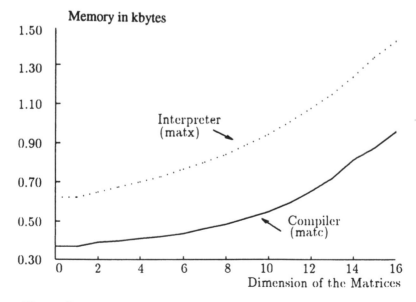

Figure 8 Memory requirements of interpreter and compiler.

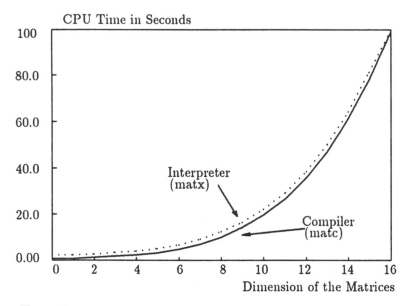

Figure 9 Computation time of interpreter and compiler.

many loops with scalar operations in the program, then the interpreter will be significantly slower than the compiled program. Incidentally, such loops with scalar operations can be written by using integers and real numbers in MaTX, which are written as a vector operation in MATLAB.

12.8.2 matx Versus matc (Comparison 2)

Here we choose for the comparison the continuous-time simulation of a linear system to which a state feedback control is applied. List 4 is the program of the simulation. We compare the execution times on some machines, such as MIPS6280, SPARC station 1, Sun4/280, Sun3/260, Sun3/60, and NEWS831 (SONY). The relative timings for both cases gave the ratios in favor of matc which is shown in Figure 10. The main reason for the good performance of the compiled program with respect to the interpreter matx is that the integrating function odesolve ( ) requires a lot of function calls which take the interpreter much more time than the compiled program.

12.9 XPLOT

Scientific and engineering data which were created in MaTX can be examined graphically in **XPLOT.** It is a very useful graphical drawing environment which runs under **X Window System.** It allows the users to perform all controls-related operations in a very consistent manner over mouse-operated menus and forms.

(CPU Time by matx) / (CPU Time by matc)

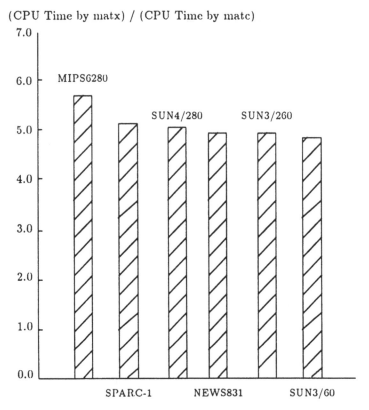

Figure 10   1D-canvas window and 2D-canvas window.

A menu hierarchy is used to group related operations together into domains. At the bottom of the menu trees, selection and action forms are used to give a highly interactive execution of most operations. XPLOT has several user-interface windows. **Canvas Windows** redraw the objects automatically whenever the hidden parts appear. The size of this window can be changed arbitrarily. **Data Information Windows** display various information in the tabular form and manage all the data in XPLOT. At the moment, the direct interface from MaTX to XPLOT is not available. We can pass the data, which are created in MaTX, to XPLOT through the file.

The fundamental object in XPLOT is the line. When the data is read from a file, the one-dimensional data information window, which manages the line data, is created. Lines are produced by putting together some one-dimensional lines. The users can select the line using a scroll-bar and change the state of the drawing with a toggle-button. Figures 11 and 12 show screen-dumps of the graphical drawing environment XPLOT. The one-dimensional data information

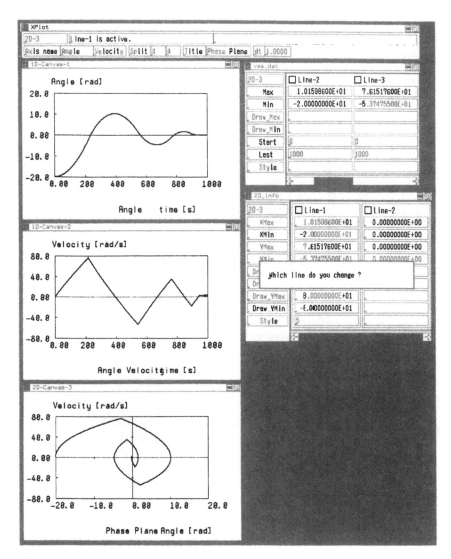

Figure 11 SemiLog-canvas window.

window and two-dimensional data information window, whose line-data were produced by putting together two one-dimensional line-data, are displayed on the right side of the screen. The one-dimensional canvas window and two-dimensional canvas window are displayed on the left side of the screen. When a toggle-button in the data information window is clicked, the corresponding line is drawn in the canvas window.

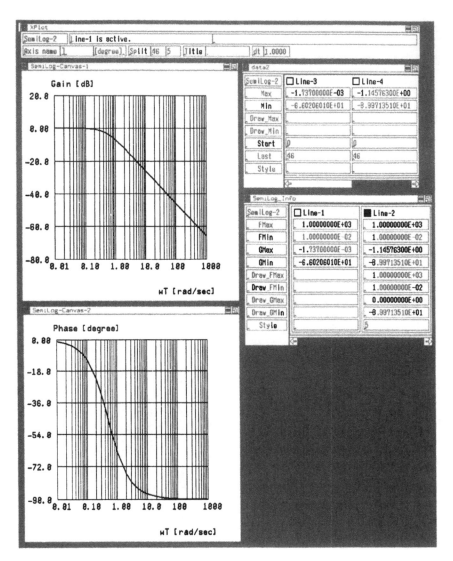

Figure 12 SemiLog-canvas window.

12.10 CONCLUDING REMARKS

A new high-performance programming language MaTX for scientific and engineering computation was presented and its features were described. It provides not only a command-driven *interpreter* but also a *compiler*. The users can extend the functionality of programs by realizing algorithms as functions in "mm-files." The interpreter (matx) enables us to test out the mm-files interactively. Furthermore, it is possible to make the executable program from the same source code using the compiler (matc). The MaTX language contains no dimension statements. Class declarations are required for compilation. Storage is allocated automatically, up to the amount available on any particular computers.

The larger the matrix, the more complicated it is to enter a matrix. The user-interface tool, matrix-editor "mated" provides us a very convenient way to enter large matrices [2, 3]. Scientific and engineering data, which were created in MaTX, can be easily examined graphically in XPLOT [2]. It is a useful graphical drawing environment which runs under X Window System. It allows the user to perform all controls-related operations in a very consistent manner over mouse-operated menus and forms.

At the moment, MaTX is checked to run on the following machines under the operating system:

Mips RC6280: UMIPS4.52C
Sun4:SunOS4.0.3, SunOS4.1.1
Sun3:SunOS4.0.3, SunOS4.1.1
NEWS-830(SONY):NEWS-OS 2.0, 3.3
J3100:UNIX System V/386 3.0, PC-DOS 3.1
PC9801(NEC):MS-DOS 3.1, 3.30

The flexibility and strength of MaTX is easily seen to be important for designers who want to combine the interactivity and the performance of optimized Fortran and C libraries. The translator which translates m-files to mm-files is under development. Further work is required to investigate and implement a higher performance for scientific and engineering computation. Work is also needed to explore and demonstrate its benefits for CACSD. The database management facilities will be needed to include version control, concurrency, and consistent data checking.

REFERENCES

1. Potter, J. E., Matrix quadratic solution, *SIAM J. Appl. Math.*, *14*, 496–501 (1966).
2. Koga, M., and Furuta, K. MaTX: A high-performance interactive software package for scientific and engineering computation, *CADCS'91*, *Swansea*, *UK*, 1991, pp. 39–44.
3. Koga, M., and Furuta, K., MaTX: A high-performance programming language (interpreter and compiler) for scientific and engineering computation, *CACSD '92*, *Napa*, *U.S.A.*, 1992.
4. Moler, C., Littel, J., and Bangert, S., *PC-MATLAB–User's Guide*, 3.1 edition, Sherborn, MA, 1987.
5. Moler, C. *MATLAB—User's Guide*. Albuquerque, NM, 1980.
6. Renes, W. A., Vanbegin, M., Van Dooren, P., and Beckers, J. W. J. J., The matlab gateway compiler. a tool for automatic linking of fortran routines to matlab, *CADCS'91*, *Swansea*, *UK*, 1991, pp. 95–100.

13

MAID: A Knowledge-Based Support System for Multivariable Control System Design

J. M. Boyle

Oxford Computer Consultants
Oxford, England

13.1 INTRODUCTION

Modern computers can play an important role as support tools for designers. This role transcends the mere utility of fast, flexible, and responsive computing; a powerful tool must also make a specialized body of knowledge easily usable and accessible and guide the designer in the use of that knowledge by providing structured design advice and explanations.

The use of a Knowledge-Based System (KBS) for control system design has drawn the interest of several researchers in recent years (e.g., Taylor and Frederick [24], James et al. [17], Birdwell et al. [1]). This chapter discusses the requirements for developing a KBS for multivariable control system design capable of providing the type of support outlined above. The object of the chapter is to examine some of the practical issues tackled during the development of MAID.

MAID (standing for Multivariable Analysis and Interactive Design) is a prototype KBS for multivariable control system design. MAID helps a control engineer to perform the design and analysis of multivariable control systems (using frequency-domain techniques). MAID also acts as an Intelligent Frontend (IFE) to the control engineering design package MATRIX$_X$ (Walker et al. [25]). (An IFE is a user interface to a software package that would otherwise be technically incomprehensible and/or too complex to be accessible to many potential users). MATRIX$_X$ stores a model of the system under design and provides a set of functions for performing design, analysis, and simulation on that model.

The chapter is divided into nine sections, organized as follows:

- Section 13.2 discusses why KBSs should be used as designers' assistants.
- Section 13.3 describes the main stages in the development of the MAID knowledge base.
- The definition of a set of systematically usable control design techniques and indicators for the analysis of designs is covered in some detail.
- Section 13.4 evaluates the environment within which MAID operates. The roles of the man and the machine are discussed and the software tools employed are evaluated.
- Section 13.5 mentions briefly how designs are performed using MAID.
- Section 13.6 provides a partial example of how MAID helps the designer to perform a design.
- Section 13.7 discusses briefly the development and implementation of the commercial version of MAID.
- Section 13.8 concludes the chapter.

13.2 A KBS APPROACH TO COMPUTER-AIDED CONTROL SYSTEM DESIGN

The use of KBSs for designing control systems is proposed for the following reasons:

- KBSs can provide a high-level design environment that is powerful, supportive, flexible, and broad in scope and readily accessible to nonexpert users [24]. They can guide a user through a design process, help him to use available Computer-Aided Control System Design (CACSD) packages and provide explanations at each stage of the design.
- KBSs can be encoded with the concepts and procedures of control engineering. (Concepts are the fundamental definitions and theory from which knowledge is derived and organized and from which such knowledge is extended, tested, and explained. Concepts are high-level organizers of knowledge which, though they may be expressed informally, are the justification of formal procedures.) A powerful environment for computer-aided control system design must have not only a user-friendly interface, reliable numerical software, interactive graphic capabilities, and good database management, but also an extensive control engineering knowledge base.

13.3 ACQUISITION OF DESIGN KNOWLEDGE

In this section, we shall describe the states in the development of MAID. The process of transferring knowledge from experts, books, and examples into a

KBS is referred to as knowledge acquisition. The knowledge acquisition life cycle has been discussed by numerous workers, notably by Buchanan et al. [8]. The main stages are considered to be:

1. **Identification** involves identifying domain characteristics, proposed users, and type of assistance to be provided by the KBS to users.
2. **Conceptualization** involves defining, in a high-level form, the main domain concepts and problem-solving strategies. This knowledge must also be partitioned and hierarchically organized.
3. **Formalization** involves developing a formal description of the concepts and strategies defined during the conceptualization phase.
4. **Implementation** involves implementing the knowledge in the form of a KBS.
5. **Testing** involves testing and revising the KBS as required.

13.3.1 Identification Phase

The identification phase establishes:

*   The scope and characteristics of the domain for which the KBS is to be developed.
*   The design and analysis techniques used in the domain. (Alternatively, a new set of techniques, more suited to a knowledge-based approach to design, may be developed.)
*   The intended class of users of the KBS and the type of support with which they should be provided.

For MAID, the following criteria were established:

(i)   Domain. MAID is to support the design of linear multivariable control systems using classical frequency-domain design and analysis techniques.
(ii) Design and analysis techniques. An expert may be able to solve complex problems but his expertise cannot be usefully encoded in a KBS if it cannot be articulated and structured into a systematically usable body of knowledge. Although many techniques exist for designing control systems, those techniques which can be clearly articulated and structured, and which are reasonably systematic to use, are best suited for implementation in a KBS. It was considered to be vitally important for the success of MAID to have a suitable set of techniques. Consequently, the major part of a research student's project [21] focused on developing such techniques.

It was also decided that the Characteristic Locus and Sequential Return Ratio design methods lend themselves well to a KBS approach. Knowledge bases for these design methods have recently been added to the commercial version of MAID described in Section 13.8.

(iii)   Class of users. Prior to developing a KBS, it is essential to have a clear model of the proposed users. This is important because the information presented by the KBS should reflect the capabilities of the intended users. For example, a design principle (i.e., the Feedback Principle) can be explained at different levels of complexity (i.e., using mathematical notation or a text-based description) and in varying amounts of detail.

MAID is intended for use by designers who have the expertise equivalent to that gained on an undergraduate control engineering course. The course taught at Cambridge University was used as a guide. A typical student receives 8 hours of lectures in multivariable control and acquires a thorough understanding of the techniques required to perform SISO designs.

(iv)   Type of support to be provided by KBS. MAID is to operate both as a **design assistant** and as an **intelligent frontend** to an engineering design package. As a design assistant, MAID's role is to help the designer to make decisions and analyze trial designs during the design process. As an intelligent frontend, MAID helps the designer to interact with a CACSD package. This package stores a model of the system under design and, therefore, allows the designer to implement trial designs and to display the results.

13.3.2   Conceptualization Phase

The aim of the conceptualization phase is to develop a high-level model of a body of knowledge. In this section, we describe the main components in the model of the MAID knowledge base.

Generalized design plan. A body of knowledge can be represented graphically as a generalized design plan. MAID's plan is shown in Figure 1 and provides:
A summary of the specialists in the knowledge base. A specialist is a module of problem-solving knowledge centered on a specific concept.
An overview of the scope of the knowledge base. Figure 1 shows that MAID has specialists for handling OBJECTIVES, DESIGN, ANALYSIS, etc.
Specialists. The generalized design plan is composed of a hierarchically organized collection of specialists. MAID is composed of eight main specialists (Figure 2).
Categories of Specialists
    1.   SUPERVISOR. The SUPERVISOR is the highest-level specialist, and, as its name implies, supervises the design process. To enable it to do this, the SUPERVISOR has a plan of the specialists it can call. Rules attached to the SUPERVISOR examine the current state of the design to determine which specialist should be called next. However, the designer can override the SUPERVISOR and call the specialists in any order.

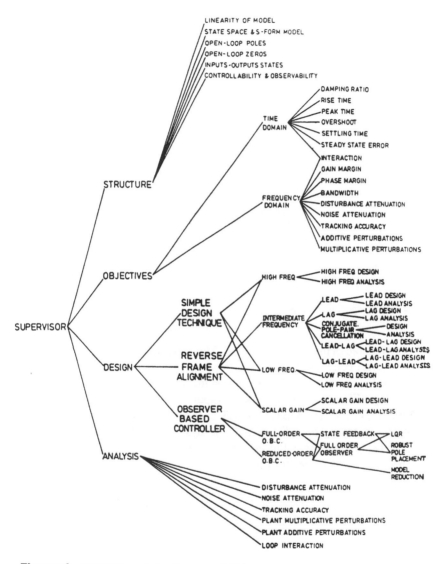

Figure 1 MAID knowledge-base specialist.

2. STRUCTURE. The STRUCTURE specialist examines the model of the artifact under design and is responsible for establishing basic information about that model. In particular, the specialist establishes the physical sources of difficulty likely to be encountered while performing control design (e.g., right half-plane zeros close to the bandwidth frequency). The specialist can take a model of a system and convert it into a form that can be handled by the design

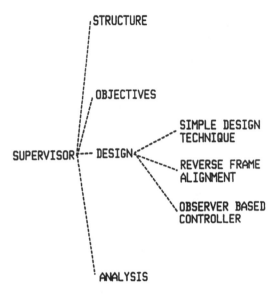

Figure 2 Main specialists in MAID.

techniques available in the knowledge base. This reworking of the model may be quite minor (e.g., converting the model from transfer-function form to a state-space description) or may involve a more significant change (e.g., measuring and controlling different plant inputs and outputs).

3.   OBJECTIVES. The OBJECTIVES specialist assists the designer (and the customer) in setting up a list of realistic design objectives. The specialist checks that the values assigned to objectives are valid and it can supply information such as default values for objectives.

4.   DESIGN

(i)   *Simple Design Technique (SDT)*. This is the simplest design technique available in the knowledge base. The SDT attempts to obtain a system with good phase properties (for stability) at high frequencies and good gain properties (for performance) at low frequencies.

(ii)   *Reverse Framed Alignment (RFA)*. This technique operates at the intermediate level of complexity. It uses the same strategies as the SDT specialist to perform the high- and low-frequency designs. It also uses dynamic compensation to obtain good gain-phase properties at intermediate frequencies.

(iii)   *Observer-Based Controller (OBC)*. This is the most complex of the three design techniques currently available in MAID. It is used when the SDT and the RFA technique fail to produce a satisfactory solution. It uses an observer to obtain information about the states of the system and uses this information to improve the gain-phase characteristics over the high- and intermediate-frequency regions. (The OBC technique is yet to be fully implemented in MAID.)

5. ANALYSIS. The ANALYSIS specialist allows the designer to perform a detailed analysis of his design. This specialist is usually called toward the end of a design session to give precise information about the behavior of the compensated system.

Levels of Abstraction of Specialists. The level of abstraction of a specialist is determined by its position in the generalized design plan. In Figure 1, the specialists at the ends of the branches of the plan are primitive specialists, whereas all the other specialists are meta-specialists.

A meta-specialist refines the design and is responsible for invoking one of its subspecialists. This may be either a further meta-specialist or a primitive specialist.

Primitive specialists are responsible for carrying out actual design activities. For example, in Figure 1, the specialist LEAD DESIGN implements a lead compensator.

Goals. Each specialist has a single goal that it attempts to achieve. The goals of the main specialist in MAID are shown in Figure 3. Goals of meta specialists deal with general aspects of the design while the goals of primitive specialists deal with specific parts of the design.

Methods. Each specialist has a set of recommendations it can make to the designer. These recommendations are referred to as the specialist's methods and are selected by the specialist in an attempt to achieve its goal.

Definition of a method. A method is any system analysis, design decision, compensator design, or other design activity which a specialist can recommend to the designer.

Examples of methods. Data-input method: input a value for the bandwidth specification; design-decision method: add integral action to the system.

Functions. A model of the system under design is stored in an engineering-design package and a designer implements trial designs and analyzes the results by executing functions in the package. Examples of functions and their use occur repeatedly in the example in Section 13.6.

Functions are executed when a specialist requires information about an attribute of the system under design or when one of the specialist's methods involves manipulating the system under design. Functions have been written in MATRIX$_X$ for performing all the design and analysis required by the techniques available in MAID.

Plans and Rules. The plans and rules of meta-specialists represent strategic (or meta-) knowledge. To refine a design, a meta-specialist calls the appropriate subspecialist(s). For example, a simple meta specialist might have the plan and rules presented in Figure 4.

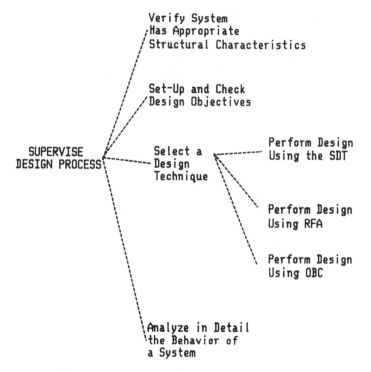

Figure 3   Goals of main specialists.

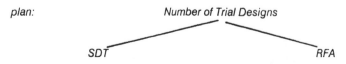

plan:

rule: Select the SDT specialist if this is the first attempt at a design, otherwise select the RFA specialist.

Figure 4   Plan of a meta-specialist.

A primitive design specialist always occurs in conjunction with a primitive analysis specialist that checks the design. For example, the design specialist HIGH FREQ (shown in Figure 1) has two primitive subspecialists:

(i)   HIGH FREQ DESIGN whose goal is to design a high-frequency subcontroller. This specialist has the design plan shown in Figure 5.

(ii)   HIGH FREQ ANALYSIS whose goal is to check any design operations performed as a result of recommendations made by the HIGH FREQ DESIGN specialist. The HIGH FREQ ANALYSIS specialist must also suggest how the design should be modified should the trial design be considered unsuccessful. This is referred to as redesign. This specialist uses the analysis plan shown in Figure 6.

Redesign Using MAID

It is important to reiterate that design is a feedback process and that trial designs, objectives, and the model of the system under design may be iteratively adjusted during the design process. Iterations occur because the design process is progressively stripping away the designer's uncertainty about what he can expect to achieve in his final design. These trial solutions are said to "fail" in some respect. Redesign is the alteration of a trial design owing to the realization that a trial design fails to meet its design objectives.

    Some of the complexity in procedural design is associated with the constructive use of information from unsuccessful trial designs. Specialists either contain their own failure-handling knowledge or they call other specialists with the required knowledge. For example, the HIGH FREQ DESIGN specialist (shown in Figure 5) has no failure-handling knowledge. Instead, it uses the HIGH FREQ ANALYSIS specialist to detect design failures and to make redesign recommendations.

    Figure 6 shows the analysis plan for the HIGH FREQ ANALYSIS specialist. Nodes 3 and 4 of the plan represent the types of design failure for which the

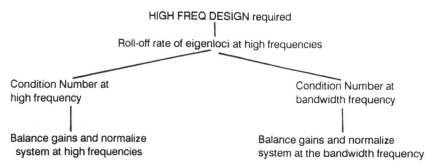

Figure 5 Plan of the HIGH FREQ DESIGN specialist.

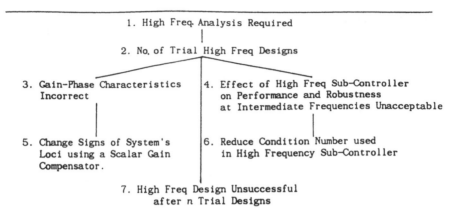

Figure 6   Design plan of HIGH FREQ ANALYSIS specialist.

specialist will look. Nodes 5 and 6 represent the redesign recommendations that the specialist can make.

MAID employs five heuristic approaches to redesign. When a design failure occurs, MAID may recommend one of the following approaches:

A.   Modify one or more parameters of a previously implemented method.
B.   Select a new method (possibly discarding the previous method).
C.   Modify the objectives causing the design failure.
D.   Select an alternative design technique.
E.   Modify the model of the plant under design.

In most cases, MAID will recommend that approach A is tried before approach B, and so on. For example, the primitive specialist HIGH FREQ ANALYSIS can recommend approaches A and B. If Node 6 in Figure 6 is selected, the redesign involves modifying one of the parameters of a previously implemented method (this involves using "redesign approach A"). On the other hand, if Node 5 in Figure 6 is selected, the redesign involves using a new method (this involves using "redesign approach B"). After a number of trial designs (usually two), the HIGH FREQ specialist (which operates at one level of abstraction lower than the HIGH FREQ ANALYSIS specialist) will be told that the HIGH FREQ ANALYSIS specialist has failed to produce an acceptable design after *n* trials. The HIGH FREQ specialist uses different approaches to redesign than HIGH FREQ ANALYSIS. HIGH FREQ will try to get the designer to modify his design objectives (this involves using "redesign approach C"). Alternatively, HIGH FREQ may inform SIMPLE DESIGN TECHNIQUE that it (i.e., HIGH FREQ) has failed. This may lead to a new design technique to be chosen (this involves using "redesign approach D").

A suggestion for future work is to improve MAID's approach to redesign. In particular, Brown's [6] work on redesign should be investigated.

Types of Failure. MAID detects a number of different types of failure. In particular,

- structural failures (e.g., if the system is not linear)
- design-objective failure (e.g., when an invalid design objective is input by the designer)
- design failures (e.g., when either some part of a trial design violates one of the design objectives or when a design method has had an unexpected or undesired effect on the system).

Partitioning and Organization of Design Knowledge

A further aim of the conceptualization phase is to partition the model of a body of knowledge into components and to organize those components into a hierarchy. The MAID knowledge base (Figure 1) is partitioned and organized as follows:

(i) The main stages of the design process are represented by separate branches of the generalized plan.
(ii) The design techniques are partitioned according to their complexity.
(iii) Each design technique is divided into high-, intermediate-, and low-frequency design specialists.
(iv) For each stage of the design process, the knowledge base is partitioned into specialists operating at different levels of abstraction.

Stages of the Design Process. The main stages of the procedural design process are:

STRUCTURE. Structural analysis which involves analyzing the model of the plant under design.

OBJECTIVES. Definition of design objectives. During this stage, the designer defines a realistic and consistent set of design specifications and constraints.

DESIGN. Selecting and applying the steps of a design technique.

ANALYSIS. Involves analyzing the behavior of a proposed design solution.

The main stages are shown as branches of the MAID knowledge base (Figure 1).

Complexity of Design Techniques. Each top-level DESIGN specialist represents one of the levels of design complexity. MAID supports three design techniques which are, in order of increasing complexity, SIMPLE DESIGN TECHNIQUE, REVERSE FRAME ALIGNMENT, and OBSERVER-BASED CONTROLLER.

Frequency Regions. Top-level design specialists have subspecialists for frequency regions.

Levels of Abstraction.   Abstraction refers to the concept of working from something general through to something more detailed. Specialists operating at
higher levels of abstraction perform rough design [6], whereas specialists operating at a lower level of abstraction perform detailed design. In Figure 1, HIGH
FREQ operates at a lower level of abstraction than SIMPLE DESIGN TECH-
NIQUE which, in turn, operates at a lower level of abstraction than DESIGN.

13.3.3   Formalization Phase

The formalization phase involves mapping the model of a body of knowledge
into a formal knowledge-representation language.

The formalization of the prototype version of MAID knowledge base was
built entirely on production rules (i.e., IF . . . THEN . . . ELSE statements).
Although these rules provided a simple means of representing knowledge and
made possible the rapid development of a prototype system, they were concluded to be a restrictive way of trying to represent formally a body of design
knowledge. A new version of MAID is being designed using a frame-based
knowledge-representation language [3].

13.3.4   Implementation Phase

The purpose of developing an early prototype was twofold. First, we wanted to
test out our model of the design process developed during the conceptualization
and formalization phases of the knowledge-acquisition process. Second, we
wanted to build a prototype system to identify whether further research was required.

The implementation of MAID has changed significantly since the research
work described in this chapter was completed in 1989. A description of the current implementation is given in Section 13.8.

13.3.5   Testing Phase

Once a prototype system has been developed, "guinea pigs" (for example, students) can be released onto the system. An intelligent student is ideal for evaluating and debugging a prototype system because:

*   He/she has the right level of ability for the class of intended users.
*   He/she can be motivated (by making the evaluation of the prototype an as-
    sessed project).
*   He/she will not only detect deficiencies in the KBS but will also recommend
    suitable modifications.
*   Students are readily available in a university environment.
*   He/she can take first-hand experience of the KBS into industry and thus gen-
    erate interest from industrial organizations.

13.4 MAID ENVIRONMENT

13.4.1 Components of the MAID Environment

MAID is part of a design environment. This environment is shown in Figure 7. Its basic components are:

- A Knowledge-Based System (KBS), MAID, which provides a knowledge framework.
- A Computer-Aided Control System (CACSD) package. This provides a manipulative framework.
- A designer who provides the environment's conceptual framework.

  (i) *Knowledge-Based System.* The KBS guides the designer through the design process using the appropriate specialists and their plans and rules. At any stage, the KBS is able to help the designer to *understand* the design process by explaining the procedures being used and the concepts on which those procedures are based.

 (ii) *CACSD package.* The package performs all the mathematical manipulations of a system required during the design process. The package must:

    - Perform all the calculations required to add compensators to the original system

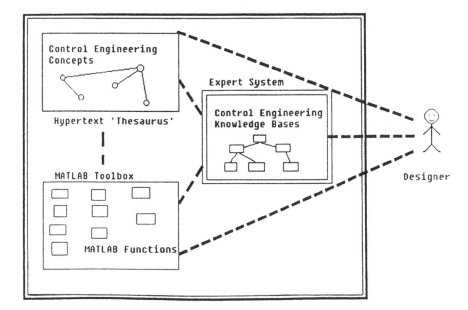

Figure 7   Components of the MAID environment.

- Perform all the calculations required to analyze the attributes of the system
- Plot data in graph form (i.e., indicators)
- Enable experienced designers to perform a range of operations on a system above and beyond those specifically required by the techniques encoded in the KBS.

(iii) *Designer.* The designer is responsible for:

- Making decisions in the light of recommendations from the KBS
- Performing qualitative analysis and interpreting of the behavior of the system using indicators recommended by the KBS
- Incorporating "real-world" knowledge into the design process. (For example, MAID takes no account of whether it would be possible to implement a particular design. Such information must be "incorporated" by the designer.)

13.5 EXECUTION OF DESIGNS IN MAID

The design process involves a constructive dialogue between the designer and the KBS. A detailed example of part of a design session is given in Section 13.6.

The key specialist is the SUPERVISOR. When any of the high-level specialists (shown in Figure 2) has completed its work, the SUPERVISOR selects the specialist required next. The SUPERVISOR has a default mode of operation. In this mode, the SUPERVISOR selects the subsequent specialist on the basis of the default sequence shown in Figure 8.

The sequence can be modified by direct intervention on the part of the designer. The SUPERVISOR only "recommends" which specialist should be used next. The modularity and the flexibility of the KBS allow the designer to execute any specialist in any sequence and at any phase of the design process. After imposing his own design strategies, the designer can always return to MAID's default sequence.

The default sequence plays an important role in tutoring designers. In particular, a novice designer or a new user of MAID can use the default design sequence to develop an understanding of (i) the design process and (ii) the features of MAID.

13.6 AUTOMOBILE GAS-TURBINE DESIGN EXAMPLE

The engine considered in this example is a gas turbine. The model used here was developed by Winterbone et al. [27]. The transfer-function matrix of the model of the system is

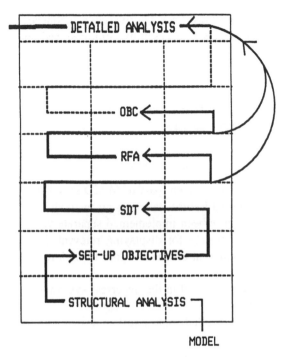

Figure 8 MAID's default design sequence.

$$
G(s) = \begin{bmatrix} \dfrac{0.806s + 0.264}{s^2 + 1.15s + 0.202} & \dfrac{-(15.0s + 1.42)}{s^3 + 12.8s^2 + 13.4s + 2.36} \\[2ex] \dfrac{1.95s^2 + 2.12s + 0.49}{s^3 + 9.15s^2 + 9.39s + 1.62} & \dfrac{7.14s^2 + 25.8s + 9.35}{s^4 + s^3 + 23.94s^2 + 191.4s + 510.3} \end{bmatrix}
$$

The presentation of the example is divided into three columns. The interaction between the designer and MAID is given in the right-hand column, whereas in the center column, the interaction between the designer and the CACSD package, MATRIX$_X$, is shown. This includes the commands executed in MATRIX$_X$ and the indicators displayed. Finally, in the left-hand column, we have added comments about the design session.

13.7   CURRENT IMPLEMENTATION OF MAID

MAID currently handles multivariable frequency-domain control engineering design. It is now an interactive design environment that guides control engineers through the multivariable design process. To the best of our knowledge, it is the only Knowledge-Based Support System (KBSS) for Multivariable Frequency-Domain Control System Design that has been successfully commissioned and used in industry.

The environment consists of:

- An expert system
- A dictionary/thesaurus of control engineering concepts
- A computer-aided control system design package
- A human designer

The main changes to MAID over the past 2 years are as follows:

- The main knowledge bases in the expert system part of MAID now handle the Characteristic Locus and Sequential Return Ratio design methods.
- The design procedure and concepts have been implemented separately. Previously, the expert system was responsible for displaying relevant design concepts. Now a hypertext tool is used to allow the designer to explore control concepts.
- Networks of concepts have been implemented in the hypertext system to allow the designer to explore the definition of related concepts.
- MATLAB is now used as the computer-aided control system design package instead of MATRIX$_X$.

A. Expert System. The expert system guides a designer through the procedural aspects of a design problem. It helps the designer select and use appropriate modules of knowledge. The modules of the two design methods currently supported by the KBSS are shown in Figures 9 and 10.

The expert system makes recommendations to the designer. These may include statements about which part of the design process to perform next, or statements about design operations that should be executed on the model of the system under design. The latter operations involve running MATLAB functions. The expert system tells the user which functions to run and sets up the required parameters.

B. Dictionary/Thesaurus of Control Engineering Concepts. This part of the KBSS incorporates definitions of concepts and the relationships between concepts. The concept thesaurus is a high-speed, automated way of browsing through concepts. Concepts are linked together in a dynamic and interactive document. Each concept points to other related concepts and, with the click of a button, the user can focus his attention on a new concept without having to work through a traditional linear document. Information is also stored at several levels of detail which the user can explore as required. Each concept is defined in terms of a set of properties:

Name: The name of the concept.
Type: The type of concept (active, object, or attribute).

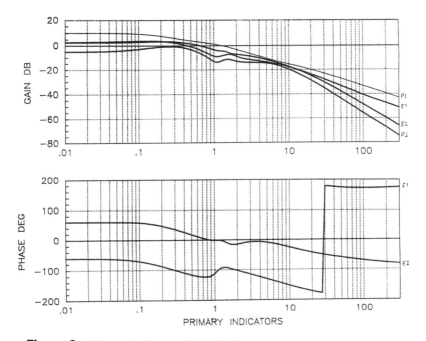

Figure 9 Primary indicators of original system.

Author:	The author of the concept.
Date:	The date the concept was written.
Definition:	Brief definition of concept.
Description:	A description of the concept.
Examples:	Examples of the concept.
References:	References to texts containing further information on the concept.
Remarks:	Important additional information (e.g., special cases).
Thesaurus:	The type of links between the defined concept and other concepts.

To represent the relationships between concepts, we have implemented links between concepts. These links can be of the following types: *Abstraction*; *Specialization*: *Part of*; *Property of*; *Representation of*; *Affects*: *Uses*; *Indicator of*; *Relates*.

C. Computer-Aided Control System Design Package. The package (MAT-LAB is used in our KBS) performs the mathematical manipulations of a system model required during the design process. The package:

- Performs the calculations required to manipulate systems
- Performs the calculations required to analyze systems
- Produces plots of data

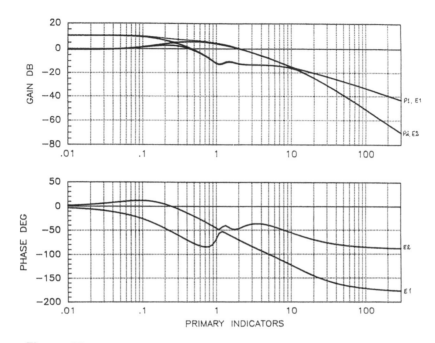

Figure 10 Primary indicators after high-frequency design.

- Enables experienced designers to perform a range of operations on a system model beyond those specifically required by the procedures encoded in the expert system

Most of the functions that need to be run during the design process are selected by cutting function names from the expert system output and pasting them into MATLAB. The functions then run automatically, and return the required values or display the required plots.

D. Designer. The designer forms an integral part of the KBSS design environment. The designer is responsible for:

- Making decisions in the light of recommendations from the expert system
- Performing qualitative analysis and interpretation of data
- Incorporating "experience" and "real-world" knowledge into the design process

13.7.1 Implementation of the KBSS

The KBSS has been implemented on an IBM compatible 386 PC with 4MB of RAM. The following software has been used:

1. The KBSS runs under Microsoft Windows which is an extension of the DOS operating system. Windows allows several tasks to be integrated. With Windows, the user can run several packages at the same time and can switch between packages at the click of a mouse. Windows also allows for an easy transfer of information between packages. For example, MATLAB commands can be copied directly from the expert system into MATLAB.
2. The expert system has been developed using an expert system shell, X_i Plus, in which all the procedural knowledge can be implemented.
3. The thesaurus of concepts framework has been developed using a hypertext tool called Guide from Owl International Ltd. Using hypertext, concepts and the links between them can be easily defined and modified by users.
4. A toolbox for multivariable control design, built on top of MATLAB, provides the manipulation framework [4].

13.7.2 Testing

The current version of the KBSS contains knowledge bases for the Sequential Return Ratio and Characteristic Locus design methods (Figures 9 and 10). The main concepts on which these two design methods are based have been fully encoded in the concept thesaurus. MATLAB's multivariable toolbox has been extended and contains the functions for these design methods.

The design of several multivariable control design problems has been successfully completed by control engineers using the developed software. The users were able to complete multivariable designs using only the software and their background in single-input single-output control theory.

13.8 CONCLUSIONS

This chapter concludes with a complete worked example. The aim of the example is to show that the technique that have been researched and developed by the author successfully help control engineers to learn about and use the control design methods on design problems.

The development of KBSSs for control engineering is an important step forward. Using these tools, learning takes place in a flexible manner controlled by the user. Once the user is comfortable with the design method, he/she can directly use the tools to assist on real design problems. As a result of such KBSSs a wider range of design methods will become accessible to a greater number of control engineers.

REFERENCES

1. Birdwell, J. D., et al. Expert systems techniques in a computer based control system analysis and design environment, *Proc. 3rd IFAC/IFIP Int. Symp.*, *CADCE '85*, 1985.

2. Boyle, J.-M., and Pang, G. K. H., The implementation of an expert system for use in control system design, *Inst. MC Workshop on Computer-Aided Control System Design, University of Salford*, July, 1986.
3. Boyle, J.- M., Knowledge Based Techniques for Multivariable Control System Design, *Ph.D.* Thesis, Cambridge University, UK, 1987.
4. Boyle, J.- M., Ford, M. P., and Maciejowski, J. P., A Multivariable toolbox for use with MATLAB, *IEEE Control Syst. Mag.*, *9*, (No. 1) (1989).
5. Bradbury, C., An investigation into the use of expert systems in design and how this technology meets the requirements of industry, Third Year Project Report, Engineering Dept., University of Cambridge, Cambridge, UK.
6. Brown, D. C., Failure handling in a design expert system, in Computer-Aided Design, F. Gero, (ed.), Butterworth & Co, London, 1985.
7. Brownston, L, Farrell, R., Kant, E., and Martin, N., *Programming Expert Systems in OPS5*, Addison-Wesley, Reading, MA, 1985.
8. Buchanan, B. G., Barstow, D., Bechtel, R., Bennett, J., Clancey, W., Kulikowski, C., Mitchell, T., and Waterman, D. A., Constructing an Expert System, Building Expert Systems, in F. Hayes-Roth, D. A. Waterman, F., and D. B. Lenat, (eds), Addison-Wesley, Reading, MA, 1983.
9. Clocksin, W., and Mellish, C., *Programming in Prolog*, Springer-Verlag, New York, 1981.
10. Croall, I. F., Ishizuka, M., Look H. E., and Waterman, D. A. (eds), Expert systems and the personal computer, *Expert Systems: The International Journal of Knowledge Engineering*, *2*(4) (1985).
11. Daniel, R. W., and Kouvaritakis, B., Analysis and design of linear multivariable feedback systems in the presence of additive perturbations, *Int. J Control*, *39*, 551–580 (1984).
12. Doyle, J. C., and Stein, G., Multivariable feedback design: concepts for a classical/ model synthesis, *IEEE Trans. Auto. Control*, *AC-26*, 4–16 (1981).
13. *Expertech Xi—User manual*, Expertech Ltd, 1986.
14. Ford, M. P., Boyle, J.-M., and Maciejowski, J. M., Multivariable frequency domain toolbox (for use with MATLAB), *Tutorial and Reference Manuals*, Cambridge Control & GEC Research, UK.
15. Hung, Y. S., and MacFarlane, A. G. J., *Multivariable feedback: A Quasi-classical Approach*, Springer-Verlag, Berlin, 1982.
16. Intellicorp. 1984. The Knowledge Engineering Environment, 707 Laurel Street, Menlo Park, California 94025-3445.
17. James, J. R., Frederick, D. K., Bonissone, P. P., and Taylor, J. H., A retrospective view of CACEIII: considerations in co-ordinating symbolic and numeric computation in a rule-based expert system, *Proc. Second Conf. on A. I. Applications*, Miami Beach, Florida, 1986.
18. MacFarlane, A. G. J., and Kouvaritakis, B., A design technique for linear multivariable feedback systems, *Int. J. Control*, *25*, 837–874 (1977).
19. MacFarlane, A. G. J., and Hung, Y. S., Indicators for multivariable feedback systems, *Int. Conf Cont 85*, *Cambridge*, 9–11 July, 1985.
20. MacFarlane, A. G. J., Gruebel, G., and Ackermann, J., Future design environments for control engineering, *Preprints of the 10th IFAC World Congress on Automatic Control. Munich*, 27–31 July 1987.

21. Pang, G. K. H., An expert systems approach to computer-aided control system design, Ph. D. Dissertation, Engineering Depart; Cambridge University, Cambridge, 1986.
22. Pang, G. K. H., and MacFarlane, A. G. J., *An Expert Systems Approach to Computer-Aided Design of Multivariable Systems*, Springer-Verlag, Berlin, 1987.
23. Shortliffe, E. H., *Computer Medical Consultations: MYCIN*, Elsevier, Amsterdam, 1976.
24. Taylor, J. H., and Frederick, D. K., An expert system architecture for computer aided control engineering, *Proc IEEE* (Dec. 1984).
25. Walker, R., Gregory, C., Jr., and Shah, S., MATRIX-x: A data analysis, system identification, control design and simulation package, *IEEE Cont. Syst. Mag.*, 2(4), 30–36 (1982).
26. Winston, P. H., and Horn, B. K. P., *LISP*, Addison-Wesley, Reading, MA, 1984.
27. Winterbone, D. E., Munro, N., and Nuske, D. J., A Multivariable controller for an automotive gas turbine, Contributed by the Gas Turbine Division of The American Society of Mechanical Engineers for presentation at the *Gas Turbine Conference*, *San Diego, California*, 12–15 March 1979.
28. Yass, D., An introduction to expert systems in control system design, Third Year Project Report, Engineering Dept., Univ. of Cambridge, Cambridge, 1986.

14

Optimization for Control Design

David Q. Mayne

*University of California
Davis, California*

Hannah Michalska*

*Imperial College of Science, Technology and Medicine
London, England*

14.1 INTRODUCTION

Optimization is not yet used as a matter of course in design, but its eventual employment for this purpose appears inevitable. To justify this assertion, which may appear perverse in view of the continuing development of sophisticated techniques provided by H_2 and H_∞ control theory, several points may be made. Rosenbrock [1] makes a useful distinction between synthesis and design. In the former, a single, simple objective, such as minimum quadratic error, is sought, whereas the latter is concerned with multiple, possibly conflicting, objectives. The former can usually be stated as a (relatively) simple problem for which a unique solution may be obtained by a (relatively) straightforward procedure, such as solving a matrix Riccati equation. The synthesis problem may, for example, be that of minimizing, over the class of stabilizing controllers, a norm which is a measure of the performance of the closed-loop system, determining a stabilizing controller which satisfies $\|WS\|_\infty < 1$, where S is the sensitivity function and W a weighting transfer function chosen by the designer, or choosing a controller such that the closed-loop system has specified eigenvalues. In each case, a well-defined theoretical problem is posed. Design, by contrast, is concerned with many issues, is iterative, and requires judgment in trading off some

---

Current affiliation: McGill University, Montreal, Canada

objectives against others. Typically, a design study involves modeling (this may consume 90% of the total effort), choice of actuators and sensors and their location, analysis, formulation of performance specifications, choice of controller structure, design of controller parameters to meet the specifications (including, if necessary, modification of sensors, actuators, and controller structure), simulation of the controlled system, controller implementation (choice of hardware/software), monitoring of the controlled system and on-line tuning of the controller.

It is not the purpose of this chapter to argue that optimization can be used to solve the design problem completely. But a more modest claim can usefully be made: that the specifications arising in a typical control design problem are so complex and widely ranging that the choice of parameters in a controller (whose structure is specified) is greatly facilitated by the use of appropriate optimization tools. It must be admitted at the outset that, in the context of linear systems, the H_2 and H_∞ synthesis techniques enable controllers to be designed which satisfy a surprisingly wide range of specifications by the appropriate choice of weighting transfer functions (such as W). Nevertheless, optimization enables one to add, for example, to the typical menu of frequency-domain constraints for robust performance and stability, hard constraints on the time responses of controller, and state variables to a specified range of test signals. It permits, albeit at appreciable computational cost, the incorporation of a wide variety of structured plant variations; these may take the form of envelopes on time or frequency responses or specified sets in which selected plant parameters take their values.

The case for optimization becomes stronger when one moves outside the class of linear systems. Even if the plant is linear, successful satisfaction of the inevitable hard constraints on controls and states often requires nonlinear controllers. Because of the lack of available theory, these nonlinear controllers are designed on an ad hoc basis. Using existing synthesis techniques, a linear controller is first designed, the controller nonlinearities are then added, and the controller returned, if necessary, using simulation. At the very least, optimization could usefully be employed for the second stage. The remarkable developments in nonlinear control theory [2,3] assume that controls and states are unconstrained; control saturation, for example, destroys exact feedback linearization and, therefore, the validity of any nonlinear controller designed using this interesting technique. Apart from a few cases when frequency-domain methods (for example, the circle criterion) can be employed, the design specifications for nonlinear systems (including linear systems with nonlinear controllers) will take the form of constraints on time responses and/or constraints on nonlinear functions of the state. This type of constraint, as will be shown in the sequel, can be handled by modern optimization techniques.

There is a caveat which should be stated at this stage. In control design, there is one constraint, the requirement that the closed-loop system is (internally) stable which must be satisfied. The remaining constraints (on performance) may be infeasible (cannot all be satisfied) as initially stated, necessitating *trade off* at some stage in the design. The problem is essentially one of *multicriterion* optimization; an appropriate test is that the design by Pareto optimal, that is, that no competing design is uniformly superior. Determining a Pareto optimal point may be formulated as a min-max problem, that is minimizing the maximum of the *weighted* objectives (see Ref. 4). Different Pareto optimal points may be obtained by changing the weighting, thus permitting, in principle, tradeoffs to be made.

To demonstrate the relevance of optimization to control design the chapter is organized as follows. A few preliminary definitions are made in Section 14.2. In Section 14.3, typical specifications for the design of linear control systems are presented. The intent of the section is to be illustrative, rather than exhaustive, showing that design constraints are typically semi-infinite in nature, resulting in an optimization problem that is nondifferentiable. In Section 14.4, a similar approach is taken in presenting design specifications for nonlinear systems. The specifications presented in Sections 14.3 and 14.4 show that many design problems may be approximated by a *canonical* semi-infinite optimization problem. Algorithms for solving this canonical problem are presented in Section 14.5. When the problem is convex, which occurs when the system being controlled is linear and Q-parameterization of the controller is employed, global solutions to the canonical problem may be obtained; in the more general situation, it can be shown under certain hypotheses that accumulation points of infinite sequences generated by the algorithms presented in this section satisfy the constraints and a necessary condition of optimality. In the final section of the chapter, a brief discussion of how these techniques may be incorporated in a suitable design environment is presented.

14.2 PRELIMINARIES

14.2.1 System Description

Consider a system described by

$$\dot{x} = f(x, u, w) \tag{1}$$

$$y = c_1(x, w) \tag{2}$$

and

$$z = c_2(x, u, w) \tag{3}$$

where $x \in \mathbb{R}^n$ is the state, $u \in \mathbb{R}^m$ the control, $w \in \mathbb{R}^p$ the exogenous signal (reference, disturbance), $y \in \mathbb{R}^k$ the measured output, and $z \in \mathbb{R}^l$ the "output" it is desired to control or limit (z may include components of u, for example). In the linear case, Eqs. (1)–(3) become

$$\dot{x} = Ax + Bu + Gw \tag{4}$$

$$y = C_1x + D_1w \tag{5}$$

and

$$z = C_2x + D_2u + E_2w \tag{6}$$

The exogenous variable is sometimes assumed to be generated by a dynamic system

$$\dot{w} = s(w) \tag{7}$$

In the linear case

$$\dot{w} = Sw \tag{8}$$

The control is also generated by a dynamic system described by

$$\dot{v} = \eta(v, y, \alpha) \tag{9}$$

$$u = c_3(v, y, \alpha) \tag{10}$$

which, in the linear case, becomes

$$\dot{v} = A_C(\alpha)v + B_C(\alpha)y \tag{11}$$

and

$$u = C_C(\alpha)v + D_C(\alpha)y \tag{12}$$

where the matrices A_C, B_C, . . . depend on the controller parameter $\alpha \in \mathbb{R}^a$ which is chosen to satisfy, if possible, the design specifications. The closed-loop system is described by

$$\dot{x}_{cl} = f_{cl}(x_{cl}, w, \alpha) \tag{13}$$

with output

$$z = c_{cl}(x_{cl}, w, \alpha) \tag{14}$$

where $x_{cl} \triangleq (x, v)$ is the state of the closed-loop system and f_{cl}, c_{cl} are obtained from Eqs. (1)–(3), (9), and (10) so that

$$f_{cl}(x_{cl}, w, \alpha) = \begin{bmatrix} f(x, c_3(v, c_1(x, w), \alpha), w) \\ \eta(v, c_1(x, w), \alpha) \end{bmatrix} \tag{15}$$

and

$$c_{cl}(x_{cl}, w, \alpha) = c_2(x, c_3(v, c_1(x, w), \alpha), w) \tag{16}$$

In the linear case, the closed-loop system is described by

$$\dot{x}_{cl} = A_{cl}(\alpha)x_{cl} + B_{cl}(\alpha)w \tag{17}$$

$$z = C_{cl}(\alpha)x_{cl} + D_{cl}(\alpha)w \tag{18}$$

where

$$A_{cl}(\alpha) \triangleq \begin{bmatrix} A + BD_C(\alpha)C_1 & BC_C(\alpha) \\ B_C(\alpha)C_1 & A_C(\alpha) \end{bmatrix} \tag{19}$$

$$B_{cl}(\alpha) \triangleq \begin{bmatrix} BD_C(\alpha)D_1 + G \\ B_C(\alpha)D_1 \end{bmatrix} \tag{20}$$

$$C_{cl}(\alpha) \triangleq [C_2 + D_2D_C(\alpha)C_1, \ D_2C_C(\alpha)] \tag{21}$$

and

$$D_{cl}(\alpha) \triangleq D_2D_C(\alpha)D_1 + E_2 \tag{22}$$

The closed-loop system, described by Eqs. (17) and (18), has input w, state $x_{cl} = (x, v)$, and output z. Let $H(s, \alpha)$ denote the transfer function from w to z. This transfer function is easily computed from Eqs. (17) and (18).

The system described by Eqs. (1)–(3) [or (4)–(6)] will be referred to as the "plant"; however, the description includes possible connections between the exogenous variable w and the controller, permitting a variety of controller configurations (see Ref. 4 for an elaboration of this point).

14.2.2 Controller Parameterization

The definitions in Section 14.2.1 cover the classical case when the controller structure is prespecified and a parameter α is sought to satisfy the design specifications; a proportional plus integral controller, for example, fits the format of Eqs. (11) and (12). An advantage of this formulation is that it yields a simple controller in those cases when a simple controller suffices to satisfy the design specifications. A disadvantage is that the transfer function $H(s, \alpha)$ is, in general, nonlinear in α which makes the associated design optimization problem more difficult than it would be if H were affine in α.

It is an interesting fact [4,5] that H can be expressed affinely in terms of the transfer function Q of a stable system S_Q which parameterizes the controller. To see this, suppose that the linear controller, defined by Eqs. (11) and (12), is observer-based so that

$$u = K\hat{x} + S_Q(\tilde{y}) \tag{23}$$

where, with some abuse of notation, $S_Q(\tilde{y})$ denotes the output of the linear, time-invariant system S_Q with input $\tilde{y}$, $\hat{x}$ is the state of a conventional observer, and $\tilde{y} \triangleq y - \hat{y}$, where $\hat{y} \triangleq C_1\hat{x}$. This controller can easily be expressed in the form of Eqs. (11) and (12), where $v = (\hat{x}, x_Q)$, x_Q being the state of S_Q. It follows that

$$H(s) = T_1(s) + T_2(s)Q(s)T_3(s). \tag{24}$$

If the observer is stable and K is chosen so that $A + BK$ is stable, then T_1, T_2, and T_3 are all stable, so that H is stable if Q is stable. Any stabilizing controller can be realized by Eq. (23) for some stable S_Q. Clearly, H is affine in Q [but not in the parameters (A_Q, B_Q, C_Q, D_Q) of S_Q].

14.3 DESIGN SPECIFICATIONS FOR LINEAR SYSTEMS

An excellent discussion of design specifications for linear control systems is given in Ref. 4; the presentation here is restricted to a few illustrative examples, chosen to emphasize certain points rather than to cover the subject comprehensively. An essential constraint is that the closed-loop system be stable; formulations of this constraint, suitable for use in optimization algorithms, are discussed next.

14.3.1 Stability

A system matrix (transfer function) is stable if all its eigenvalues (poles) lie in the open left half of the complex plane $\mathbb{C}$. A matrix (transfer function) is γ-stable if each eigenvalue λ of the matrix (each pole λ of the transfer function) lies in the set $\Gamma \subset \mathbb{C}$ defined by

$$\Gamma \triangleq \{z \in \mathbb{C} \mid \gamma(z) \leq 0\} \tag{25}$$

where $\gamma : \mathbb{C} \to \mathbb{R}$ is any function satisfying $\gamma(z) = \text{Re}(z) + f(\text{Im}(z))$ for all $z \in \mathbb{C}$, where $f : \mathbb{R} \to \mathbb{R}_+$ is even, positive-valued, and continuously differentiable and Re and Im denote, respectively, "real part of" and "imaginary part of." An example of such a function is $\gamma(z) = \max\{\text{Re}(z) + \alpha, \text{Re}(z) + \rho \, \text{Im}(z), \text{Re}(z) - \rho \, \text{Im}(z)\}$; any eigenvalue satisfying $\gamma(z) \leq 0$ has a real part less than $-\alpha$ and a damping factor which is not less than $1/(1 + \rho^2)^{1/2}$. Let D_γ denote the "Nyquist" contour consisting of the set $I_\gamma \triangleq \{z \in \mathbb{C} \mid \gamma(z) = 0\}$ (this set replaces the imaginary axis I) and an arc of infinite radius, joining the "end" points of I_γ [points corresponding to $\text{Im}(z) = \pm\infty$] and passing through the right half-plane.

The constraint that a matrix A be Γ-stable may, therefore, be expressed as

$$\gamma(\phi(A)) \leq 0 \tag{26}$$

where

$$\phi(A) \triangleq \max\{\text{Re}(\lambda) \mid \lambda \text{ is an eigenvalue of } A\} \tag{27}$$

but this is not useful since the function $A \mapsto \gamma(\phi(A))$ is not even Lipschitz continuous, making the newly developed algorithms for nondifferentiable optimization problems inapplicable.

Likewise, the standard Nyquist criterion is not amenable to the use of optimization. Provided there are no hidden modes, the necessary and sufficient condition for stability of the closed-loop system is that the multivariable Nyquist plot *encircles* the origin n times counterclockwise, where n is the number of eigenvalues of the open-loop system [zeros of $\det(sI - A)\det(sI - A_C)$] in the closed right half-plane; the multivariable Nyquist plot is the curve traced out by $\det(I + L(s, \alpha))$ as s traverses the standard Nyquist contour D, where the loop transfer function L is defined by

$$L(s, \alpha) \triangleq P(s)C(s, \alpha) \tag{28}$$

$$P(s) \triangleq C(sI - A)^{-1}B \tag{29}$$

and

$$C(s, \alpha) \triangleq D_c + C_c(sI - A_c)^{-1}B_c \tag{30}$$

A necessary and sufficient condition for γ-stability can be obtained by replacing the Nyquist contour D by a suitably modified contour D_γ which includes the curve $I_\gamma = \{z \in \mathbb{C} \mid \gamma(z) = 0\}$.

Encirclement does not fit well with optimization. However, a modification of the Nyquist criterion may be used.

The linear closed-loop system $S(\alpha) \triangleq \{A_{cl}(\alpha), B_{cl}(\alpha), C_{cl}(\alpha), D_{cl}(\alpha)\}$, whose transfer function from w to z is $H(\alpha, s)$, is γ-stable if and only if the eigenvalues of $A_{cl}(\alpha)$ lie in Γ. Let the closed-loop characteristic polynomial be defined by

$$\rho_{cl}(s, \alpha) \triangleq \det(sI - A_{cl}(\alpha)) \tag{31}$$

and let $d(s, \beta)$ denote a monic, Hurwitz polynomial in s, whose degree is the dimension of n_{cl} of x_{cl}, defined as follows:

$$d(s, \beta) \triangleq (s + \beta_1)(s^2 + \beta_2 s + \beta_3) \cdots (s^2 + \beta_{n_{cl}-1}s + \beta_{n_{cl}}) \tag{32}$$

if n_{cl} is odd, and

$$d(s, \beta) \triangleq (s^2 + \beta_1 s + \beta_2)(s^2 + \beta_3 s + \beta_4) \cdots (s^2 + \beta_{n_{cl}-1}s + \beta_{n_{cl}}) \tag{33}$$

if n_{cl} is even. Clearly, $d(s, \beta)$ is Hurwitz if and only if $\beta \geq 0$ (i.e., each component of β is greater than or equal to zero). A suitable stability test is provided by the following result, due to Polak [6]: $A_{cl}(\alpha)$ is γ-stable if and only if there exists a $\beta \geq 0$ such that the γ-Nyquist plot of $\rho_{cl}(s, \alpha)/d(s, \beta)$ [the curve traced out by $\rho_{cl}(s, \alpha)/d(s, \beta)$ as s traverses D_γ] lies in the open right half-plane. For design purposes, this condition may be reformulated as a semi-infinite inequality constraint

$$\mathrm{Re}\left(\frac{\rho_{\mathrm{cl}}(s, \alpha)}{d(\beta, s)}\right) \geq \epsilon \quad \text{for all } s \in I_\gamma \tag{34}$$

for some $\epsilon > 0$.

This stability test illustrates two points. First, stability may be expressed as a semi-infinite constraint. Second, it may be necessary to reformulate an existing condition into an equivalent, computationally useful form. From the computational point of view, utilizing the Nyquist plot of $\det(I + L(s, \alpha))$ [instead of $\rho_{\mathrm{cl}}(s, \alpha)/d(s, \beta)$] has the disadvantage of introducing encirclements into the stability test.

The stability test (34) is suitable for use with conventionally parameterized controllers. It is, however, not convex in the parameters α and β which must be determined via optimization.

To avoid a nonconvex stability constraint, Q-parameterization can be employed. In this parameterization, the controller consists of an observer with state $\hat{x}$, a feedback matrix K, and a subsystem S_Q [see Eq. (23)]. Suppose that the observer is γ-stable and that K is such that $A + BK$ is γ-stable; then $A_{\mathrm{cl}}(\alpha)$ is γ-stable if and only if the subsystem $S_Q = \{A_Q, B_Q, C_Q, D_Q\}$ [whose transfer function is $Q(s)$] is γ-stable; this is a constraint on the eigenvalues of A_Q which are nonconvex functions of the elements of A_Q. To avoid the problem of nonconvexity, S_Q may be constrained to be a weighted parallel combination $(\Sigma \alpha_i S_i)$ of a set of *prespecified* subsystems $\{S_i \mid i \in \mathbb{N}\}$, $S_i \triangleq (A_i, B_i, C_i, D_i)$, where, for all $i \in \mathbb{N}$, the eigenvalues of A_i lie in Γ so that S_Q is, by construction, γ-stable. Clearly, the eigenvalues of A_Q also lie in Γ and the transfer function of S_Q is

$$Q(s, \alpha) = \sum_{i=1}^{N} \alpha_i Q_i(s) \tag{35}$$

where $Q_i(s) = C_i(sI - A_i)^{-1}B_i + D_i$ and N is the total number of terms. The controller realization (35) will be referred to as the "Q-linear" parameterization in the sequel. Under reasonable conditions, any γ-stable transfer function can be arbitrarily closely approximated by choosing N sufficiently large.

The price to be paid for this approach is that the dimension of the controller is equal to the sum of the dimensions of the plant, observer, and S_Q; this can be considerably larger than that of a conventional controller if N is large. The advantages are, of course, that γ-stability is automatically ensured and that the closed-loop transfer function $H(s, \alpha)$, in terms of which the performance objectives are specified, as affine in α.

14.3.2 Performance

The plant description has been chosen so that all variables (including, if desired, the control u) which it is desired to regulate appear in the output z; also, w, the exogenous signal, includes all reference signals and all external disturbances. Thus, all performance objectives, such as rapid tracking of reference signals and attenuation of disturbances, can be expressed in terms of the closed-loop transfer function H from w to z or the corresponding time response.

For example, the requirement that output z_i is not excessively affected by disturbance w_j in the frequency range $[\omega_1, \omega_2]$ could be expressed by the semi-infinite constraint

$$|H_{ij}(j\omega, \alpha)| \leq \epsilon \quad \text{for all } \omega \in [\omega_1, \omega_2]. \tag{36}$$

Many such constraints may be involved in a design study. It is, of course, possible to group a collection of such constraints into a single inequality. This can be done by taking the maximum of a set of constraints; thus, the single semi-infinite constraint

$$\Psi(j\omega, \alpha) \triangleq \max\{ \ |H_{ij}(j\omega, \alpha)| \ |(i, j) \in I_d \ \} - \epsilon \leq 0 \tag{37}$$

is equivalent to the set of constraints $| H_{ij}(j\omega, \alpha) | \leq \epsilon$, $(i, j) \in I_d$. Here I_d indexes the appropriate (disturbance) input-output pairs. Other measures may, of course, be employed. A common form of the "disturbance constraint" is the semi-infinite inequality

$$\sigma_{\max}(H^*(j\omega, \alpha)) \leq \epsilon \quad \text{for all } \omega \in [\omega_1, \omega_2] \tag{38}$$

where $\sigma_{\max}(A)$ denotes the maximum singular value of the matrix A and H^* denotes an appropriate submatrix, corresponding to row-column pairs $(i, j) \in I_d$, constructed from H. This formulation beings analytic convenience but also a degree of approximation if the actual objective is $\Psi(j\omega, \alpha) \leq 0$. The maximum singular value of a matrix is, unlike an eigenvalue, Lipschitz continuous in its entries.

Another example of a performance objective is tracking; this often takes the form that an appropriate response lies in a given envelope which enforces objectives such as maximum overshoot, maximum rise time, and maximum settling time. Let $h_{ij}(t, \alpha)$ denote the step response from input w_j to output z_i. Then, a typical tracking objective takes the form

$$h_{ij}(t, \alpha) \in [s_l(t), s_u(t)] \quad \text{for all } t \in [0, t] \text{ and for all } (i, j) \in I_r \tag{39}$$

where I_r indexes the appropriate (reference) input-output pairs. The functions s_l and s_u define the envelope in which the step response is constrained to lie. Hard constraints on state variables may be similarly expressed.

To obtain decoupling, appropriate frequency responses (or step responses) can be bounded in magnitude. Thus, if $I_r = \{(1, 1), (2, 2), \ldots, (r, r)\}$, then the following semi-infinite inequality ensures a degree of decoupling:

$$|h_{ij}(t, \alpha)| \leq \epsilon(t) \quad \text{for all } t \in [0, t_1] \text{ and for all } i, j \in \underline{r}, i \neq j \tag{40}$$

where $\underline{r} \triangleq \{ 1 \cdot \cdot \cdot r \}$. Hard constraints on control signals are similar [the function $\epsilon(t)$ is replaced by a constant]. Enough has, perhaps, been said to indicate that many of the performance objectives arising in control design are semi-infinite.

Do these constraints have any structure which can be exploited? This question is relevant because the relative complexity of semi-infinite constraints makes it important to introduce simplification wherever possible. It turns out [4,5] that most of the conventional design objectives are convex, or quasi-convex, in the closed-loop transfer function H, but *not* necessarily in the design parameter α *unless* Q-linear parameterization of the controller is employed.

Many of the performance constraints have the form

$$\phi(\alpha, \gamma) \leq 0 \quad \text{for all } \gamma \in \Gamma \tag{41}$$

where γ denotes time and/or frequency and/or an uncertain plant parameter, and Γ is, correspondingly, a time interval, a frequency interval, a subset of parameter space, or an appropriate cartesian product of these spaces. These constraints may be expressed as

$$\Psi(\alpha) \triangleq \max \{ \phi(\alpha, \gamma) \mid \gamma \in \Gamma \} \leq 0 \tag{42}$$

Maximization over a convex set preserves convexity so that Ψ is convex if ϕ is convex in α for each γ in Γ. Also, any norm is a convex function. Because an induced norm may be expressed as the maximum of a norm over a convex set, it follows that induced norms are convex functions. For example, $\sigma_{\max}(A)$, the maximum singular value of a matrix $A \in \mathbb{C}^{m \times m}$, is the induced 2-norm and may be expressed as

$$\sigma_{\max}(A) = \max \{\|Ad\| \mid d \in D\} \tag{43}$$

where D is the unit sphere in $\mathbb{C}^m$. Hence constraints of the form (37)–(40) are all convex in H. They are also convex in α if Q-linear parameterization of the controller is employed. This convexity can be used, to advantage, in the optimization algorithms employed to solve the design problem; see Section 14.5.

In the above discussion of performance constraints, the envelopes, or upper bounds, of time and frequency responses were assumed fixed. Parameters of the envelopes, or upper bounds, such as rise time, or bandwidth, are *not* convex in H. However, as shown in Ref. 4, many performance constraints of this type (e.g., the rise time of a step response should not exceed T) are *quasi-convex* in H (a function is quasi-convex if all its level sets are convex). For example, rise time, appropriately defined, is quasi-convex in H since, for all T, the set of transfer functions with rise time not exceeding T is convex. Other examples are given in Ref. 4. These constraints are quasi-convex in α if Q-linear parameterization is employed. Quasi-convexity may also be exploited in optimization.

14.3.3 Robustness

The model of the plant, employed in the design process, is never exact. Typically, many simplifications are employed in modeling to obtain a linear, time-invariant model of moderate complexity which matches the behavior of the possibly nonlinear, distributed parameter system reasonably accurately over the operating range. To allow for this in the design stage, model error must be quantified. Abstractly, the plant p is assumed to be a member of a set P_p of possible plants. The closed-loop system (corresponding to a particular controller) is *robustly stable* if it is stable for all $p \in P_p$. A particular performance specification of the closed-loop system is robust if it is satisfied for all $p \in P_p$. To quantify this, the actual plant is often modeled by a multiplicative or additive perturbation of the nominal plant; in either case, the actual plant may be described as in Eqs. (4)–(6) but with the state dimension appropriately increased (by an amount equal to the state dimension of the perturbation) and with the (actual) plant parameters A, B, etc., now functions of a parameter $\delta \in \Delta$ chosen to ensure that p ranges over P_p (or a superset of P_p) as δ ranges over Δ. It follows that A_{cl}, B_{cl}, C_{cl} and D_{cl} in (29) and (30) and the characteristic polynomial ρ_{cl} are all functions of α and δ. The perturbed closed-loop system is

$$S(\alpha, \delta) \triangleq \{A_{cl}(\alpha, \delta), B_{cl}(\alpha, \delta), C_{cl}(\alpha, \delta), D_{cl}(\alpha, \delta)\}.$$

Hence, ρ_{cl} is a function of s, α, and δ. It then follows that the closed-loop system is robustly γ-stable [i.e., $S(\alpha, \delta)$ is γ-stable for all $\delta \in \Delta$) if, for every $\delta \in \Delta$, there exists a $\beta \geq 0$ satisfying

$$\text{Re}\left(\frac{\rho_{cl}(s, \alpha, \delta)}{d(\beta, s)}\right) \geq \epsilon \qquad (44)$$

for all $s \in I_\gamma$. A simpler, but conservative, robust stability constraint is the existence of a $\beta \geq 0$ such that Inequality (44) is satisfied for all $s \in I_\gamma$, $\delta \in \Delta$, a semi-infinite constraint. Obtaining a robustly stable system requires the determination of an α such that Inequality (44) is satisfied for all $s \in I_\gamma$, $\delta \in \Delta$, a semi-infinite feasibility problem.

The transfer function from w to z of the perturbed closed-loop system is $H^p(s, \alpha, \delta)$. A particular performance criterion is robust if it is satisfied for all $\delta \in \Delta$. For example, a robust version of (37) is

$$|H^p_{ij}(j\omega, \alpha, \delta)| \leq \epsilon \quad \text{for all } \omega \in [\omega_1, \omega_2] \text{ and for all } \delta \in \Delta \qquad (45)$$

another semi-infinite feasibility constraint. Many robust performance specifications can similarly be expressed as semi-infinite constraints, whose evaluation requires maximization with respect to time (or frequency) and a perturbation parameter δ.

Although, as pointed out earlier, most design specifications, when the model error is ignored, are convex, or quasi-convex, in the nominal transfer function H, this is not necessarily the case when plant perturbations are considered. However, in many cases the perturbed closed-loop system $S(\alpha, \delta)$ can be represented as a (feedback) interconnection of the (stable) nominal system $S_n(\alpha)$ and a perturbation system $S_p(\delta)$; in such cases, the small gain theorem can be employed to obtain sufficient conditions for robust stability and performance in the form of convex inequalities on the *nominal* closed-loop transfer function H; see Refs. 4 and 5. The perturbation system is usually characterized by disk bounds on its frequency response and the requirement that it be γ-stable. A typical specification for robust stability, which is clearly convex in H, is

$$\sigma_{\max}(H(j\omega, \alpha)) \leq l(\omega) \tag{46}$$

for all ω in a specified frequency interval; here $l(\omega)$ is a measure of multiplicative model error and usually increases with frequency, enforcing "roll-off" in the nominal closed-loop transfer function H. Inequalities such as (45) are clearly convex in α if Q-linear parameterization of the controller is employed.

14.4   DESIGN SPECIFICATIONS FOR NONLINEAR SYSTEMS

Design of controllers for nonlinear systems (or nonlinear controllers for linear systems) is at a less advanced state despite recent spectacular progress due to the complexity and variety of the systems encountered in practice. In practice, of course, controllers are nearly always nonlinear, even if designed using linear theory. Practical requirements, such as necessity to restrict control magnitude, control rate, or certain states (temperature, for example), enforce nonlinearities in the controller. The controlled system is, therefore, nonlinear, even if the plant being controlled is linear. The lack of techniques for analyzing general nonlinear systems necessitates the use of extensive simulation to ensure that performance and stability objectives are met. Because design by simulation is relatively inefficient (it is difficult for a person to make an optimal choice of more than two or three variables), it appears almost certain that optimization procedures will eventually be extensively employed in nonlinear controller design. Since frequency- domain techniques are irrelevant, except in a restricted class of problems, performance specifications are expressed in the time domain. An optimization- based design environment would, therefore, consist of a nonlinear simulator to which is coupled a suitable optimizer whose purpose is to adjust the parameters of a controller to ensure stability and satisfaction of time domain performance specifications. As in linear systems, it is important to exploit appropriate system-theoretic advances. Among these is the interesting work on input-output linearization [2] and its use in "inverse" control to make the input-output map linear with a specified input-output transfer function. This approach has been used to good effect in control of vertical takeoff and landing (VTOL) air-craft and control of robot arms (the "computed control torque" method). How-ever, a controller of this type cancels the zero dynamics and can, therefore, only be employed when the zero dynamics are stable (the minimum phase case when the system is linear). In Ref. 6, approximate methods to deal with this problem have been developed. More recently, Isidori and Byrnes [3] have developed a nonlinear analogue of linear regulator theory [7], yielding necessary and suffi-cient conditions for the local solvability of the state and output feedback regu-lator problems. This work is an important advance, indicating the possibility of regulation even when the zero dynamics are unstable and providing useful guid-ance on the structure of the controller. It does, nonetheless, rely on the absence of state and control constraints.

14.4.1 Stability

Stability remains the overiding objective. Unfortunately, there is no simple method for determining whether a particular nonlinear system is stable or not. There exist certain systems, usually of the form of a feedback interconnection of a linear, time-invariant, and a memoryless, sector-bounded nonlinearity, for which frequency response techniques such as the circle and Popov criteria can be employed to assess stability. In the single-input single-output case, the Popov criterion (and the circle criterion when the transfer function of the linear system is minimum phase) may be expressed as a semi-infinite inequality in the fre-quency domain. For more general nonlinear systems, it appears necessary to employ Lyapunov theory, even though the computational cost in doing so may be high.

The work involved in constructing a Lyapunov function for a given autono-mous system is comparable to that of determining a nonlinear feedback law. Be-cause the need to assess stability may recur frequently in the design process, it appears necessary to avoid repeated modification of the Lyapunov function. To show how this may be done, consider, for this subsection only, a simplified problem in which the system to be controlled is described by $\dot{x} = f(x, u)$, where $f(0, 0) = 0$ (f may include some controller dynamics if required) and the design objective is regulation to the origin. State feedback is used so that the controller is $u = c(x, \alpha)$, where $c(0, \alpha) = 0$ for all α. The closed-loop system is de-scribed by $\dot{x} = f_{cl}(x, \alpha)$, where $f_{cl}(x, \alpha) \triangleq f(x, c(x, \alpha))$.

One method to avoid repeated modification of the Lyapunov function is to use a Lyapunov function for the linearized system. Suppose that $\dot{x} = Ax + Bu$, the linearization of the nonlinear system at the origin, is stabilizable and that a linear controller $u = Kx$ has been chosen so that $A_c = A + BK$ is stable. This defines a Lyapunov function $x \mapsto V_L(x) = \frac{1}{2}x^T P x$ for the linear closed-loop system, where P is the symmetric positive definite solution of $A_c^T P +$

$PA_c = -Q$ *for some positive definite matrix Q.* Suppose X, a level set of V_L, is the domain of interest. Then a sufficient condition that the closed-loop system $\dot{x} = f_{cl}(x, \alpha)$ is asymptotically stable, with a region of attraction X, is that the inequality

$$\dot{V}_L(x) = x^T P f_{cl}(x, \alpha) \leq -\epsilon V_L(x) \tag{47}$$

is satisfied for all $x \in X$ and some $\epsilon > 0$. However, the region of attraction X may be small, especially if control constraints have to be satisfied.

A possibly larger region of attraction can be obtained using a nonquadratic Lyapunov function which can be constructed as shown below provided that the original system is state-feedback equivalent to a linear system. More precisely, suppose that $\dot{x} = f(x, u)$ can be transformed via a (nonlinear) diffeomorphic state transformation $x \mapsto \pi = \Pi(x)$ and a nonlinear feedback $u = \kappa(x)$, $\kappa(0) = 0$, into a linear system

$$\dot{\pi} = A_c \pi \tag{48}$$

for which

$$V_L(\pi) = \frac{1}{2}\pi^T P \pi \tag{49}$$

is a positive definite Lyapunov function with derivative $\dot{V}_L(\pi) = -\frac{1}{2}\pi^T Q \pi$, where Q is positive definite. Then, as is easily shown [8], the function $V:\mathbb{R}^n \rightarrow \mathbb{R}$ defined by

$$V(x) \triangleq V_L(\Pi(x)) \tag{50}$$

is a positive definite Lyapunov function for the closed-loop system

$$\dot{x} = f(x, \kappa(x)) \tag{51}$$

and decreases along any trajectory of Eq. (51). The nonlinear control law κ may, of course, transgress performance specifications and control or state constraints because these were not considered in its derivation. However, the actual controller employed is $u = c(x, \alpha)$ resulting in a closed-loop system $\dot{x} = f_{cl}(x, \alpha)$. The control parameter α is chosen, if possible, to satisfy the performance specifications and constraints, and the following stability constraint:

$$\dot{V}(x, \alpha) = \Pi(x)^T P \Pi_x(x) f_{cl}(x, \alpha) \leq -\epsilon \Pi(x)^T Q \Pi(x) \tag{52}$$

for all $x \in X$, where X is a level set of V. If the stability constraint (52) is satisfied for all $x \in X$, then the closed-loop system is asymptotically stable with a region of attraction X.

These two methods avoid the problem of computing Lyapunov functions at each stage of the design (i.e., at each value of the controller parameter value α)

by using a Lyapunov function obtained using a linear model of the system or, in the second case, ignoring all constraints. It is, therefore, not certain that the semi-infinite inequality in Eq. (47) or (52) can be satisfied for all $x \in X$. For this reason, a relaxed condition was introduced in Ref 11; the requirement that Lyapunov function V be reduced continuously along controlled trajectories is replaced by the condition that the sequence $\{V(x(t_i), i \in \mathbb{N}\}$, constructed by sampling the Lyapunov function at t_i, $i \in \mathbb{N}$, decreases monotonically to zero. This permits the Lyapunov function to increase temporarily in intersample periods. More precisely, if V is a continuous, positive definite Lyapunov function, X is a compact level set of V, $T \in (0, \infty)$, $\delta_1 \in (1, \infty)$, and $\delta_2 \in (0, 1)$, then a sufficient condition for the closed-loop system $\dot{x} = f_{cl}(x, \alpha)$ to be asymptotically stable, with a region of attraction X, is that the following inequalities are satisfied:

$$V(x(t; x_0, \alpha)) \leq \delta_1 V(x_0) \tag{53}$$

for all $x_0 \in X$, all $t \in T$, and

$$V(x(T; x_0, \alpha)) \leq \delta_2 V(x_0) \tag{54}$$

for all $x_0 \in X$. In Inequalities (53) and (54), $x(t; x_0, \alpha)$ denotes the solution at time t of $\dot{x} = f_{cl}(x, \alpha)$ with initial state x_0 at time 0. It is possible, as shown in Ref. 9, to make T, δ_1, and δ_2 functions of x, which permits extra flexibility in the design of controllers for nonlinear systems. [It may be desirable, at any state x such that $V(x)$ is large, to permit a longer period T in which V is to be reduced.]

Hence, stability of nonlinear systems may also be expressed as a semi-infinite inequality which, in this case, requires global maximization over X, a subset of $\mathbb{R}^{n_{cl}}$. This very severe requirement seems unavoidable if general nonlinear systems with state and control constraints are considered and is a reflection of current practice in which extensive simulation is employed in the final stages of the design of a nonlinear system to ensure that all specifications, including stability, are satisfied.

In the above discussion, a simplified control problem has been considered. Sufficient conditions for stability with respect to a set-point, or to a periodic orbit (corresponding to tracking a periodic reference), may also be expressed as semi-infinite inequalities. The conditions may be generalized to the case when output, rather than state, feedback is employed.

14.4.2 Performance

Performance constraints are naturally expressed in the time domain and will have a form similar to (39) or (40). Because nonlinear systems are involved, properties of the input (its magnitude, for example) become relevant. Let $\dot{x} = f_{cl}(x, w, \alpha, \delta)$ denote the perturbed form of Eq. (13) with $\delta = 0$ corresponding, as before, to the nominal system. Then the nonlinear version of (39) is

$z_i(t, \alpha, w, x) \in [s_l^{w,x}(t), s_u^{w,x}(t)]$ for all $t \in [0, t_1]$, all $(w, x) \in T_r$,
and all $i \in I_r$ (55)

where $z_i(t, \alpha, w, x)$ denotes the response of the ith output to (w, x), where w is the test input, x is the initial state of the closed-loop system, and the controller parameter is α; T_r is the set of test input, initial state pairs, and I_r indexes the appropriate outputs. State and control constraints can be similarly expressed. Again, semi-infinite inequalities are involved.

14.4.3 Robustness

For robust stability, consider again the simpler problem considered in Section 14.4.1 and suppose that the real plant is described by

$$\dot{x} = f(x, u, \delta) \tag{56}$$

rather than by $\dot{x} = f(x, u)$; here $\delta \in \Delta$ is a perturbation parameter and $\delta = 0$ specifies the nominal system. The corresponding closed-loop system is described by

$$\dot{x} = f_{cl}(x, \alpha, \delta) \tag{57}$$

Suppose V and X are defined as above [immediately prior to Inequality (53)] and that $w = 0$. Then the closed-loop system is robustly stable (stable for all $\delta \in \Delta$) with a region of attraction X if the following inequalities are satisfied:

$$V(x(t; x_0, \alpha, \delta)) \le \delta_1 V(x_0) \tag{58}$$

for all $x_0 \in X$ and all $t \in T$, and

$$V(x(T; x_0, \alpha, \delta)) \le \delta_2 V(x_0) \tag{59}$$

for all $t \in [0, \cdot t_1]$, all $x_0 \in X$, and all $\delta \in \Delta$. In Inequalities (58) and (59), $x(\cdot; x_0, \alpha, \delta)$ denotes the solution of Eq. (57) with initial state x_0.

Robust performance specifications can similarly be expressed, for the general case, as inequalities which must be satisfied for all $t \in [0, t_1]$, $\delta \in \Delta$, and all $(w, x) \in T_r$ where these variables have the meaning defined above [see (55)] and the closed-loop system is $\dot{x} = f_{cl}(x, w, \alpha, \delta)$, a perturbation of Eq. (13).

14.5 ALGORITHMS

It has been shown above that design specifications are often semi-infinite. Thus, a significant part of the control design process can often be expressed as finding an α solving the following *canonical* problem:

$$\min \{f(\alpha) \mid \Psi(\alpha) \le 0\} \tag{60}$$

or finding a feasible point, that is, a point α satisfying $\Psi(\alpha) \leq 0$, where

$$\Psi(\alpha) \triangleq \max \{g^i(\alpha), i \in \underline{k}; \phi(\alpha, \gamma_j), \gamma_j \in \Gamma_j, j \in \underline{m} \} \tag{61}$$

$\underline{k} \triangleq \{ 1, \ldots, k \}$, $\underline{m} \triangleq \{ 1, \ldots, m \}$, $f: \mathbb{R}^a \to \mathbb{R}$ is a performance index, $g^i: \mathbb{R}^a \to \mathbb{R}$ a conventional constraint, and $\phi^j: \mathbb{R}^a \times \mathbb{R}^{p_j} \to \mathbb{R}$ a semi-infinite constraint. It is assumed, for simplicity, that the functions f, g^i, ϕ^j are all continuously differentiable. The constraint function Ψ is *non-smooth* (is not continuously differentiable, but is Lipschitz continuous). The sets Γ_j may represent time or frequency intervals, or subsets of parameter or state space, and are assumed to be compact. The problem formulation may include weights to permit trade off between constraints. The problem when all the sets Γ_j are intervals on the real line is appreciably simpler since global optimization [required to evaluate $\Psi(\alpha)$] is much simpler in this case. Two classes of algorithms have been developed for semi-infinite problems, namely, *descent* and *outer approximation* algorithms; a brief description of each class is given below. When the design problem is convex, algorithms which exploit convexity can be profitably used. This case is discussed first.

14.5.1   Convex and Quasi-Convex Problems

Consider, for simplicity, the problem

$$P_D: \min_{\alpha \in A} \Psi(\alpha) \tag{62}$$

where

$$\Psi(\alpha) \triangleq \max \{ \phi(\alpha, \gamma) \mid \gamma \in \Gamma \} \tag{63}$$

and A is a compact "box" in $\mathbb{R}^a$.

Cutting Plane Algorithms
Assume that ϕ is convex in α for each γ. Then Ψ is convex so that, at each point α_0 in A, there exists a subgradient $g(\alpha_0)$ satisfying

$$\Psi(\alpha) \geq \Psi(\alpha_0) + \langle g(\alpha_0), \alpha - \alpha_0 \rangle \quad \text{for all } \alpha \in A \tag{64}$$

The subdifferential $\partial\Psi(\alpha_0)$ of Ψ at α_0 is the set of all subgradients at α_0. Because ϕ is continuously differentiable, $\nabla_\alpha \phi(\alpha_0, \gamma_0)$, where γ_0 is *any* maximizer of $\phi(\alpha_0, \cdot)$, is a subgradient of Ψ at α_0 so that subgradients of Ψ are easily obtained. If $A_i \triangleq \{ \alpha_0, \ldots, \alpha_{i-1} \}$ is a subset A, then the piecewise linear, convex function

$$\Psi_{A_i}(\alpha) \triangleq \max \{\Psi(\alpha_j) + \langle g(\alpha_j, \alpha - \alpha_j \rangle \mid j \in \{0, \cdots, i - 1\}\} \tag{65}$$

is a lower bound for Ψ, hence satisfying

$$\Psi(\alpha) \geq \Psi_{A_i}(\alpha) \quad \text{for all } \alpha \in A \tag{66}$$

The minimizer

$$\alpha_i = \arg \min \{\Psi_{A_i}(\alpha) \mid \alpha \in A\} \tag{67}$$

of Ψ_{A_i} is easily obtained since the optimization problem in Eq. (67) is equivalent to the linear program

$$P_i: \min \{w \mid \Psi(\alpha_j) + \langle g(\alpha_j), \alpha - \alpha_j\rangle \leq w, j \in$$
$$\{0, \cdots, i - 1\}, \alpha \in A, w \geq 0\} \tag{68}$$

Clearly, $\Psi_{A_{i+1}}$, where $A_{i+1} \triangleq A_i \cup \{\alpha_i\}$ is a better approximation to Ψ than Ψ_{A_i}. This method of successively improving the lower bound function Ψ_{A_i} by replacing A_i by a larger subset of A leads to the construction of a family of simple and effective convex programming algorithms of which Kelley's algorithm [10] is a prominent representative. Assume that an initial set $A_1 = \{\alpha_0\}$ is available.

Kelley's Cutting Plane Algorithm for Solving Problem P_D.   For all $i \in \mathbb{N}$,

(a)   Determine α_i, the minimizer of

$$\Psi_i = \min \{\Psi_{A_i}(\alpha) \mid \alpha \in A\}$$

(b)   Set $A_{i+1} = A_i \cup \{\alpha_i\}$

It is easily shown that Ψ_i increases monotonically and converges to $\Psi^* \triangleq \min \{\Psi(\alpha) \mid \alpha \in A\}$ as $i \rightarrow \infty$. If Ψ is strictly convex, α_i converges to the unique solution of P_D; else, any accumulation point $\hat{\alpha}$ of an infinite sequence generated by the algorithm is a solution to P_D. At iteration i, Ψ_i is a lower bound, and $\Psi(\alpha_i)$ an upper bound, to Ψ^*. The number of constraints in the linear program in Step (a) increases at every iteration, increasing its complexity. However, there exists schemes for dropping constraints, without prejuducing convergence; these limit the complexity of the subproblem (a). The cutting plane algorithm is a member of a larger family of closely related algorithms, the outer approximations algorithms for semi-infinite programming [11–14].

Ellipsoid Algorithms

Another family of convex programming algorithms, which also relies on the construction of the supporting hyperplanes to the level sets of the function Ψ, are the ellipsoid algorithms [15]. The ellipsoid algorithms merely require that the level sets of Ψ are convex, (i.e., that Ψ is quasi-convex) and that, at each α in A, a quasi-gradient $g(\alpha)$ which defines the supporting hyperplane to the level set is available, so that, for all α,

$$L_\alpha \subset \{\alpha' \mid \langle g(\alpha), \alpha' - \alpha\rangle \leq 0\} \tag{69}$$

where the level set $L_\alpha \triangleq \{ \alpha' \mid \Psi(\alpha') \leq \Psi(\alpha) \}$. A quasi-gradient of Ψ at α is any vector $g(\alpha)$ satisfying

$$\Psi(\alpha') \geq \Psi(\alpha) \qquad \text{whenever } \langle g(\alpha), \alpha' - \alpha \rangle \geq 0 \tag{70}$$

The ellipsoid algorithm computes a sequence $\{E_i\}$ of ellipsoids of decreasing volume, each containing the minimizer. Each ellipsoid E_i is defined by its center α_i and the positive definite matrix M_i, that is,

$$E_i \triangleq \{\alpha \in \mathbb{R}^a \mid (\alpha - \alpha_i)^T M_i (\alpha - \alpha_i) \leq 1, M_i = M_i^T > 0\}$$

Thus, α_i is the estimate, at iteration i, of the solution P_D. Suppose an initial ellipsoid E_0 containing the minimizer is available.

Basic Ellipsoid Algorithm for Solving Problem P_D. For all $i \in \mathbb{N}$, compute (α_i, M_i) which defines the ellipsoid E_i of *minimum volume* containing the set

$$E_{i-1} \cap \{\alpha \mid \langle g(\alpha_{i-1}), \alpha - \alpha_{i-1} \rangle \leq 0\}$$

Since the initial ellipsoid E_0 and each level set of Ψ contain the minimizer, each of the ellipsoids subsequently constructed must also contain the minimizer. It can be shown (see, e.g., Ref. 4) that the volume of E_{i+1} is less than that of E_i by a factor $\exp(-1/2a)$, so that the volume of E_i decreases to zero as $i \to \infty$. Hence, α_i converges to a solution of P_D. Because $\exp(-1/2a) \approx 1$ when the dimension a of α is large, convergence can be extremely slow. Formulas for updating α_i and M_i are given in the literature (see, e.g., Ref. 4).

14.5.2 Nonconvex Problems

Descent Algorithms

These algorithms, as the name suggests, reduce the constraint $\Psi(\alpha)$ when α is infeasible [$\Psi(\alpha) > 0$] and reduce the performance index $f(\alpha)$, maintaining feasibility, when α is feasible. The principles underlying these algorithms can be more easily understood in the context of a simpler problem P (which replaces A in P_D by $\mathbb{R}^a$).

$$P: \min_{\alpha \in \mathbb{R}^a} \Psi(\alpha) \tag{71}$$

where ϕ is continuously differentiable and Γ is compact. Although the function Ψ is *not* differentiable, it does have a *generalized derivative* [16]

$$\partial \Psi(\alpha) = \text{co}\{ \nabla_\alpha \phi(\alpha, \gamma) \mid \gamma \in \hat{\Gamma}(\alpha)\} \tag{72}$$

where

$$\hat{\Gamma}(\alpha) \triangleq \{\alpha \mid \phi(\alpha, \gamma) = \Psi(\alpha)\} \tag{73}$$

The function $\partial\Psi$ is *set-valued* and is *not* continuous (in the set-valued sense). A necessary condition of optimality for P is $0 \in \partial\Psi(\alpha)$. Consider the search direction

$$h(\alpha) \triangleq - \arg\min \{ \text{½} \|h\|^2 \mid h \in \partial\Psi(\alpha) \} \tag{74}$$

which may be regarded as the direction of the steepest descent in the nonsmooth case. Because $\partial\Psi$ is not continuous, a steepest descent direction algorithm employing h can jam up at a point $\hat{\alpha}$ which does not satisfy $0 \in \partial\Psi(\hat{\alpha})$. Hence, research is directed to finding *smooth* alternatives. Polak [17] has proposed various alternatives which may be regarded as developments of techniques employed in feasible directions algorithms. One of these is the set $\overline{G}\Psi$ defined by

$$\overline{G}\Psi(\alpha) \triangleq \text{co} \left\{ \begin{bmatrix} \Psi(\alpha) - \phi(\alpha, \gamma) \\ \nabla_\alpha\phi(\alpha, \gamma) \end{bmatrix} \,\middle|\, \gamma \in \Gamma \right\} \tag{75}$$

Polak [17] has shown that $\overline{G}\Psi$ is a convergent direction finding map, that is, $\overline{G}\Psi$ is continuous, $\overline{g} = (g^0, g) \in \overline{G}\Psi(\alpha)$ implies $g^0 \geq 0$, and $\overline{g} = (0, g) \in \overline{G}\Psi$ (a) and only if $g \in \partial\Psi(\alpha)$.

The functions $\theta : \mathbb{R}^n \rightarrow \mathbb{R}$ and $\overline{h} : \mathbb{R}^n \rightarrow \mathbb{R}^{n+1}$ of Ref. 17 are defined by

$$\theta(\alpha) \triangleq \min \{ \text{½} \|\overline{g}\|^2 \mid \overline{g} \in \overline{G}\Psi(\alpha) \} \tag{76}$$

$$\overline{h}(\alpha) = (h^0(\alpha), h(\alpha)) \triangleq - \arg\min \{ \text{½} \|\overline{g}\|^2 \mid \overline{g} \in \overline{G}\Psi(\alpha) \} \tag{77}$$

The term $\tfrac{1}{2} \|\overline{g}\|^2$ in (76) and (77) may be replaced by, for example, $g_0 + \tfrac{1}{2} \|\overline{g}\|^2$; see Ref. 17 for further discussion of this point.

It can be shown that $0 \in \partial\Psi(\alpha)$ if and only if $0 \in \overline{G}\Psi(\alpha)$, $\theta :$ $\mathbb{R}^n \rightarrow \mathbb{R}$, and $\overline{h} : \mathbb{R}^n \rightarrow^{n+1}$ are continuous, $\theta(\alpha) \leq 0$ for all α, and $\theta(\alpha) = 0$ if and only if $0 \in \partial\Psi(\alpha)$. Hence, θ and h are, respectively, suitable optimality and search directions functions to be employed in the following nondifferentiable steepest descent algorithm for P. Assume that an initial point α_1 is available.

Steepest Descent Algorithm for Problem P.   For all $i \in \mathbb{N}$,

(a)   Determine the search direction: $h_i = h(\alpha_i)$.
(b)   Determine the step length: $\lambda_i \in \arg\min \{ \Psi(\alpha_i + \lambda h_i \mid \lambda \leq 0 \}$.
(c)   Update: $\alpha_{i+1} = \alpha_i + \lambda_i h_i$.

If $\hat{\alpha}$ is an accumulation point of an infinite sequence $\{\alpha_i\}$ generated by the algorithm, then $\hat{\alpha}$ satisfies the necessary condition of optimality $0 \in \partial\Psi(\hat{\alpha})$.

An obvious concern is the practicality of computing a solution to the problem defined by Eq. (77) to obtain the search direction h. In fact, Eq. (77) is a convex problem for which proximity algorithms (such as those due to Gilbert and Wolfe) may be employed. A practical algorithm employs an Armijo procedure to compute step length, truncates the computation of $\overline{h}$, and uses discretization in the evaluation of ψ so that all computations at iteration i of the algorithm require only a finite number of steps. Full details are given in Ref. 17. He and Polak [18] have shown how the discretization process may be optimized. There exist further possibilities for simplifying the algorithm. For example, $\overline{G}\Psi$, the *smooth* modification of $\partial\Psi$, may be replaced by a simpler object $\overline{G}_\epsilon\Psi$, defined by

$$\overline{G}_\epsilon\Psi(\alpha) \triangleq \mathrm{co}\left\{ \begin{bmatrix} \Psi(\alpha)-\phi(\alpha, \gamma) \\ \nabla_\alpha\phi(\alpha, \gamma) \end{bmatrix} \middle| \gamma \in \Gamma_\epsilon(\alpha) \right\} \tag{78}$$

where

$$\Gamma_\epsilon(\alpha) \triangleq \{\gamma \in \Gamma \mid \Psi(\alpha) - \phi(\alpha, \gamma) \leq \epsilon\} \tag{79}$$

Outer Approximation Algorithms

Consider, for simplicity, the (feasibility) problem of determining a $\alpha \in \mathbb{R}^n$ satisfying the semi-infinite constraint

$$\Psi(\alpha) \leq 0 \tag{80}$$

that is, satisfying (in the simple case considered here)

$$\phi(\alpha, \gamma) \leq 0 \quad \text{for all } \gamma \in \Gamma \tag{81}$$

Assume that an initial set $\Gamma_1 = \{\alpha_0\}$ is available.

Basic Outer Approximations Algorithm. For all $i \in \mathbb{N}$,

(a) Determine α_1 satisfying $\phi(\alpha_i, \gamma) \leq 0$ for all $\gamma \in \Gamma_i$.
(b) Determine $\gamma_i = \arg \max \{\phi(\alpha_i, \gamma) \mid \gamma \in \Gamma\}$.
(c) Set $\Gamma_{i+1} = \Gamma_i \cup \{\gamma_i\}$.

The algorithm has the following features. The semi-infinite feasibility problem is decomposed into an infinite sequence of conventional (finite-dimensional) feasibility problems, each *required* to be solvable. One *global* optimization is required per iteration of the algorithm. The set $F_{\Gamma_i} \triangleq \{ \alpha \in \mathbb{R}^n \mid \phi(\alpha, \gamma) \leq 0$ for all $\gamma \in \Gamma_i$ is an *outer approximation* of the feasible set F (i.e., $F \subset F_{\Gamma_i}$), where F is defined by

$$F \triangleq \{\alpha \in \mathbb{R}^n \mid \phi(\alpha, \gamma) \leq 0 \text{ for all } \gamma \in \Gamma\} = \{\alpha \in \mathbb{R}^n \mid \Psi(\alpha) \leq 0\} \tag{82}$$

If the outer approximations algorithm generates an infinite sequence $\{\alpha_i\}$, then any accumulation point $\hat{\alpha}$ of $\{\alpha_i\}$ is feasible [i.e., satisfies $\Psi(\hat{\alpha}) \leq 0$].

A conceptual outer approximation algorithm for solving P_D [see (62)] can be obtained by replacing Step (a) in the algorithm by

(a) Determine α_i, the minimizer of

$$\min_\alpha \max_\gamma \{\phi(\alpha, \gamma) \mid \gamma \in \Gamma_i\}$$

A practical algorithm, which solves the problems in (a) and (b) approximately, increasing precision as necessary, and drops elements from Γ_i to reduce the complexity of these problems, is presented in Ref. 14. Recent research is aimed at reducing the number of (approximate) global optimizations required.

The great virtue of the outer approximations algorithm is its simplicity. Its practical version requires, in Step (a), approximate solution of a conventional feasibility or constrained optimization problem for which well-tested software is readily available. Step (b) requires approximate global optimization but this is not too difficult when time and/or frequency responses are involved. Expressing singular values as a max function [see Eq. (43)] yields a relatively simple algorithm for solving design problems with constraints on singular values [24].

14.6   DESIGN ENVIRONMENT

The first requirement for a design environment is a facility for simulating the system and its controller, together with a suitable friendly interface for inputting the system descriptions and modifying them. The simulation should yield time responses to specified inputs, and, in the case of linear systems, frequency responses. This facility is readily available in commercially available software (MATLAB, $MATRIX_X$, etc.).

The second requirement is a facility for inputting design specifications, such as envelopes for time and frequency responses, and for specifying appropriate test signals.

A third requirement is the ability to compute sensitivities, for example, the gradient $\nabla_\alpha \phi(\alpha, \gamma)$ at specified values of α and γ when one of the design specifications is $\phi(\alpha, \gamma) \leq 0$ for all $\gamma \in \Gamma$. This facility is *not* available in commercial software although the means for computing these derivatives has been developed. For linear systems, the derivatives have been evaluated in design software developed at Imperial College and University of California, Berkeley using diagonal and Schur decompositions of $A_{cl}(\alpha)$ (see Ref. 19). An additional, useful feature of this software was its ability, using a special-purpose symbolic language, to permit the controller to be specified algebraically in terms of the parameter α. For nonlinear systems, it is necessary to employ the adjoint, preferably the adjoint of the discrete-time system corresponding to an integration procedure used for simulating the system. This procedure reduces errors and has been employed, for example, in SPEEDUP, software developed at Imperial College for the design of chemical processes. In some design software, finite differences are employed; because derivatives can be computed without excessive additional overhead, it seems desirable to include this facility in new design environments. This is an area which could benefit from the same careful attention that has been devoted to developing sound mathematical procedures for evaluating, for example, the Riccati and Lyapunov equations and frequency responses.

A fourth requirement is the coupling of optimization algorithms to the simulator (which yields the values of the specifications) and the sensitivity program (which yields the gradients of the specifications at specified values of the arguments). Because the optimization problems involved in design are essentially semi-infinite, the algorithms are more complex than those normally employed, and this has raised the fear that they are difficult to use and require optimization expertise. Whereas difficulties should not be minimized, it should be emphasized that there are available well-tested programs for solving conventional optimization problems (problems with a finite number of constraints) and that the outer approximations algorithms, in particular, are very simple and merely solve, or approximately solve, a sequence of such problems. All the outer approximations algorithm does is specify, at each iteration, a few points in Γ at which the design specifications must be satisfied. To do this, it must approximately solve, at each iteration, a global optimization problem of the form max $\{\phi(\alpha, \gamma) \mid \gamma \in \Gamma\}$. This not prohibitively difficult when Γ is a time or frequency interval, which is the common situation, because it involves scanning time and frequency responses which the designer would, in any case, like to do. When the design problem is convex, the cutting plane algorithm solves a sequence of finite- dimensional linear program for which excellent software is readily available.

Software involving most of the above features has been developed at the following universities: Imperial College, University of California, Berkeley (DELIGHT), University of Maryland (CONSOLE), and Stanford University (QDES). DELIGHT was particularly ambitious, providing an optimization environment for many applications (design of control systems, structures, electronic circuits). CONSOLE is simpler, and, therefore, more robust, limiting itself to one optimization routine FSQP, a well-tested program for solving conventional constrained optimization problems using sequential quadratic programming. It has the useful feature of maintaining feasibility with respect to hard constraints (e.g., stability constraints). CONSOLE has been used in some challenging aerospace design problems; it is currently the most powerful optimization-based package generally available. QDES solves convex control design problems employing the cutting plane or ellipsoid algorithms. The fact that commercial packages (e.g., MATLAB and X_{MATH}) now have optimization facilities, encourages the hope that the essential features of a design environment, listed above, may soon be available commercially.

14.7 CONCLUSION

Early formulations of controller design as an optimization, or feasibility, problem appear in Refs. 20 and 21. The semi-infinite nature of the design problem is emphasized in Refs. 14, 17, and 22–26.

One of the earliest descent algorithms for nonlinear semi-infinite optimization problems appears in Ref. 22. An improved version appears in Ref. 27. These algorithms, developed before the advent of the nonsmooth calculus of

Clarke [16], are developments of ideas inherent in feasible directions algorithms. Further developments are excellently reviewed in Ref. 17. Alternative descent algorithms are presented in Ref. 28.

An approach to solving nonlinear semi-infinite optimization problems, using outer approximations, is presented in Ref. 14, that introduces a new technique for dropping constraints to reduce the dimensionality of subproblems solved at each iteration. Improved techniques for constraint dropping are presented in Ref. 29.

The fact that many control design problems may be cast as a convex optimization problem is discussed in Refs. 4 and 30. The resultant controller can be complex. Convex optimization problem can, of course, be solved by the algorithms presented in Section 14.5.2 for nonconvex problems. The ellipsoid algorithms can be very slow when the number of design parameters (the dimension of α) is large.

DELIGHT-MIMO, the control design package using DELIGHT, is described in Ref. 25; CONSOLE is described in a technical research report of the System Research Center at the University of Maryland and QDES in Ref. 4.

ACKNOWLEDGEMENT

This work was supported by the National Science Foundation under grant ECS-9024944.

REFERENCES

1. Rosenbrock, H.H., Design of multivariable control systems using the inverse Nyquist array, *Proc. IEE, 116*, 1929–1936 (1969).
2. Isidori, A., *Nonlinear Control Systems*, Springer-Verlag, New York, 1989.
3. Isidori, A., and Byrnes, C. I., Output regulation of nonlinear systems, *IEEE Trans. Auto. Control, TAC, 35*, 131–140 (1990).
4. Boyd, S. P., and Barratt, C. H., *Linear Controller Design, Limits of Performance*, Prentice-Hall, Englewood Cliffs, NJ, 1991.
5. Doyle, J. C., Francis, B. A., and Tannenbaum, A. R., *Feedback Control Theory*, Macmillan, New York, 1992.
6. Polak, E., and Wuu, T-L., On the design of stabilizing compensators via semi-infinite optimization, *IEEE Trans. Auto. Control, TAC, 34*, 196–200 (1989).
7. Hauser, J., Sastry, S., and Meyer, G., Nonlinear controller design for flight control systems, University of California, Berkeley ERL Memo, M88/76 (1988).
8. Michalska, H., and Mayne, D. Q., Approximate global linearization of nonlinear systems via online optimization, *Proceedings First European Control Conference, Grenoble*, 1991, Vol. 1, pp. 182–187.
9. Polak, E., and Mayne, D. Q., Design of nonlinear feedback controllers, *IEEE Trans. Auto. Control, TAC, 26*, 730–733 (1981).
10. Kelley, J. E., The cutting-plane method for solving convex problems, *J. Soc. Ind. Math., 8*, 703–712 (1960).

11. Levitin, E. S., and Polyak, B. T., Constrained optimization methods, *Zhurnal Vychislitelnoi Matematiki Matematicheskoi Fiziki*, 5, 529–542 (1966).
12. Eaves, B. C., and Zangwill, W. I., Generalized cutting plane algorithms, *SIAM J. Control*, 9, 529–542 (1971).
13. Blankenship, J. W., and Falk, J. E., Infinitely constrained optimization problems, George Washington University, Institute for Management Science and Engineering Report, T-301 (1974).
14. Mayne, D. Q., Polak, E., and Trahan, R., An outer approximations algorithm for computer-aided design problems, *J. Optim. Theory Appl.*, 28, 331–352 (1979).
15. Bland, R. G., Goldfarb, G. D., and Todd, M. J., The ellipsoid method, a survey, *Oper. Res.*, 26, 1039–1091 (1981).
16. Clarke, F., *Optimization and Nonsmooth Analysis*, Wiley, New York, 1983.
17. Polak, E., On the mathematical foundations of nondifferentiable optimization in engineering design, *SIAM Rev.*, 29, 21–89 (1987).
18. He, L. and Polak, E., Optimal diagonalization strategies for the solution of a class of optimal design problems, *IEEE Trans. Auto. Control, TAC*, 35, 258–267 (1990).
19. Wuu, T. L., Becker, R. G., and Polak, E., On the computation of sensitivity functions of linear time-invariant system responses via diagonalization, *IEEE Trans. Auto. Control, TAC*, 31, 1141–1143 (1986).
20. Fegley, K. A., and Hsu, M. I., Optimal discrete control by linear programming, *IEEE Trans. Auto. Control, TAC*, 9, 114–115, (1965).
21. Zakian, V., and Al-Naib, U., Design of dynamical and control systems by the method of inequalities, *Proc. IEE*, 120, 1421–1427 (1973).
22. Polak, E., and Mayne, D. Q., An algorithm for optimization problems with functional constraints, *IEEE Trans. Auto. Control, TAC*, 23, 184–193 (1976).
23. Becker, R. G., Heunis, A. J., and Mayne, D. Q., Computer aided design of control systems via optimization, *Proc. IEE, TAC*, 126, 573–578 (1979).
24. Mayne, D. Q., and Polak, E., Algorithms for the design of control systems subject to singular value inequalities, *Math. Programming Study*, 18, 112–134 (1982).
25. Polak, E., Siegel, P., Wuu, T., Nye, W. T., and Mayne, D. Q., DELIGHT-MIMO: an interactive optimization-based multivariable control systems design package, *Control Syst. Mag.*, 2, 9–14 (1982).
26. Polak, E., Mayne, D. Q., and Stimmler, D. M., Control system design via semi-infinite optimization: a review, *Proc. IEEE*, 72, 1777–1794 (1984).
27. Gonzaga, C., Polak, E., and Trahan, R., An improved algorithm for optimization problems with functional inequality constraints, *IEEE Trans. Auto. Control, TAC*, 25, 49–54 (1980).
28. Kiwiel, K. C., *Methods of Descent for Nondifferentiable Optimization*, Springer-Verlag, New York, 1985.
29. Gonzaga, C., and Polak, E., On constraint dropping schemes and optimality functions for a class of outer approximations algorithms, *J. SIAM Control Optim.*, 17, 477–493 (1979).
30. Polak, E., and Salcudean, S., On the design of linear multivariable feedback systems via constrained nondifferentiable optimization in H_∞ spaces, *IEEE Trans. Auto. Control, TAC*, 34, 268–276 (1989).

15

Analysis and Design of Relay Control Systems

D. P. Atherton

The University of Sussex
Brighton, England

15.1 INTRODUCTION

In contrast to the situation for linear feedback systems, very little software has been written for the study of nonlinear feedback systems. The reason for this is that the only general approach to the study of nonlinear systems is simulation and this is the only nonlinear facility contained in commercially available computer-aided control system design software. Although simulation is a particularly important facility, because all systems contain some nonlinearity, it is primarily an analysis facility and sheds little, if any, light on design. Further, there are undoubtedly nonlinear problems whose simulation in some languages must be done with care and some solutions provided, such as the evaluation of limit cycles with sliding, discussed later, may prove inaccurate. There is, therefore, definitely a need for analytically based methods for the study of nonlinear systems, but unfortunately, no general-purpose approach exists. The most widely known and used approach for the study of nonlinear systems is the describing function method [1,2], which is an approximate procedure and has some restrictions on its applicability. It can, however, prove useful both for synthesis and design. The describing function method is a procedure for estimating limit cycles and it is based on the assumption that the input to any nonlinear element will be approximately sinusoidal. The Tsypkin-based approach [1,3,4] for continuous systems, which is discussed in this chapter, is a method for finding limit cycles, also makes an assumption to start the analysis, namely, that the form of the nonlinearity output waveform is known. This is possible when the nonlinearity is a relay because the output from a relay does not depend on the input at all instants of time but only on those instants of time which cause the relay to switch from one output level to another. For example, if the nonlinearity in Figure 1 is an ideal relay which switches between its output level of ±1 when the input passes through zero; the usual limit cycle found, if the oscillation is odd symmetric, will be a square wave with a one-to-one mark-to-space ratio at the relay output. What is not known is the fundamental frequency of this limit cycle and this is determined by the Tsypkin method. This chapter gives the basic background material required to understand the application of the Tsypkin method and describes a computer program developed to assist in applying the method, which involves the solution of one or more nonlinear algebraic equations. The program is used to investigate several examples which have been carefully chosen not only to show the variety of problems which can be examined using the method but also to illustrate a range of phenomena which can occur in the simple nonlinear feedback loop of Figure 1.

15.2 THE TSYPKIN METHOD

As mentioned in Section 15.1, the Tsypkin method can be used to determine exactly any limit cycles which may exist in the system of Figure 1 when the nonlinearity is a relay-type element. The method has a further advantage that when a limit cycle is found, its stability, provided the loop transfer function $G(s)$ is a ratio of polynominals in s, can also be ascertained exactly. The starting point of the method is to assume a form for the periodic waveform at the relay output, which typically for a relay with a dead zone shown in Fig. 2a will have the form shown in Figure 2b. It is, in fact, possible to have more complicated pulse-type oscillation waveforms, but all can be considered as a summation of several periodic pulse trains of the form of $y_i(t)$ shown in Figure 3.

Consider the Fourier series for $y_i(t)$ which can be written

$$y_i(t) = \left(\frac{h\Delta t_i}{T}\right) + \left(\frac{h}{\pi}\right) \sum_{n=1}^{\infty} \left(\frac{1}{n}\right) \{\sin n\omega\Delta t_i \cos n\omega(t - t_i)$$
$$+ (1 - \cos n\omega\Delta t_i)\sin n\omega(t - t_i)\} \tag{1}$$

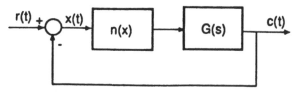

Figure 1 Simple nonlinear feedback system.

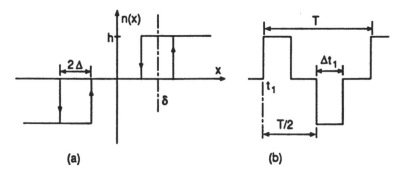

Figure 2 Relay with (a) Dead zone and (b) output waveform.

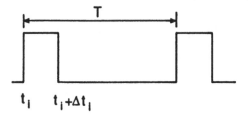

Figure 3 Periodic pulse train.

If $y_i(t)$ is the input to a transfer function $G(s)$ which, for ease of presentation, we assume is such that $\lim_{s \to \infty} sG(s) = 0$, then the output $c_i(t)$ of $G(s)$ is given by

$$c_i(t) = \left(\frac{h\Delta t_i G(0)}{T}\right) + \left(\frac{h}{\pi}\right) \sum_{n=1}^{\infty} \left(\frac{g_n}{n}\right)\{\sin n\omega\Delta t_i \cos(n\omega t - n\omega t_i + \phi_n)$$
$$+ (1 - \cos n\omega\Delta t_i)\sin(n\omega t - n\omega t_i + \phi_n)\} \qquad (2)$$

and its derivative by

$$\dot{c}_i(t) = \left(\frac{\omega h}{\pi}\right) \sum_{n=1}^{\infty} g_n\{-\sin n\omega\Delta t_i \sin(n\omega t - n\omega t_i + \phi_n)$$
$$+ (1 - \cos n\omega\Delta t_i)\cos(n\omega t - n\omega t_i + \phi_n)\} \qquad (3)$$

where

$$G(jn\omega) = g_n \exp(j\phi_n) = U_G(n\omega) + jV_G(n\omega) \qquad (4)$$

If we define an A locus of a transfer function $G(s)$ by

$$A_G(\theta,\omega) = \text{Re } A_G(\theta,\omega) + j \text{ Im } A_G(\theta,\omega) \qquad (5)$$

with

$$\text{Re } A_G(\theta,\omega) = \sum_{n=1}^{\infty} V_G(n\omega)\sin n\theta + U_G(n\omega)\cos n\theta \tag{6}$$

and

$$\text{Im } A_G(\theta,\omega) = \sum_{n=1}^{\infty} \left(\frac{1}{n}\right)\{V_G(n\omega)\cos n\theta - U_G(n\omega)\sin n\theta\} \tag{7}$$

then $c_i(t)$ is expressible in terms of Im A_G and $\dot{c}_i(t)$ in terms of Re A_G, where Re and Im denote real and imaginary parts, respectively. For the relay with a dead zone having the output waveform of Figure 2b,

$$y(t) = \sum_{i=1}^{2} y_i(t) \quad \text{with } y_2(t) = -y_1(t + 0.5T) \tag{8}$$

and the resultant output of $G(s)$, $c(t)$, is given by

$$c(t) = \left(\frac{2h}{\pi}\right)\{\text{Im } A_G^o(-\omega t + \omega t_1,\omega) - \text{Im } A_G^o(-\omega t + \omega t_1 + \omega\Delta t_1,\omega)\} \tag{9}$$

and

$$\dot{c}(t) = \left(\frac{2\omega h}{\pi}\right)\{\text{Re } A_G^o(-\omega t + \omega t_1,\omega) - \text{Re } A_G^o(-\omega t + \omega t_1 + \omega\Delta t_1,\omega)\} \tag{10}$$

where t_1 is the instant at which $y_1(t)$ switches positive, Δt_1 is the pulse duration, and A^o denotes the A locus with odd values of n only in Eqs. (6) and (7). Because the relay input $x(t) = -c(t)$, necessary conditions for the limit cycle to exist are

$$x(t_1) = -c(t_1) = \delta + \Delta \tag{11}$$
$$\dot{x}(t_1) = -\dot{c}(t_1) > 0$$

and

$$x(t_1 + \Delta t_1) = -c(t_1 + \Delta t_1) = \delta - \Delta \tag{12}$$
$$\dot{x}(t_1 + \Delta t_1) = -\dot{c}(t_1 + \Delta t_1) < 0$$

These conditions, which can be written taking $t_1 = 0$ and $\Delta t_1 = \Delta t$ without loss of generality, become

$A_G^o(0,\omega) - A_G^o(\omega\Delta t,\omega)$ must have

$$\text{I.P.} = -\frac{\pi(\delta + \Delta)}{2h}, \quad \text{R.P.} < 0 \tag{13}$$

$A_G^o(0,\omega) - A_G^o(-\omega\Delta t,\omega)$ must have

$$\text{I.P.} = \frac{\pi(\delta - \Delta)}{2h}, \quad \text{R.P.} < 0 \tag{14}$$

where R.P. and I.P. denote real part and imaginary part, respectively.

Because closed-form expressions for the A_G loci can be found for any given transfer function $G(s)$, necessary conditions for the assumed limit cycle are positive solutions for ω and Δt, with $\omega \Delta t < \pi$, for Eqs. (13) and (14) for the I.P. and satisfaction of the inequalities for the R.P.

Sufficiency can be determined once a solution has been found by checking that the waveform $x(t)$ does not pass through the relay switching levels such as to give additional switchings to those assumed. This is easily done computationally because with a graphical display, the solution waveform $x(t)$ equal to $-c(t)$, given by Eq. (9), can be displayed when ω and Δt are known.

15.2.1 Stability of a Limit Cycle Solution

For the simple autonomous feedback loop of Figure 1 with nonlinear element, $n(x)$, and linear transfer function, $G(s) = p(s)/q(s)$, the differential equation for the system can be written

$$q(D)x + p(D)n(x) = 0 \tag{15}$$

where D denotes the differential operator d/dt. If $x^*(t)$ is the limit cycle solution and it is perturbed by a small amount $\Delta x(t)$, then the differential equation for the perturbation is

$$q(D)\Delta x(t) + p(D)n'(x^*(t))\Delta x(t) = 0 \tag{16}$$

when $n'(x) = dn(x)/dx$. The limit cycle is stable if this equation, which has periodically time varying coefficients, has a stable solution. If this equation is written in the state-space form

$$\Delta \dot{x}*(t) = A(t)\Delta x*(t) \tag{17}$$

then the periodically time varying matrix $A(t)$ can be shown to be given by

$$A(t) = A + Bn'(x^*(t))C \tag{18}$$

where (A, B, C) is a state-space description of $G(s)$. The matrix $A(t)$ is piecewise constant; being equal to A at all times apart from the infinitesimal time intervals, it takes the relay to switch. It is known [5] that for this situation of $A(t)$ piecewise constant, Eq. (17) has asymptotic orbital stability if and only if the eigenvalues of the matrix

$$Q = \sum_{i=1}^{m} \exp(A_i t_i) \tag{19}$$

have magnitude less than unity, apart from one which will have unit magnitude. Here A_i are the constant values of the A matrix in the time intervals t_i and the period T is given by

$$T = \sum_{i=1}^{m} t_i \tag{20}$$

Because the relay output waveform of Figure 2b possesses odd symmetry, the matrix Q may be evaluated over the half-period $T/2$ and is given by

$$Q = \exp\left[A\left(\frac{T}{2} - \Delta t\right)\right]\exp\left[\frac{hBC}{|\dot{x}(\Delta t)|}\right]\exp[A\Delta t]\exp\left[\frac{hBC}{|\dot{x}(0)|}\right] \tag{21}$$

where Δt is the width of the pulse and $\dot{x}(0)$ and $\dot{x}(\Delta t)$ are the derivatives of the solution waveform at the relay switching instants. Thus, once the limit cycle solution is found by the Tsypkin method, Q can be evaluated and the limit cycle stability ascertained.

15.2.2 The A Loci

It is easily shown from the definitions given in Eqs. (6) and (7) that the A loci have the following properties:

1. The A locus for $\theta = 0$ is identical, apart from a constant factor, with the Tsypkin locus $\Lambda(\omega)$ if

$$\lim_{s\to\infty} sG(s) = 0 \tag{22}$$

   The relationship is

$$\Lambda(\omega) = \left(\frac{4h}{\pi}\right)A^o(0,\omega) \tag{23}$$

   and, thus, the A locus can be regarded as a generalized Tsypkin locus.
2. The A locus satisfies the superposition property, that is, if the linear plant $G(s) = G_1(s) + G_2(s)$, then

$$A_G(\theta,\omega) = A_{G_1}(\theta,\omega) + A_{G_2}(\theta,\omega) \tag{24}$$

3. If $G'(s) = G(s)\exp(-sT_d)$, then

$$A_{G'}(\theta,\omega) = A_G(\theta + \omega T_d,\omega) \tag{25}$$

4. The loci are periodic in θ with period, 2π, that is,

$$A_G(\theta,\omega) = A_G(\theta + 2\pi,\omega) \tag{26}$$

5. For the loci with odd terms only, that is, A^o, the periodicity is odd with

$$A_G^o(\pi - \theta,\omega) = -A_G^o(-\theta,\omega) \tag{27}$$

Because any plant transfer function can be put into partial fractions in terms of transfer functions having real or complex pairs of poles, use of Eq. (24) allows

A loci for any transfer function to be obtained in terms of the *A* loci of a few basic transfer functions such as those given in Table 1. In addition, the use of Eq. (25) allows *A* loci for transfer functions with a time delay to be found.

Substitution of the expressions for the basic transfer functions in Eqs. (6) and (7) yields an infinite series which can be evaluated in closed form for both odd and all values of *n*. For example, if $G(s) = 1/(1 + s\tau)$, then

$$\text{Im } A_G(\theta,\omega) = \sum_{n=1}^{\infty} -\frac{n\lambda \sin n\theta}{1 + n^2\lambda^2} + \frac{\cos n\theta}{1 + n^2\lambda^2} \tag{28}$$

where $\lambda = \omega\tau$. This gives

$$\text{Im } A_G(\theta,\omega) = -\lambda S_{1,1}(\theta,\lambda) + C_{0,1}(\theta,\lambda) \tag{29}$$

or for odd terms only in the series

$$\text{Im } A_G^o(\theta,\omega) = -\lambda S_{1,1}^o(\theta,\lambda) + C_{0,1}^o(\theta,\lambda) \tag{30}$$

Here

$$S_{j,k}(\theta,\lambda) = \sum_{n=1}^{\infty} \frac{n^j \sin n\theta}{(1 + n^2\lambda^2)^k} \quad \text{for } 0 < \theta < 2\pi \tag{31}$$

and

$$C_{j,k}(\theta,\lambda) = \sum_{n=1}^{\infty} \frac{n^j \cos n\theta}{(1 + n^2\theta^2)^k} \quad \text{for } 0 \le \theta < 2\pi \tag{32}$$

All the other *A* loci for the transfer functions in Table 1 can be written in terms of these series apart from those numbered 7 and 8 for complex poles. A complete listing of the expressions for the *A* loci of these transfer functions can be found in Ref. 4.

15.2.3 The Computer Program

The basic computer program for the determination of limit cycles in the relay system of Figure 1 is menu driven and asks the user for the transfer function

Table 1 Basic transfer function types

Type No.	Transfer Function	Type No.	Transfer Function
1	$1/s\tau$	5	$1/s^3\tau^3$
2	$1/(1 + s\tau)$	6	$1/(1 + s\tau)^3$
3	$1/s^2\tau^2$	7	$1/(s^2 + 2\zeta\omega_0 s + \omega_0^2)$
4	$1/(1 + s\tau)^2$	8	$s/(s^2 + 2\zeta\omega_0 s + \omega_0^2)$

$G(s)$, which may include a time delay, the characteristics of the relay, and initial guesses for the limit cycle solution parameters which for a relay with a dead zone are the limit cycle frequency ω and the pulse width $\omega\Delta t$ in radians. These values are used in a nonlinear algebraic equation solver which searches for solutions to Eqs. (13) and (14). The A loci are normally evaluated from the stored closed-form expressions described in the previous section, but optionally may be evaluated approximately by simply summing a user-defined number of terms in the infinite series. This procedure is of interest because it enables the user to compare the exact solution with either the describing function solution, obtained using the first term only in the series, or a solution using just a few harmonics. The limitation used in the theoretical development, namely, that $\lim_{s\to\infty} sG(s) = 0$ is not required, the only requirement being that $G(s)$ is proper. The limit cycle solution waveform is displayed and this enables the user to check that the relay switching levels are not crossed at other than the assumed times, thus verifying the correctness of the solution. If the system has no time delay, that is, $G(s)$ is a rational function of s, the eigenvalues of the Q matrix can be displayed to show whether the limit cycle is stable or unstable.

15.3   EXTENSIONS OF THE METHOD

For ease of presentation, the detailed theoretical analysis given in the previous section has only been done for the determination of symmetrical oscillations in a single-loop system. Several extensions are available in the software and these will be described briefly below. Some examples of these situations are given in Section 15.4

15.3.1   Saturation Nonlinearity

It is relatively straightforward to show [6] that the procedure with minor modifications can be extended to systems containing the ideal saturation characteristic of Figure 4a if the saturation output is approximated by the linear

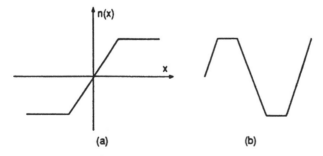

Figure 4   (a) Ideal saturation and (b) approximate output.

segmented waveform shown in Figure 4b. An approximate test for evaluating the stability of the limit cycle can also be obtained.

15.3.2 A Quantization Characteristic

A quantization characteristic is shown in Figure 5 and it is obvious that it can be modeled by a number of relays with dead zone placed in parallel. It is, therefore, straightforward to formulate a Tsypkin solution for a quantization characteristic as the nonlinearity in Figure 1.

This is a useful approach for solving for limit cycles in the system of Figure 1 with any nonlinear characteristic because by increasing the number of quantization levels any nonlinearity can be approximated as closely as required. The difficulty, of course, is that the more quantization levels used, the larger the number of equations which have to be solved. Two nonlinear algebraic equations, the same as for a relay with dead zone, have to be solved for every quantization level. One factor which makes the solution of the problem tractable, however, is that one can start with a course quantization, that is, one or two quantization levels, and gradually increase the number. It is then possible to use the information obtained from the limit cycle solution for a few quantization levels to obtain approximate initial parameter guesses for the solution for a higher number of quantization levels. This procedure has been adopted in a computer program [7] which solves for a limit cycle in the feedback system of Figure 1 with a quantization characteristic.

15.3.3 Limit Cycle with Sliding

An interesting phenomenon which can take place in relay systems is the occurrence of a motion which is referred to as sliding. This is a rapid switching of the

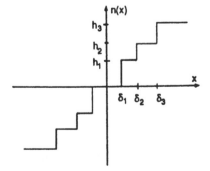

Figure 5 Quantization characteristic.

relay from one position to another. The result is that the output waveform from the relay during sliding has a value which varies between the two switching levels. To obtain solutions for limit cycles with a sliding mode, the relay output is approximated by a linear segmented waveform while sliding is taking place. As explained in section 15.3.1 an approximate solution can be obtained for this type of nonlinearity output waveform.

15.3.4   System Compensation

When the closed-loop system contains a relay with a dead zone, then a stable equilibrium may exist with zero output from the relay. In most systems, stability is obtained as the gain is reduced; one feature of the program is that it has a facility for determining the value of gain when the limit cycle disappears. This is, of course, simply done by calculating the limit cycle and using an algorithm to change the gain until no limit cycle solution exists. This feature is particularly useful because it enables a proportional controller to be designed to give a given gain margin. It is also possible to obtain the phase margin of a stable system if an additional time delay is entered into the transfer function, and the value of phase shift across the time delay is determined when the limit cycle is just about to cease.

15.3.5   Asymmetrical Oscillations

It is obviously relatively easy to extend the Tsypkin method to determine limit cycles which do not have odd symmetry. The difficulty, of course, is that the number of nonlinear algebraic equations to be solved increases. For example, for an asymmetrical oscillation in a relay system where the relay has no dead zone, two rather than one nonlinear algebraic equations need to be solved. The normal situation is that the number of nonlinear algebraic equations to be solved doubles when one considers the possibility of asymmetrical oscillations. As outlined in the theory, the only other difference is that the *A* loci stored series for all harmonics must be used in the software.

15.3.6   Multiple Input-Output Systems

Although, in principle, the theory presented allows limit cycles to be evaluated for any combination of relays and linear plant transfer functions, it is quite difficult to write a general program which allows this to be done for a completely free structure. A program has, therefore, been written [8] for the multivariable configuration shown in Figure 6, where any element in the nonlinear block can be either a relay or a saturation characteristic, and in the linear block, one can have any linear transfer function for each element.

Although the software has been written to handle systems with five inputs and five outputs, difficulties arise in estimating initial conditions for solutions to

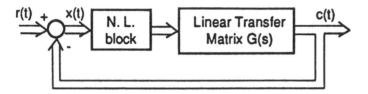

Figure 6 Nonlinear multivariate system.

the resulting large number of nonlinear algebraic equations. It has, therefore, typically been used for two-input–two-output systems and, on occasion, three-input–three-output systems. The number of nonlinear, algebraic equations to be solved is two or one, depending on whether the relay has a dead zone, for each relay in the system.

15.4 ILLUSTRATIVE EXAMPLES

To illustrate applications of the software and to show various aspects of nonlinear system behavior, several examples are considered.

15.4.1 Relay with No Dead Zone

Here the system of Figure 1 is considered with $G(s) = K/s(1 + s\tau)$, and the nonlinearity is a relay with outputs $\pm h$ and hysteresis $\pm\Delta$. For $r(t) = 0$, the limit cycle at the relay output will be a symmetrical square wave and the limit cycle solution is given by $A_G^o(0,\omega)$ must have

$$\text{I.P.} = -\frac{\pi\Delta}{4h} \qquad \text{R.P.} < 0 \tag{33}$$

For the given $G(s)$, it can be shown that

$$\text{Im } A_G^o(0,\omega) = \left(\frac{K\pi\tau}{4}\right)\left[\left(-\frac{\pi}{2\lambda}\right) + \tanh\left(\frac{\pi}{2\lambda}\right)\right] \tag{34}$$

where $\lambda = \omega\tau$, so the limit cycle frequency is given by

$$\left(\frac{\pi}{2\lambda}\right) - \tanh\left(\frac{\pi}{2\lambda}\right) = \frac{\Delta}{hK\tau} \tag{35}$$

Use of the describing function method for this problem yields

$$\lambda(1 + \lambda^2) = \frac{4hK\tau}{\pi\Delta} \tag{36}$$

for the limit cycle solution. The exact Eq. (35) for the limit cycle solution has been obtained easily in this instance because of the simple expression for $G(s)$, and, because the relay has no dead zone, only one nonlinear algebraic equation has to be solved. It is interesting to note for the relay with no dead zone, because the imaginary part of $-1/N(a)$, where $N(a)$ is the relay describing function, is constant and equal to $-\pi\Delta/4h$, that graphical solutions for the describing function method and the exact method are the frequencies on $G(j\omega)$ and $A_G^o(0,\omega)$, respectively, when they intersect the straight line parallel to the negative real axis with imaginary part $-\pi\Delta/4h$. The graphical solution is shown in Figure 7 for the above example with $K = 1$, $\tau = 1$, and $h/\Delta = 3$. The exact solution frequency is 1.365 rads/sec and the describing function solution frequency is 1.352 rads/sec.

15.4.2 Multiple Limit Cycles

In this example, an ideal relay with output levels ± 1 is taken as the nonlinearity, and the linear transfer function is

$$G(s) = \frac{12(s + 1)^2}{s^3(s^2 + 1.2s + 16)} \tag{37}$$

The Nyquist locus $G(j\omega)$ is shown in Figure 8 from which it is seen that there are two possible limit cycle solutions. The software gives the solution frequencies as 1.067 and 3.657 rads/sec with the former an unstable limit cycle, shown in Figure 9, and the later a stable limit cycle, shown in Figure 10. If the more general transfer function

$$G(s) = \frac{12(s + 1)^2}{s^3(s^2 + 0.3\lambda s + \lambda^2)} \tag{38}$$

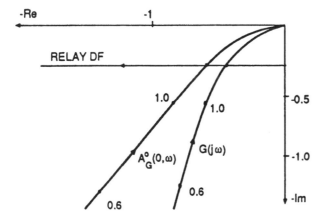

Figure 7 Graphical solution for limit cycle frequency.

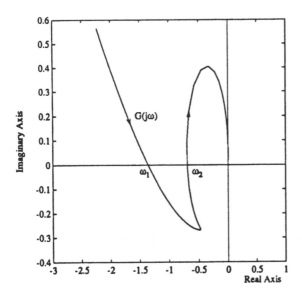

Figure 8 Nyquist locus for $G(s)$ of Eq. (37).

is considered and λ varied, then the intersections of $G(j\omega)$ with the negative real axis change both in position and frequency. As λ decreases, the ratio $|G(j\omega_1)|/|G(j\omega_2)|$ decreases and the limit cycle at the higher frequency, ω_1, becomes unstable when $\lambda < 3.73$. For $\lambda < 3.73$ and when two limit cycle frequency intersections ω_1 and ω_2 still exist, simulations show the existence of a combined mode oscillation consisting primarily of these two frequencies. The existence of this mode can also be predicted using the dual input describing function [1,9].

15.4.3 Invalid Solutions

In Section 15.2, the requirement to check that the solution waveform does not violate the assumed relay switching conditions was mentioned. To illustrate this situation, a system with an ideal relay with outputs ± 1 and linear plant $G(s) = (1 + 2s)/(s^2 + 1)(s + 1)$ is considered. A transformed Nyquist locus of $G(j\omega)$, where $|G(j\omega)|$ is replaced by $(2/\pi)\tan^{-1}|G(j\omega)|$, so that the infinite circle becomes the unit circle, is shown in Figure 11. Because the locus does not cross the negative real axis, no limit cycles are predicted by the describing function method. Again, because the transfer function is not too complex, it is possible to show by the Tsypkin method that an infinite number of limit cycle solutions exist with frequencies

$$\omega = \frac{1}{2n + 0.5} \quad \text{for } n = 1, 2, \ldots, \infty. \tag{39}$$

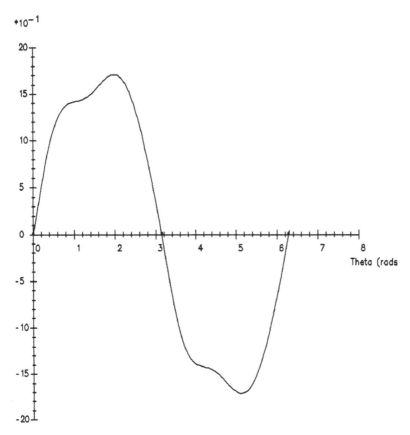

Figure 9   Unstable limit cycle.

The lowest frequency solution is ω = 0.4 rads/sec and using this value in the software gives the waveform shown in Figure 12 for the limit cycle at the relay input.

It can be seen that this waveform violates the condition that it should not pass through the relay switching, levels at other than the assumed times and is, therefore, an invalid solution. This is also the situation with the other frequencies given by Eq. (39) and is due to the resonant nature of $G(s)$, which has two poles on the imaginary axis.

15.4.4   Design of Relay Systems

A relay system with a plant of transfer function $G(s) = 10/s(s + 1)(s + 2)$ and relay with dead zone ±1 and output ±1 is given. It is required to design a compensator to give a step response with an overshoot of less than 20% for all input

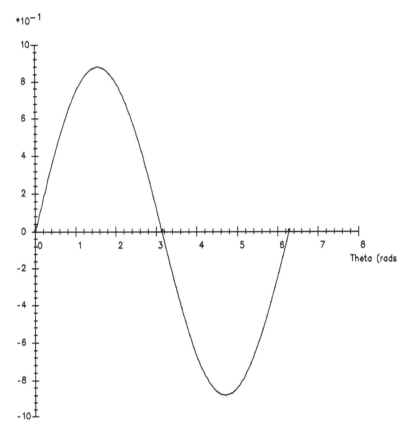

$\ast10^{-1}$

Theta (rads

Figure 10 Stable limit cycle.

step magnitudes. First, if the system is examined with no compensator, it is found to be unstable, having an unstable limit cycle at a frequency of 1.236 rads/ sec, as shown in Figure 13, and a stable limit cycle of frequency 1.404 rads/sec, shown in Figure 14. The program indicates that the maximum value of gain for stable operation is 9.0. Examination of the system behavior using the describing function approach suggests that the system can be stabilized using a phase advance compensator. A transfer function of $(1 + s)/(1 + 0.1s)$ is, therefore, chosen for the compensator. Analyzing this system with the compensator added together with a time delay, T_d and gain K yields the results that for $T_d = 0$, the system goes unstable with $K = 3.5$, and with $K = 1$, it goes unstable with $T_d = 0.226$ sec at a frequency of 2.08 rads/sec, which gives a phase shift through the time delay of 27°. The system, thus, has a gain margin of 10.9 dB and a phase margin of 27°, which seem reasonable to meet the specifications.

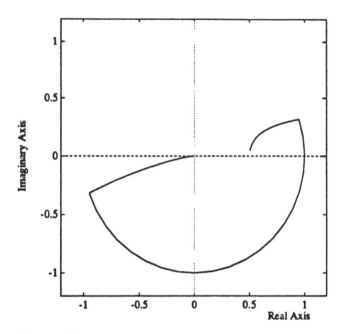

Figure 11   Transformed Nyquist locus.

Simulation of the resulting system showed this to be true with little overshoot in the step response for any input amplitude. Figure 15 shows the step response for a step input of 2, which gave approximately the maximum overshoot with the chosen compensator and also the response for the same input when the compensator zero frequency gain is increased to 2.

15.4.5   Limit Cycle with a Sliding Regime

The system of Figure 1 is considered with the nonlinearity a relay with a dead zone ± 1, output levels ± 1, and linear transfer function $G(s) = 8(s - 0.5)/ (s^2 - 0.5s + 1)$. In using the program to search for a limit cycle, the waveform of Figure 16 was found.

This solution is invalid because the switching at point A is such that the relay input, instead of continuing to increase, immediately reverses. Because when sliding takes place the relay switches rapidly between two levels, it is argued that the immediate reversal of this waveform at the switching instant is an indication of sliding and the program gives a message to this effect. The relay output is then assumed to be of the form shown in Figure 17 where the sloping line from time 0 to t_s is the assumed average output in the sliding region. To

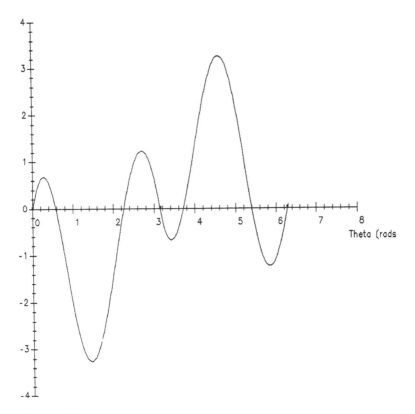

Figure 12 Invalid limit cycle solution.

obtain a solution waveform of this type, the user is asked to provide estimates for ω, ωt_s, $\omega \Delta t$, and h. These are estimated from the incorrect solution of Figure 16 by typically taking the same value of ω, ωt_s approximately equal to AC, $\omega \Delta t$ approximately equal to its value in Figure 16, and a value for h between 0 and 1. The new solution obtained is shown in Figure 18, where it can be seen, as required, that the input remains approximately constant at the switching level during sliding. The program allows the solution to be further improved by using more linear segments to approximate the relay output during sliding. The resulting solution is shown in Figure 19 for a three linear segment approximation, which can be seen to be of the required form.

15.4.6 Use of Quantization Characteristics

In this section, the evaluation of a limit cycle is considered in a feedback system where the nonlinear element is approximated by a quantization characteristic.

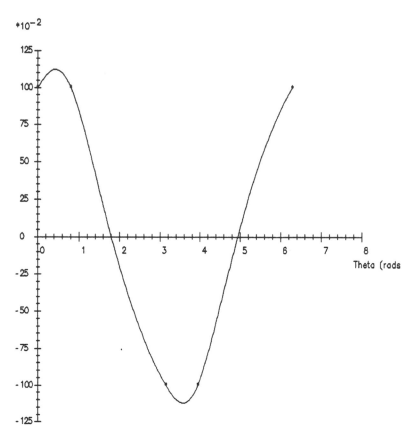

$\bullet 10^{-2}$

Figure 13   Unstable limit cycle (Section 15.4.4).

This procedure is easier to apply when the nonlinearity is of the saturation type since, if the input amplitude to the quantizer is larger than anticipated, the error in matching the nonlinearity by the quantizer will usually not be significantly increased. Here, however, to illustrate the general applicability of the procedure, the method is applied to a feedback loop with $G(s) = 6(s^2 + 2s + 1)/s^3(s^2 + 1.2s + 16)$ and the nonlinearity equal to $x|x|$, an odd square law characteristic. The program [7] requires the user to estimate the maximum input amplitude to the nonlinearity and then offers several options for providing a good fit to the nonlinearity by a quantizer over this amplitude range, with the user selecting the number of quantization levels. As mentioned previously, the procedure usually starts with a low value, say between 2 and 4, and can be built up to a maximum of 10 to improve accuracy. The program allows the user to display the nonlinearity and its quantized approximation.

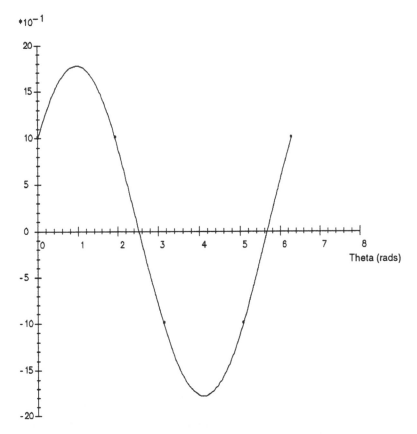

Figure 14 Stable limit cycle (Section 15.4.4).

Figure 20 shows the quantized approximation for the nonlinearity and Figure 21 the resulting limit cycle solution. This limit cycle agrees very closely with the solution obtained by digital simulation, the limit cycle frequency, for example, being in error by less than 0.1%.

15.4.7 Chaotic Motion

Although the Tsypkin method cannot be used to determine chaotic motion in a relay system it may provide some indication of when this behavior may exist. A simple illustration of this fact is provided here, but for further details the reader is referred to the appropriate references [10–13]. A system with an ideal relay and $G(s) = -1/(s^3 + \alpha_2 s^2 + \alpha_1 s + \alpha_0)$ is considered. Using the Tsypkin method, it was found that for $\alpha_0 = 4$, $\alpha_1 = 1.25$, and $0 < \alpha_2 \leq 2.8$, the system had a stable limit cycle typically with more than one oscillation

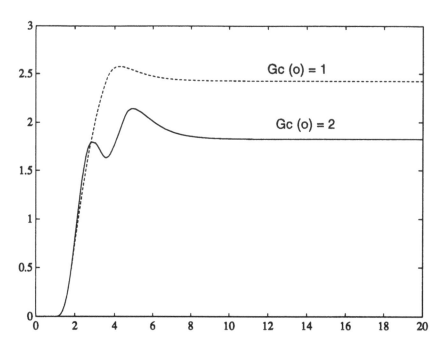

Figure 15   Responses to step input of two units.

in each half-period. The stable limit cycle for $\alpha_2 = 2.8$ is shown in Figure 22, which, for obvious reasons, is usually referred to as a spiral limit cycle. For $2.8 < \alpha_2 < 3.2$, several limit cycles were found by the Tsypkin method, all of which were unstable. They included one limit cycle which was sinusoidal and others of the spiral type having different numbers of oscillations per half-period and a maximum amplitude less than that of the sinusoidal limit cycle. Simulations starting with an initial condition inside the unstable sinusoidal limit cycle resulted in chaotic motion of the form shown in Figure 23, where the chaotic motion basically consists of "jumps" among the several unstable spiral limit cycles. Thus, one possible outcome, provided certain other conditions are satisfied, when multiple unstable limit cycles are found in a nonlinear feedback system may be the occurrence of chaotic motion.

15.5   A MULTIVARIABLE RELAY SYSTEM

As mentioned in Section 15.3, it is possible to write equations using the Tsypkin approach which yield limit cycle solutions for any combination of relay and linear elements. To write a computer program for such a general approach is not easy, but the task is considerably simplified if the structure is restricted to the multivariable configuration of Figure 6. This configuration offers more

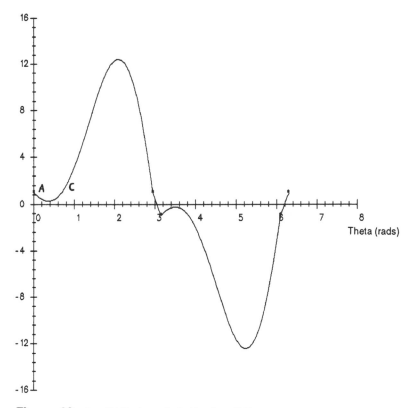

Figure 16 Invalid limit cycle indicating sliding.

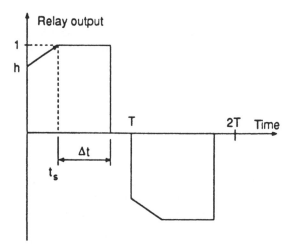

Figure 17 Assumed relay output waveform.

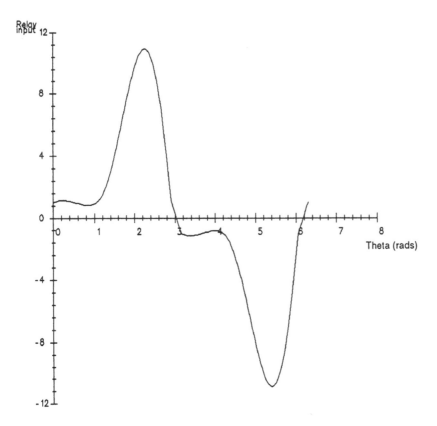

Figure 18 Approximate sliding limit cycle solution.

flexibility than is perhaps apparent initially since, for example, a single-loop system with nonlinearities and linear elements in series can be obtained by making several elements of the transfer-function matrix zero. A major difficulty in analyzing multiloop systems is that they may possess limit cycles with a high harmonic content which may cause one or more relays to have several switchings in the period of the lowest limit cycle frequency. This, for example, is easily seen if two feedback loops which have limit cycles are coupled together. If the coupling is weak, then both limit cycles will exist in both loops, but as the coupling is strengthened the limit cycles may synchronize either to the same basic frequency or with one a subharmonic of the other. The Tsypkin approach can be used to find limit cycles in the latter case but not in the former. Synchronization at a subharmonic frequency may result in one or more relays switching several times during the basic period of the oscillation, thus increasing the number of equations to be solved by the Tsypkin method to determine the limit cycle

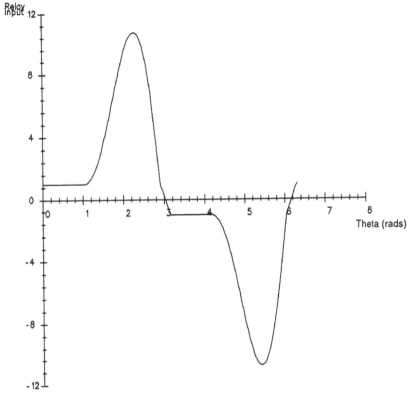

Figure 19 More-accurate sliding limit cycle solution.

parameters. Unless reasonable initial parameter guesses for the limit cycle are available, difficulty may be obtained in finding any solutions to the increased number of nonlinear algebraic equations. When the coupled loops oscillate with the same basic frequency, more than one stable limit cycle solution may exist.

To illustrate the complexity of the problem, a system with

$$G(s) = \frac{1 - 0.3s}{(1 + s)^2} \begin{pmatrix} \dfrac{1}{s} & \dfrac{\mu}{1} \\ \mu & \dfrac{1}{s} \end{pmatrix}$$

and two ideal relays with ± 1 output levels on the diagonal of the nonlinearity matrix is considered. This system has been studied in some detail [14,15] to determine the behavior as μ varies. Here, the only case considered is for $\mu = 0.3$ and seven limit cycle solutions are obtained, three of which are stable. One of the three stable solutions corresponds to the limit cycles in phase and the waveform of this limit cycle is shown in Figure 24.

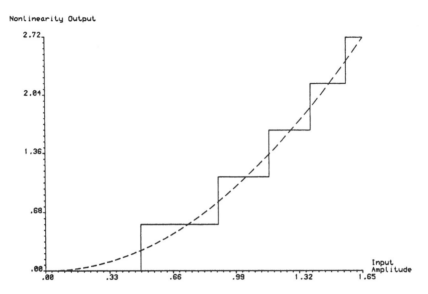

Figure 20   Quantized approximation to the nonlinearity.

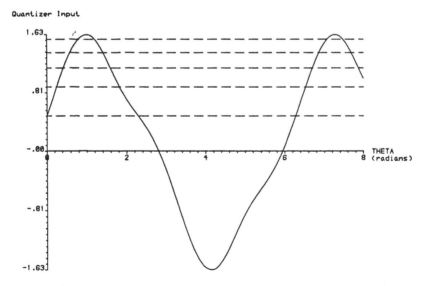

Figure 21   Limit cycle in quantized system.

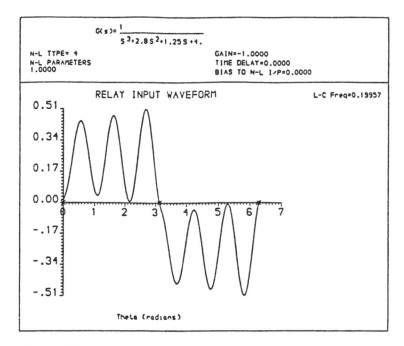

Figure 22 Stable limit cycles for $\alpha_2 = 2.8$.

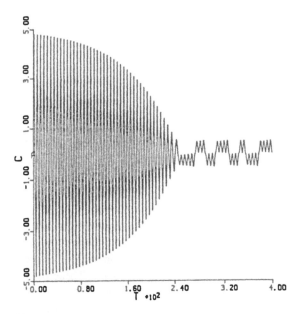

Figure 23 Response with $\alpha_2 = 3.0$ leading to chaotic motion.

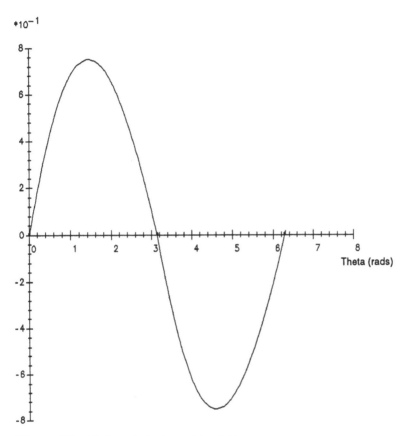

Figure 24   Limit cycle in the multivariable system.

15.6 CONCLUSIONS

In this chapter, the Tsypkin method for investigating limit cycles in relay control systems has been presented and details of a program to implement the method have been described. Although the initial theoretical background for the approach was presented over three decades ago by Tsypkin, its successful application is not possible, as has been seen from the examples considered, without modern computer-aided design facilities. Several problems showing the application of the method have been included and these have been chosen to some extent to show the variety of situations for which the method can be used either to determine limit cycles or shed light on unusual system behavior. These have included, among others, applications to systems with other than relay nonlinearities using a quantization approximation, the determination of limit cycles with sliding, and the indication of chaotic motion. Some other topics, such as the determination of limit cycles where the relay switches several times in an oscillation period [7], more detailed studies of applications to systems with asymmetrical limit cycles [7] and detailed investigations of multivariable systems [7,8,14,15], have not been covered and the interested reader is referred to the appropriate references. It is also worth commenting on the fact that a similar Tsypkin-type procedure can be used for evaluating limit cycles at subharmonics of the sampling frequency in nonlinear sampled data systems [16,17]. Here the output of a zero order hold is a pulse-type waveform where "the switching times" in this case, at the sampling period, are known but the unknowns are the amplitude levels. Nonlinear algebraic equations can be obtained to solve for these amplitudes in a similar manner to that described for the continuous system with relays in this chapter.

REFERENCES

1. Atherton, D. P., *Nonlinear Control Engineering, Student Edition*, Van Nostrand-Reinhold, New York, 1982.
2. Gelb, A., and Vander Velde, W. E., *Multiple-input Describing Functions and Nonlinear System Design*, McGraw-Hill, New York, 1968.
3. Atherton, D. P., *Stability of Nonlinear Systems*, Wiley, New York, 1981.
4. Atherton, D. P., Oscillations in relay systems, *Trans. Inst. MC*, *3*(No. 4), 171–184 (1982).
5. Willems, J. L., *Stability Theory of Dynamical Systems*, Nelson, London, 1970.
6. Atherton, D. P., and Ramani, N., A method for the evaluation of limit cycles, *Int. J. Control*, *21*, 375–384 (1975).
7. Wadey, M. D., Extensions of Tsypkins' Method for Computer Aided Control System Design, D. Phil. thesis, University of Sussex, 1984.
8. Atherton, D. P., and Wadey, M., Periodic modes in multivariable systems with relay and saturation elements. *Inst. MC Symposium on Applications of Multivariable System Techniques*, Plymouth, 1984.
9. Atherton, D. P., and Choudhury, S. K., An exact solution for oscillations in relay systems with complex poles, *Int. J. Control*, *21*, 1015–1020 (1975).
10. Ogorzalek, M. J., Some observations on chaotic motion in single loop feedback systems, *Proceedings, 25th CDC, Athens*, 1986, Vol. 1, pp. 588–589.
11. Amrani, D., and Atherton, D. P., Designing autonomous relay systems with chaotic motion, *Proceedings, 28th CDC, Tampa*, Florida, 1989, Vol. 1, pp. 513–517.
12. Genesio, R., and Tesi, A., Chaos prediction in nonlinear feedback systems, Report 18/89, Dipartimento di Sistemi e Informatica, Universita di Firenze, 1989.
13. Genesio, R., and Tesi, A., Chaos prediction in a third order relay system, Memorandum GT-01-90 Dipartimento di Sistemi e Informatica, Universita di Firenze, 1989.
14. Balasubramanian, R., and Atherton, D. P., Limit cycles in coupled biological systems, *IFAC Symposium on Theory and Application of Digital Control*, New Delhi, 1982.

15. Atherton, D. P., and Linkens, D. A., *Analysis of Limit-Cycle Conditions in Inter-coupled Relay Oscillators with Reference to Gastrointestinal Modelling*, Bulletin of Mathematical Biology, *45*, 239–257 (1983).
16. Goucem, A., and Atherton, D. P., Limit cycles in nonlinear discrete systems, *Proceedings 25th CDC*, Athens, 1986, Vol. 1, pp. 573–577.
17. Goucem, A., Computer Aided Design for Nonlinear Sampled Data Systems, D. Phil. thesis, University of Sussex, 1988.

16

MATRIX$_X$

Naren K. Gupta, Dan Groshans, and Steve P. Houtchens

Integrated Systems, Inc.
Santa Clara, California

16.1 INTRODUCTION

The MATRIX$_X$ products line is an integrated set of CAE/CASE tools that aid engineers in the design, simulation, and implementation of control systems for a wide range of aerospace, automotive, computer peripheral, and industrial applications. The use of these products both improves the design and accelerates the development of products using real-time control logic. Introduced in 1983, MATRIX$_X$ was the first commercial product line to address this application area and still provides the most comprehensive set of capabilities for the control system designer, the system simulation engineer, and the software developer. The MATRIX$_X$ product family is developed, marketed, and supported by Integrated Systems Inc.

Xmath: The foundation of the MATRIX$_X$ Family is Xmath. Written in C++ and fully object oriented, Xmath provides a high-level fourth-generation language for mathematical analysis and engineering computations. It allows rapid design of control systems and prototyping of new algorithms. Xmath represents the next generation of MATLAB-type technology for control system analysis and design software.

SystemBuild: Systembuild is a graphical modeling and simulation software package for building block diagrams. SystemBuild allows an engineer to describe a system graphically and to evaluate its behavior by simulation. Both

395

linear and nonlinear multirate systems can be described. Early simulation can be used to compare competing designs, aid selection among available options, and identify errors prior to the implementation phase. The graphical description also creates a database for the automatic code generation family of products. It was first introduced in 1985.

AutoCode: AutoCode generates real-time source code from graphical models described in SystemBuild. The generated code can be in C, Ada, or Fortran.

AC-100: AC-100 Rapid Prototyping System is used to test real-time control logic developed by the other product families. AC-100 can be used to evaluate the performance of real-time software in conjunction with electronic and mechanical components of a system.

pSOS +: pSOS + accelerates the implementation of the real-time software on embedded processors. (This product line will not be discussed further in this chapter.)

The MATRIX$_X$ Family of Products gives Integrated Systems, Inc. (ISI) a unique offering in the control system development market. ISI's tools are unique in that they support the engineering process from requirements analysis through design and simulation to implementation and testing. All of these functions are available within a single design environment backed by a common datastructure. The following sections describe the first four product families. A detailed description of pSOS + can be obtained from the authors.

16.2 Xmath

The past decade has seen remarkable gains in the quality and number of software tools available to control engineers. Looking to the future, however, we see that the current generation of MATLAB-based software (MATRIX$_X$, Ctrl-C, Pro-MATLAB, etc.) have reached the limits of their original architecture: Iterative improvements are harder to come by and inherent limitations are nearly impossible to remove. Meanwhile, the demand for new features and capabilities has grown as more and more people use ISI software in new and different ways.

For these reasons, ISI created Xmath. Xmath is a completely new product which is intended to replace the MATLAB generation of software tools. It provides substantial capabilities in the areas of mathematical analysis, user interaction, programming, and graphics. Many of its features were specifically requested by MATRIX$_X$ users and many others originated within ISI.

16.2.1 Overall Architecture

Xmath includes the following major elements:

• A graphical user interface built on the Motif X-Window toolkit.

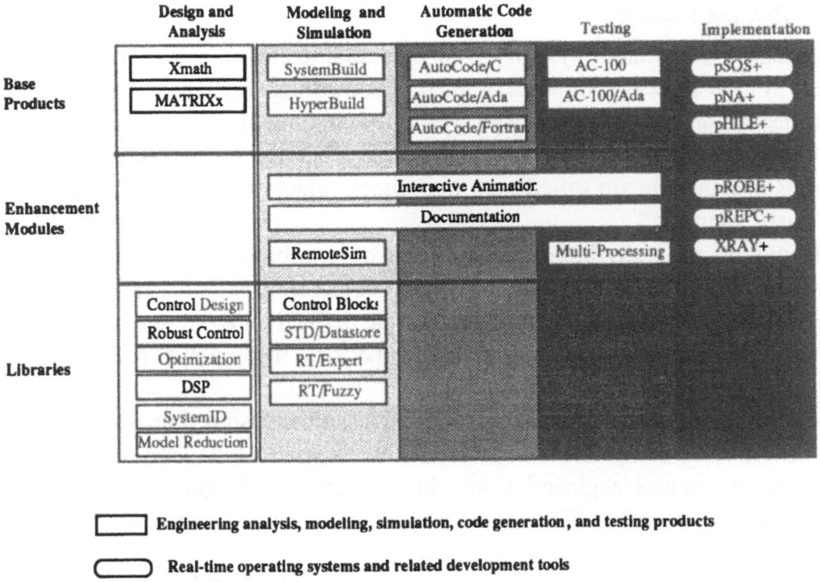

	Design and Analysis	Modeling and Simulation	Automatic Code Generation	Testing	Implementation
Base Products	Xmath	SystemBuild	AutoCode/C	AC-100	pSOS+
	MATRIXx	HyperBuild	AutoCode/Ada	AC-100/Ada	pNA+
			AutoCode/Fortran		pHILE+
Enhancement Modules		Interactive Animation			pROBE+
		Documentation			pREPC+
		RemoteSim		Multi-Processing	XRAY+
Libraries	Control Design	Control Blocks			
	Robust Control	STD/Datastore			
	Optimization	RT/Expert			
	DSP	RT/Fuzzy			
	SystemID				
	Model Reduction				

☐ Engineering analysis, modeling, simulation, code generation, and testing products

⬭ Real-time operating systems and related development tools

Figure 1 Overview of the MATRIX$_X$ family.

- A high performance interpreted language called MathScript that parses and executes user commands and external functions. The scripting environment includes a graphical, source-level debugger.
- A hierarchy of both numerical and non-numerical objects, including control-specific objects.
- An interactive two-dimensional and three-dimensional plotting environment.
- A hypertext help system.

Most importantly, the Xmath architecture is extremely open. Functions can easily be added, either in MathScript or using C++ or C routines (which are dynamically linked). The user interface is highly customizable: It can even be localized for foreign languages without recompiling! Source code is also provided for all functions written in MathScript.

16.2.2 The Xmath Object Hierarchy

The lack of advanced data structures is one of the most significant limitations of MATLAB-based languages. Of course, Xmath supports MATLAB's basic complex matrix type, but it also includes many special matrix types as well as objects which are particularly well-suited for control engineers. The various objects in the hierarchy are given below.

Control-Related Objects

The concept of a "dynamic system" is fundamental to control engineering. The dynamic system object in Xmath is defined as something which has inputs, outputs, and a sample rate (which is exactly zero for continuous systems). Dynamic systems can exist in either state-space or transfer-function form. The names of the inputs and outputs are stored with the object, which makes documentation particularly simple.

Polynomials are really basic mathematical objects but they are also of importance to control engineers. The polynomial object in Xmath makes it easy to define and use polynomials. Polynomials can be defined either in terms of their roots or their coefficients, allowing Xmath to perform many polynomial operations with exceptional speed and accuracy.

The parameter-dependent matrix (PDM) in Xmath is designed to store a matrix function of an independent variable. Although this is a general object, its applications to control engineering are direct. Time and frequency responses are conveniently stored as PDMs with time and frequency as the independent variable.

For a dynamic system with m outputs and n inputs, the PDM used to store a frequency response would be made of $m \times n$ matrices (one row per output and one column per input). The PDM used to store the results of a simulation would just be made of $m \times 1$ matrices (one row per output).

Numeric Objects

The general matrix data type in MATLAB-generation software is just that: general. In the world of linear algebra, however, there are many special types of matrices which can be handled with efficiency and elegance by numerical software if the special nature of the matrix is known a priori. This results in substantial speed and accuracy improvements over the MATLAB-generation packages.

In Xmath, there are more than eight special kinds of matrices. Some, such as square, rectangular, vector, or scalar, are distinguished by their shape, whereas others, such as symmetric, upper or lower triangular, Hessenberg, or diagonal, are distinguished by patterns in their data.

Supporting Objects

The numeric and control-related objects in Xmath are complemented with various supporting objects. These include strings, tables (an $m \times n$ matrix where each element is a string of any length), lists (a list of other objects), and index lists (basically, a vector of row–column indices).

16.2.3 MathScript Language Features

The MathScript language has been designed to be object-independent. Thus the grammar rules are not dependent on the definition of the objects on which the

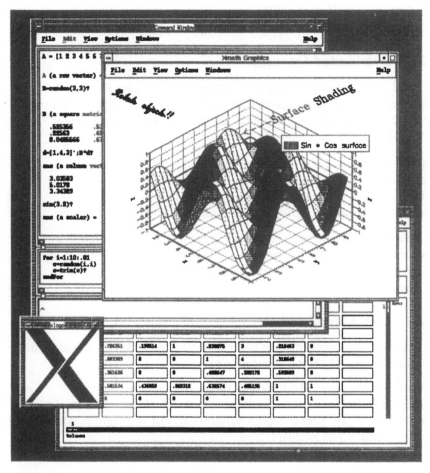

Figure 2 Xmath combines numerical functions, graphics, and interactive scripting with a friendly motif interface.

language operates. This is an extremely important consideration, since it will permit the Xmath environment to evolve far beyond its original design objectives without requiring continued changes to the language grammar.

16.2.4 Functions Take Advantage of Objects

In Xmath, the same function may perform different operations for different types of objects. It may be defined for one object but not for another. Sometimes, the meaning is the same but a different algorithm is used. Consider the standard EIG (eigenvalue) function for example: it is only defined for square matrices. Giving it a vector or a string would return an error. For full matrices, it uses a general eigensolver. However, if the matrix is an upper triangular matrix object, it automatically chooses an upper triangular matrix eigensolver.

16.2.5 Overloaded Operators

MATLAB made linear algebra easy by including matrix operators $(+, -, *, /)$ in the basic language. Xmath extends this flexibility to other types of objects by "overloading" operators. For example the basic multiplication operator * has the following meanings:

$matrix_1$ * $matrix_2$	is matrix multiplication
$poly_1$ * $poly_2$	convolves two polynomials
$system_1$ * $system_2$	connects two systems in series
system * timeResponse	performs a time-domain simulation

The definition of operators for dynamic systems have been designed with extreme care to make their usage very "familiar" with conventional written notation. This ensures, for example, that $(sys_1*sys_2)*response == sys_1*(sys_2*response)$

16.2.6 Data Management

To allow one to effectively manage large amounts of data, Xmath provides three key advances from MATLAB-generation software:

- Partitions—these are like folders or directories of data.
- Comments—you can attach comments (up to 255 characters) to variables.
- Time-stamps—are maintained on variables so that one can see when they were created or last modified.

16.2.7 User Interface

The user interface in Xmath is built on the Motif X-Windows toolkit. It is mouse-driven, intuitive, flexible, and extremely configurable. It includes:

- A command window for textual input and output.
- A graphics window supports interactive two-dimensional and three-dimensional graphics. Once a plot has been made, its look can be quickly changed with a few clicks of the mouse. Curve styles and colors can be changed, comments can be added, grids can be turned on and off and their spacing can be changed, and three-dimensional plots can be rotated. Three-dimensional surfaces are shaded using a point light source.
- A variables window (or data dictionary) which lists the names, sizes, types, and comments of variables in all your partitions. Menus allow you to rapidly load, save, delete, rename, and copy variables.

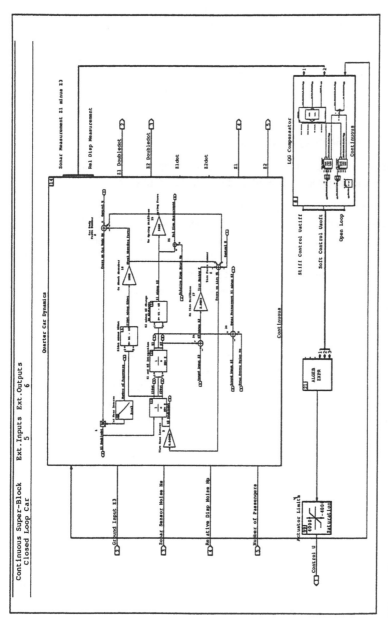

Figure 3 SystemBuild provides a user-friendly editor for graphical modeling with a powerful simulator.

- A matrix-editor window. This window acts as a kind of scientific spreadsheet. It allows you to easily create, browse, and edit large matrices. Typing an Xmath expression into a cell causes it to be immediately evaluated and then substituted.
- A hypertext help window gives you information on hundreds of topics in Xmath. Clicking on a word and then pressing a button causes the help system to jump to a new topic. The syntax of Xmath functions is also easily isolated. It can then be pasted directly into the command window's input area.

16.2.8 Extensibility Features

Xmath has been designed from the outset to be open and programmable. Advanced scripting support is provided through the following:

- User-defined functions and commands.
- A source-level debugger: Xmath provides a source-level debugger which makes the debugging and verification of scripts much easier. You can step through scripts one line at a time, set breakpoints, and watch key variables.
- Keyword arguments: MathScript supports both positional and keyworded arguments for functions and commands. This makes both the writing and use of functions easier because you do not always have to remember the order of arguments.
- Error checking and reporting utilities.
- "Smart" evaluation of logical expressions. Xmath short circuits logic expressions when sufficiency is established.
- Dynamically linked user functions: Although the MathScript interpreter is extremely fast, users who want maximum possible performance (or who just want to integrate existing code) can dynamically link functions written in Fortran, C, or C++ to Xmath.
- A callable interface: The overall architecture of Xmath has two main parts: the user interface and the interpreter. The user interface calls the interpreter through a callable interface. The specifications to this interface are published and the interpreter binary is separable so that users or third parties can embed the MathScript interpreter into their applications.

16.3 SYSTEMBUILD

SystemBuild™ is an integrated graphical model editor and simulator. SystemBuild accelerates the design and debugging of both linear and nonlinear dynamic systems. Functional block diagrams are used routinely by designers for system descriptions. Using the built-in function library plus user-defined functions, the SystemBuild hierarchical block editor quickly captures both system topology and behavior.

From the block diagram description, SystemBuild automatically builds a simulation model, eliminating programming and code debugging and allowing design engineers to concentrate on creative design issues. The entire design is simulated directly from the block diagram picture—truly a What-You-See-Is-What-You-Simulate design environment. SystemBuild is linked to the engineering analysis and design tools of Xmath through a common database. Results from nonlinear systems built and simulated with SystemBuild are transparently available to Xmath for further analysis, plotting, or manipulation. In addition, SystemBuild models can be linearized and brought into Xmath to utilize the powerful linear systems tools for controller design or frequency analysis.

16.3.1 Model Editor: An Intuitive Mouse-Driven Interface

SystemBuild accelerates systems design by using block diagrams to build functional models. The SystemBuild editor utilizes a user-friendly mouse interface complete with pull-down menus and pop-up forms for entering information. Each of the component blocks has an informative icon representation which identifies its function, important parameters, and timing information. The block diagrams which you create with these informative icons are "living pictures" of the system model. Components are selected from a palette and connected using simple mouse commands. They are placed on the screen, connected with signal flow paths, dynamically dragged, or resized using simple menus and the mouse. Additional features include cut, paste, zoom, multiple block selections and operations, and "rubber banding" of signals when connected blocks are moved.

The easy-to-learn interface provides the flexibility necessary to develop even very complex models. For the advanced user, special accelerator keys have also been incorporated. In their very first session, new users can take advantage of the mouse and menu interface to build models, whereas experienced users can use the accelerator keys to further expedite model building.

16.3.2 Unlimited Hierarchical Modeling

SystemBuild models are developed using a hierarchical modeling strategy which facilitates both abstract and detailed system representations. A block diagram containing many components may be grouped into a SuperBlock, given a name, and used in other models as a building block. SuperBlocks can be added to the library palette making them accessible by other engineers in the design group.

The SuperBlock approach has become an industry standard for dynamic system modeling. Hierarchical structures facilitate the construction of large system models, allowing both top-down and bottom-up design, individual testing and debugging of components, and division of tasks among large engineering teams. For example, an aircraft might be modeled at the top level with two component SuperBlocks: Aircraft Aerodynamics and Flight Control Law. Each of these two elements might, in turn, comprise many smaller subcomponents, permitting a very detailed description of the aircraft. Each subcomponent may be designed and tested separately by different engineers, and then brought together for the overall model. This is a simple two-level hierarchy. SystemBuild, however, places no limits on the hierarchical levels or on the number of shared SuperBlocks.

16.3.3 Extensive Block Library

SystemBuild includes a comprehensive block library which spans the whole range of dynamic system design. This block library includes the following:

- User-defined **Superblocks** for hierarchical model creation
- **Nonlinear Elements** such as saturation, preloads, and linear interpolation tables
- **Algebraic Equations** such as summing junctions and cross-products
- **Dynamic Systems** including PID controllers, Transfer Functions, State-Space Systems, and Gain/Pole/Zero Representations
- **Trigonometric Functions**
- **Coordinate Transformations** for cartesian, polar, and spherical coordinate frames, as well as 3-axis rotations and inverse rotations
- **Transcendental Functions** such as square roots, logarithms, and exponentiation
- **Logic Functions** including AND and OR gates, data path switches, and gain schedulers
- **Signal Generators** for sinusoid, steps, ramps, and user-defined inputs
- **Global Data Registers** for intertask communication
- **User Code Blocks** (blocks created by the user)

16.3.4 Finite-State Machine Editor

SystemBuild includes a complete facility for designing finite-state machines: the State Transition Diagram (STD) block. This block has a built-in "bubble editor" incorporating all of the same easy-to-use characteristics as the SystemBuild block editor, but with enhancements tailored to finite-state machines. The STD can be used to model the control flow of systems with discrete modes of operation and discrete transitions between these modes. It is a valuable addition to the SystemBuild library, permitting the graphical programming of such things as IF, THEN, ELSE structures, decision trees, and adaptive logic. STDs are ideal for adaptive control applications and system fault recovery analyses. Some of the features include outputs based on state conditions (Mealy Outputs), outputs based on transition occurrences (Moore Outputs), hierarchical structure, and complete integration into the overall SystemBuild block diagram environment.

16.3.5 Simulation

As scientists and engineers everywhere have discovered, our world is a very nonlinear place. Coulomb friction, gear backlash, and transistor switching are just a few examples of the real-world nonlinear problems encountered by scientists trying to model physical systems. SystemBuild is the first software solution to address nonlinear system modeling graphically. With System-Build, you can include any type of nonlinearity in your model. Standard nonlinear blocks include gain tables, deadbands, preloads, quantization, hysteresis, switches, absolute values, and many more. Logical blocks can be included in both continuous and discrete models, and a facility is provided to allow the inclusion of any user-defined or measured nonlinearity in the form of C, Fortran, or Ada source code. Models built up with these nonlinearities can then be simulated using any of the powerful integration techniques described below.

Continuous- and Discrete-Time Simulation

Continuous, discrete, hybrid, event-driven, and even multi-rate digital systems can be modeled and simulated, dramatically expanding the scope of systems which can be built in SystemBuild. Timing requirements and time skews are simply specified for the different elements in the model, and the SystemBuild simulator automatically handles all of the complicated calculations involved. In this way, extremely complex, mixed-timing models can be designed, and the effects of timing changes on the system performance can be analyzed. For example, an automobile model could include subsystems representing different continuous-time physical components such as Finite Element models, suspension subsystems, and engine dynamics. In addition, the same model could include digital engine controllers with multiple microprocessors running at different sampling rates and an interrupt-driven ignition system. This entire system could be simulated, with the results stored, plotted, and analyzed using any of the tools available with other modules in the ISI Product Family.

Choice of Integration Algorithms

Different integration algorithms are required to accurately and efficiently simulate particular systems. Six essential integration algorithms are available in SystemBuild. The SystemBuild simulator "compiles" the continuous block diagram model into its corresponding linear and nonlinear differential equations and integrates these equations with the user-specified integration technique. Integration techniques include four fixed step algorithms ranging from first to fifth order, and two variable step algorithms. Fixed step algorithms are useful for handling systems with multiple discontinuities, whereas variable step algorithms are generally faster for the case of smoothly changing functions. The variable step algorithms vary the integration step size to satisfy an error criterion which can be defined by the user. This allows the simulator to "jump" very quickly through slowly changing parts of the simulation. In addition, the Stiff System Solver algorithm incorporates the latest techniques for integrating difficult equations which contain widespread eigenvalues or even implicit algebraic loops. Discrete and event-triggered subsystems are seamlessly integrated into the simulation, providing updates at the correct times, regardless of the choice of integration step size.

Pause, Resume, and Model Trimming

Ongoing simulations may be stopped, and the intermediate simulation results and state values saved and analyzed. The simulation may then be restarted from any saved point. With this feature comes the ability to "Trim" models about their steady-state operating points to facilitate repeated analyses. For example, an aircraft model could be "flown" to 30,000 feet, the state values saved, and a series of subsequent analyses could be performed from that 30,000-foot operating point. This feature eliminates the need to "fly" the aircraft to the operating point each time a new analysis is desired. In addition, internal simulation variables may be traced, and breakpoints set to aid in the design of the system model.

User-Defined Linearization

Nonlinear models developed in SystemBuild are extremely useful for system simulation purposes. These models may be simulated, with the simulation results sent to the Xmath database. In this way, simulation results may be analyzed, displayed, and compared in the same way as any other data. Nonlinear models, however, cannot be used directly with the linear design and analysis tools which are available through linear systems theory and control design. SystemBuild solves this problem by generating equivalent linear models from nonlinear SystemBuild block diagrams. Operating points, input values, and perturbation parameters can be specified, resulting in an extremely flexible linearization capability. Linearized models can then be analyzed using any of the linear techniques available with the Xmath linear systems tool, or with advanced modules such as the Control Design Module (optional).

Frequency Response Calculations

Tools are available in the Digital Signal Processing Module (optional) to perform frequency response calculations for nonlinear, hybrid, and even multi-rate systems. These tools automatically simulate the SystemBuild model, exciting the modes of interest, and determine the frequency response based on the input/output data. In this way, no inaccuracies are introduced through linearization or resampling of the model.

Open Architecture

SystemBuild is designed to integrate easily into existing simulation environments. Existing models and components can be incorporated directly into
SystemBuild through the User Code Block (UCB) interface. With the UCB,
custom or proprietary algorithms written in C, Fortran, or Ada can be added
to the SystemBuild library. These blocks are then treated the same as any
other element in the standard SystemBuild library. Measurements and external stimuli may also be included in SystemBuild through the Xmath interface, providing the capability for extremely accurate models which have a
one-to-one mapping to your real system. A typical example is the inclusion
of actuator nonlinear characteristics in the form of an input/output mapping.
Known input values are put into the actuator, the outputs are measured using
data acquisition software, and the results are loaded directly into the System-
Build model.

Interactive Animation

Interactive Animation (IA) provides animated ''flight simulator-like'' control
panels that interact with SystemBuild models during simulation. IA uses information from the simulator to animate familiar devices such as meters, gauges,
and strip charts. The graphic display is continuously updated to monitor continuous dynamic systems. IA accelerates modeling and debugging by giving designers access to signal values and inputs.

The same IA control panels used with SystemBuild to debut control algorithms are also used to accelerate prototype testing with the AC-100^{TM} real-
time controller. The IA interface makes it easy to view actual signals from the
hardware prototype, to adjust parameters in the embedded control software, or
to select a new control algorithm from several alternatives.

Animation is an effective and versatile way to visualize dynamic systems.
Whether the system is an active suspension, a servo motor, or an aircraft engine,
IA allows design team members to see the interaction between the system and
the controller. IA and printed copies of control panels are also useful in presenting a concept or preparing a document.

Interactive Nonlinear Simulation

IA enhances the SystemBuild simulator with interactive control panels. Through
animated instruments, virtually all signals can be viewed during simulation. Using familiar input devices, any input or parameter can be changed at any point in
the simulation. IA works with continuous, discrete, hybrid, and linear or nonlinear models. Controlling a system model through an animated interface gives
the developer a better feel for the system dynamics and a realistic idea of the
effectiveness of control algorithms.

For simulating large systems, IA supports multiple control panels that are implemented as separate windows on the workstation screen. IA windows may be organized by function, actual target control panels, or by modeling hierarchy.

Because any system input can be set by the user through IA, it is easy to test "what if?" scenarios. By taking advantage of SystemBuild modeling techniques, many parameters such as transfer-function coefficients, logical conditions, and gains can also be changed during the simulation. Tradeoff analyses of alternative controllers can be performed using a model constructed to select one control strategy from among several possibilities based on user inputs.

Interactive debugging accelerates the design process. Some IA devices directly control the SystemBuild simulator. Through the PAUSE/RESUME device, it is possible to freeze a simulation, review outputs, and then continue simulating. Other IA devices help the designer to verify that signal values stay within expected ranges. Instruments such as strip charts, bar meters, and digital displays allow the user to set threshold values at which the instrument will change color if a threshold is crossed. Special alarm devices can alert the user visually if desired ranges are exceeded and store a record of the event in a file. Together, these devices make debugging a model with IA faster and easier.

Fast Control Panel Editor

Building IA control panels is simplified with a mouse-driven editor. Devices are chosen from a palette with the mouse and placed on the screen. Mouse commands let the user resize, change colors, and connect to the simulation model. The IA editor also includes menu commands to print and save/load control panels to/from files.

Extensive Animation Device Library

With the extensive IA Device Library, the user has a great deal of freedom to define the appearance and function of IA control panels. The library contains more than 100 basic devices. Many of these devices are animated to provide even greater ease of use, including turning knobs, moving sliders, scrolling strip charts, etc.

Interactive Prototype Display

With the AC-100 Rapid Prototyping System, it is no longer necessary to build control panels in hardware. Instead, the IA panels used with the SystemBuild simulator can be used for prototype debugging and testing. IA control panels are useful for observing data values, changing inputs, and evaluating control strategies. All control variables can be viewed via IA.

With the AC-100, IA is used to acquire data from the controller or from external devices connected to the controller. The acquired data can be displayed interactively. Acquired data can also be used to make subsequent SystemBuild simulations more realistic.

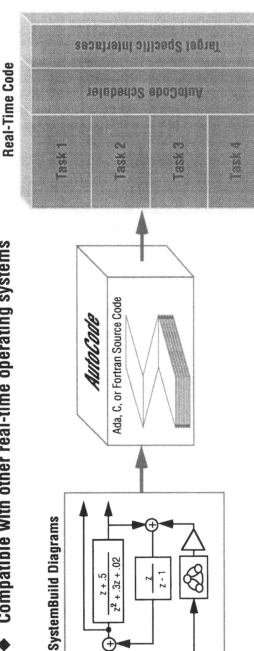

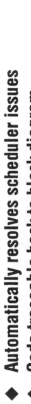

Figure 4 AutoCode—Automatic Code Generator.

16.4 AUTOCODE

Automatic code generation enables engineers to develop and maintain systems at the block diagram level, rather than in lines of cryptic programming code. This makes prototyping and experimentation much less costly and error prone. It further reduces designer frustration and configuration control headaches by ensuring that the implementation always reflects the current design documents, whether it be a satellite attitude control system, an antilock braking system, or a chemical process controller.

Engineers input their designs interactively in block diagram form using ISI's SystemBuild^{TM}, the premier graphical environment for system modeling and simulation. SystemBuild provides one interactive environment in which to create and simulate continuous, discrete, and hybrid multirate block diagrams. The combination of a rich built-in block library plus user-defined blocks provides the flexibility needed to represent almost any dynamic system. Hierarchical block modeling helps manage design complexity and encourages engineers to create their own customized block libraries for commonly used functions.

16.4.1 Breakthrough Technology

AutoCode translates SystemBuild block diagrams into real-time source code for implementation on dedicated digital computers. Any discrete-time or aperiodic algorithm in SystemBuild can be converted to standard Ada, C, or Fortransource code. The generated code is highly optimized for speed to meet the demanding needs of real-time control applications and hardware-in-the-loop simulations.

AutoCode can generate a complete, self-contained application. This code is designed to run on embedded microprocessors or minicomputers. The only external services required are basic hardware input/output drivers and timer setup. Once written, these routines provide a general framework to support AutoCode applications in the target computing environment.

Generated stand-alone applications are easily ported from one computer and/ or operating system to another by linking with different sets of hardware I/O routines. ISI supplied drivers even allow the generated real-time code to be run in batch mode on any workstation or mainframe computer. This accelerates complex discrete-time simulation and/or code verification before testing with target hardware.

AutoCode can also generate callable subprograms. These automatically generated subroutines may then be called by existing handwritten applications. Because AutoCode generated subroutines all conform to SystemBuild's UCB calling convention, it is easy to link them back into SystemBuild for simulation as User Code Blocks.

The generated real-time application is highly optimized for execution speed. To minimize overhead, loops are unrolled and complex dynamics expressions

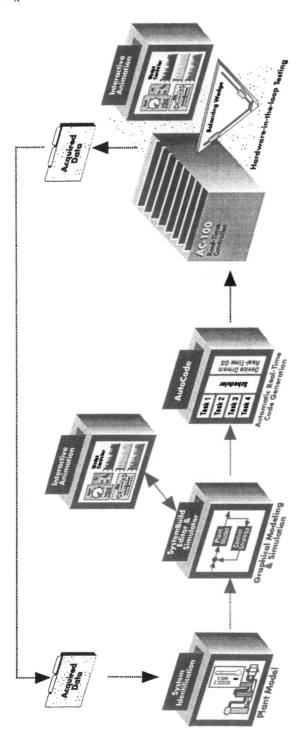

Figure 5 Tool integration accelerates system development.

are coded in-line. These optimizations are very tedious to manage manually in a large, changing application. They are usually left to the end of the development cycle. However, AutoCode automatically performs them whenever appropriate.

16.4.2 Documentation and Traceability

The generated code is not a "black box." It is heavily commented to facilitate understanding and design traceability. A variable map documents the name and use of each variable in the code. Comments in the main body trace each line back to the original SystemBlock block diagram. If a problem is discovered in an algorithm using a traditional development tool (such as a source-level debugger), these comments indicate where the block diagram should be changed to correct it. Hand patching of the generated code is certainly possible, but with AutoCode it is usually more convenient to modify the SystemBuild diagram instead. This keeps the design diagram current with the implementation even during system debugging and prototyping.

16.4.3 Multitasking

The generated code for every stand-alone AutoCode application includes a hard-deadline preemptive priority-based scheduler. This technology ensures that the processor is utilized most efficiently to meet the timing constraints indicated in the original SystemBuild block model. If a timing deadline is missed, the real-time scheduler will immediately trap the error and pass control to a user-written error handler, which takes the appropriate recovery action.

With Autocode, there is little of the overhead (and none of the cost) associated with general purpose real-time kernels or operating systems. The real-time scheduler is part of the application—tailored specifically to support just that application, rather than the needs of every application conceivable.

16.4.4 Code Tailoring

Good real-time programmers know how to optimize the performance of a given algorithm for a particular computer. AutoCode allows programmers to capture their know-how in the form of ASCII coding templets that determine the actual code generated for various SystemBuild constructs. By modifying the templet files, advanced users can systematically alter the generated code to conserve memory and/or execute as fast as possible on their intended target hardware.

16.5 AC-100 RAPID PROTOTYPING SYSTEM

The AC-100 automates the development of real-time systems by combining graphical modeling tools with a real-time controller. It can be used for control

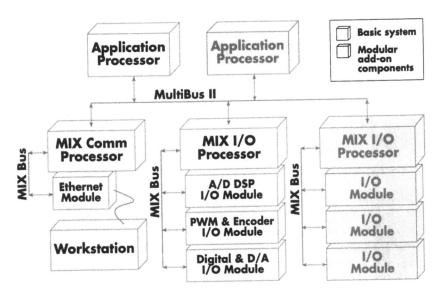

Figure 6 Distributed processing and a modular design make the AC-100 Controller powerful and flexible.

system development or as a real-time testbed for a wide range of automotive, aerospace, and process control applications.

By automating steps that are usually performed manually, the AC-100 streamlines the production of functional prototypes. Whereas the development of a working prototype usually requires considerable hardware and software engineering, the AC-100 provides a versatile hardware controller that can be programmed without writing code. Risk is reduced when a working prototype is created early in the design process. Time saved with the AC-100 can be used to improve a design, to perform more rigorous testing, or simply to complete a project sooner.

The AC-100 is a complete workbench for system development. A graphical user interface provides easy access to control design tools, a block diagram editor and simulator (SystemBuild), a code generator (AutoCode), and a hardware controller. No programming skills are required for the engineer to convert the block diagram into source code. The compiled code is then downloaded to the controller for live testing. Live tests are monitored through animated control panels and, if a modification is required, the designer simply edits the block diagram. New code is automatically generated and the modification is quickly tested with target hardware. By reducing the time required to implement an idea, the AC-100 allows designers to find the best solution.

16.5.1 Tool Integration Accelerates System Development

In the AC-100, tools for each stage of system development are integrated into a single environment. This integration allows a design to move easily from one stage to the next. Whereas many projects lose time because separate tools are used for data acquisition, control design, software engineering, and integration, the AC-100 streamlines these processes.

16.5.2 Data Acquisition Integrated with Model Synthesis Tools

Developing accurate models is often the most difficult and time-consuming aspect of system design. The AC-100 provides data-acquisition hardware and automatic model generation software to automate this step. The controller hardware can sample digital and analog signals over a wide range of frequencies. The sampled data can then be used by the System Identification Module to synthesize models of sensors, actuators, or entire systems.

16.5.3 MultiCode™ for Multiple Processors

Many applications require higher performance than a single processor can provide, so the AC-100 supports up to 11 processors working in parallel. MultiCode segments the generated code for each of the processors. It generates additional code to handle complex interprocessor communication and synchronization.

16.5.4 System Monitoring and Control

The AC-100 includes Interactive Animation (IA) control panels. IA provides animated instruments for viewing signal values and input devices to set parameters. A mouse-driven editor is used to create new panels from a library of over 100 animated devices. IA extends SystemBuild to allow interactive simulation and provides integrated, cost-effective instrumentation for the AC-100.

16.5.5 Distributed Architecture Provides Expandability

Multiple processors and a modular I/O interface make the AC-100 Controller both powerful and flexible. One or more Application Processors is dedicated to running the primary application code. A Communication Processor supports the Ethernet link to the host workstation. I/O Processors move data between the Application Processor(s) and the I/O Modules. The distribution of system functions to these dedicated processors makes it possible to scale the controller for greater system throughput.

Support for coarse-grain parallel processing is unique to the AC-100. This capability allows the user to run the primary application on just 1 or as many as

11 Application Processors. MultiCode software segments the application code for the number of processors specified by the user.

A modular I/O design makes it easy to tailor the controller interface to fit each application. Controller interfaces are built from combinations of three basic I/O Modules. Local processors on two of the modules provide signal processing on input and output signals. These smart modules improve system performance by reducing the load on the I/O Processor. Performance is further enhanced through the use of a private bus (32-bit Intel MIX standard) that connects the I/O Modules to I/O Processors and reduces traffic on the primary system bus. The basic capabilities of each I/O Module are described below.

16.5.6 I/O Modules

68332/PWM and Encoder I/O Module

* 16 Pulse Width Modulation (PWM) channels
* User specifies duty cycle and frequency
* 8 Encoder Channels with 32 bit counters

Digital and Digital-to-Analog I/O Module

* 64 bidirectional digital lines can be used as discrete lines or as 8- or 16-bit wide channels
* 16 analog output channels

Analog-to-Digital DSP Data Acquisition Module

* 46 analog input channels
* Auto calibration: gain and offset adjustments made by DSP processor
* Simultaneous sample-and-hold

Table 1 Automobile Gas-Turbine Design Example

Comments	Interaction with MATRIX$_x$	Interaction between Designer and MAID
At various stages in the design process, advice is available in the form of help texts rather than as design procedures and principles. This applies particularly to heuristics knowledge which is often difficult to express as procedures.		TOP LEVEL OF SUPERVISOR. You can: • Investigate the structural characteristics of your system • Input a set of frequency- and/or time-domain design objectives • Use one of the available design techniques • Perform a detailed analysis of the behaviour of your system • Run MAID in its default sequence. {Structure→Objectives→ Design→Detailed Analysis} Select one of • Investigate structure • Objectives • Design • Detailed analysis • Default sequence 》design Which design technique do you wish to try? • Simple design technique • Reverse frame alignment • Observer-based controller • Help with selection • Default (simple design technique) 》help with selection You are advised to use the Simple Design Technique initially. If that fails to give an acceptable design solution, the Reverse Frame Alignment technique should be tried. Particularly difficult problems can be tackled by designing an Observer-Based Controller.

Which design technique do you wish to try?

- Simple design technique
- Reverse frame alignment
- Observer-based controller
- Help with selection
- Default (simple design technique)

》simple design technique

Entering: Simple Design Technique specialist
The Simple Design Technique is broken down into two phases, dealing with the high- and low-frequency regions.

Entering: High-Frequency Design specialist
The objectives of the high-frequency design are to:
(i) align and thus normalize the system at high frequencies and reduce the interaction between loci at high frequency
(ii) balance the principal gains at high frequencies

Are the high-frequency roll-off rates the same?
    PRIMARY (s,ns);

Select one of

- Yes
- No
- Default {no}
》why
————Why————
Trying to determine: high-frequency roll-off rates are the same for use in:
rule 4

Figure 9 shows the principal gains {P1 & P2} and characteristic gains and phases {E1 & E2} of the system in the form of a primary indicator diagram. This indicator shows that the system's eigenloci roll off at different rates at high frequencies. One locus rolls off at 20 dB/decade, the other at 40 dB/decade.

Table 1 (cont.)

Comments	Interaction with MATRIX$_X$	Interaction between Designer and MAID

When the designer does not understand one of MAID's questions, he can display the procedure currently being used by MAID and/or the principles underlying the question. The principles and procedure should help the designer to understand why the question has been asked and they should also help him to formulate his answer to the question.
 	
References are attached to the principle. If the designer requires further information, he can look up the appropriate texts.
		IF high-frequency roll-off rates are the same IS no
		THEN high-frequency design IS balance gains at the bandwidth frequency
		AND CHECK the condition number at the bandwidth frequency
		which is being used to determine: high-frequency design
		
		  Principles
		
		NAME: SDT-1 STATUS: own
		ORIGIN: SDT-high-frequency design WEIGHT: 1.0
		HEADER: The principal gains can be balanced over the whole of the high-frequency region of the order of the infinite zeros are the same.
		BODY: If the principal gains roll-off at the same rate, it follows that the gain loci can be balanced (have equal values) at all high frequencies. The result is that interaction is reduced at all high frequencies. If the roll-off rates are different, the designer can only balance the gains at one frequency (usually this is done at the bandwidth frequency).
		EXAMPLE: (*not included*) REFERENCE: Hung and MacFarlane (1984) (Section 2.3.2)
		
		Are the high-frequency roll-off rates the same?
		  PRIMARY(s,ns);
		
		Select one of . . .

The HIGH-FREQUENCY DE-SIGN specialist has gathered enough information to make a design recommendation. In making design recommendations, MAID is taking on the role of a design assistant.

Having made a recommendation MAID takes on the role of an intelligent frontend. In this role, MAID tells the designer which function(s) (when executed as a command in MATRIX$_X$) will implement the required compensator. MAID also helps the designer to establish values for the parameters of the function. In this example, the function required is HFCOMPX. Note how MAID helps the designer to establish the "condition number" parameter of the function and how MAID tells the designer that the condition number is the third parameter of the function.

⟨condition no vector⟩ = HFCOND(system,no states, freq for condition no);
⟨condvec⟩ = HFCONDS (s,ns,10);
condvec = ⟨1 1.6637⟩

system after high-freq compensation,
no states after high-freq compensation,
high-freq subcontroller = HFCOMPX(system, no states, condition number vector);
hsl,nhs1,hk1 = HFCOMPX(s,ns,1 1.6637);

$$hk1 = \begin{vmatrix} 0.382 & 0.924 \\ -1.537 & 0.635 \end{vmatrix}$$

• Yes
• No
• Default {no}
⟩⟩no

What is the condition number at the closed-loop bandwidth frequency (10 rads/sec) H FCOND(s,ns,10);

Enter a number between 1 and 100

⟩⟩1.66

———Design Recommendation———

Balance the gains at the bandwidth frequency.
cs,ncs,khf HFCOMPX(s,ns, 1 1.66, 10);

In systems where the high-frequency roll-off rates are not the same, the loci can only be balanced at a single frequency. This is usually taken as the bandwidth frequency

(The compensator will also align the input and output gain directions at high frequencies.)

Table 1 (cont.)

Comments	Interaction with MATRIX$_x$	Interaction between Designer and MAID
After the implementation of a compensator, MAID calls a specialist that will help the designer to analyze the effect of the compensator on the system.		*Entering: High-Frequency Analysis specialist* Does the system have the correct gain-phase characteristics? 　　PRIMARY(s,ns); Select one of: • Yes • No • Default {yes} 〉〉yes
Figure 10 shows the primary indicator of the system following the high-frequency design. The compensator "hk1" has successfully aligned the system at high frequencies and has balanced the principal gains at the bandwidth frequency.		Has the high-frequency subcontroller caused an unacceptable deterioration in the normality of the system at intermediate frequencies? 　　PRIMARY(s,ns); MS(s,ns); Select one of: • Yes • No • Default {no}
Here, we show how the designer attempts to understand a question he has been asked by MAID. He receives two types of on-line help: (1) the procedure MAID is currently using and (2) the principles associated with the question he has been asked.		——Why—— Trying to determine: unacceptable deterioration at intermediate frequencies for use in: IF unacceptable deterioration at intermediate frequencies IS yes THEN intermediate freq after high-freq IS intermediate-freq unsatisfactory AND REPORT The high-frequency subcontroller has caused an unacceptable deterioration in robustness at intermediate frequencies. The designer should reduce the gain introduced by the high-frequency design stage to improve robustness.

AND RESET intermediate freq after high freq
AND CHECK intermediate freq after high freq

which is being used to determine: intermediate freq after high-frequency design.

—Principles—

NAME:	High Analysis Robustness	STATUS: own
ORIGIN:	HIGH-FREQUENCY DESIGN	WEIGHT: 1.0
HEADER:	The manipulation of the gains and phases in one part of the frequency spectrum may have an adverse effect on the system in another part of the spectrum.	
BODY:	The subcontroller may adversely effect another part of a system because the process of aligning the input and output frames at one frequency may misalign the frames at some other frequency.	
EXAMPLE:	(*not included*)	REFERENCE: Pang and McFarlane (1987) (Section 6.5)

Has the high-frequency subcontroller caused an unacceptable deterioration in the normality of the system at intermediate frequency?

PRIMARY(s,ns); MS(s,ns);

Select one of:

• Yes
• No
• Default {no}

》no

High-frequency design complete

The principles associated with a question always give the designer references he can look up if the on-line information is insufficient or too complicated for him.

The designer does not need ask for the primary indicators to be re-plotted unless the system is modified in some way. The designer can use Figure 10 to perform all the analysis following the high-frequency design.

The designer is expected to understand the principles of a technique. MAID protects him from the details of the theory and from the complex mathematical operations that must be performed when multivariable control design methods are used.

Source: Trans Inst MC, Vol. 11, No. 1, Jan. – Mar. 1989.

17

Computer-Aided Control Engineering Environments*

James H. Taylor

Odyssey Research Associates
Ithaca, New York

Magnus Rimvall and Hunt A. Sutherland

General Electric Corporate R & D
Schenectady, New York

17.1 INTRODUCTION

17.1.1 Motivation

Progress in many areas of technology requires better performance from embedded control systems. These controls may be in a manufacturing process where the need is for better control in the sense of more efficient use of resources and materials or tighter quality control, or the controls may be an integral part of a product and the performance of that product must be improved (e.g., the fuel efficiency of a vehicle has to be increased).

More stringent demands on the performance of technological systems require the use of advanced controls. This translates into the current trend toward integrating the control of subsystems so that beneficial subsystem coupling can be exploited and adverse coupling can be reduced or eliminated. It also promotes the use of recent advances in control theory to accommodate dynamic variability, uncertainty, component failures, and other effects and phenomena that may degrade control system performance in some sense.

---

*This contribution is based on "Computer-aided control engineering environments: architecture, user interface, data-base management, and expert aiding" by the above-named authors, which appeared in *Proc. 10th IFAC World Congress*, Tallinn, Estonia, August 1990.

In turn, the growing demand for advanced integrated control necessitates improvements in computer-aided control engineering (CACE) software, so better designs can be obtained with less cost in terms of time, effort, and errors in design or implementation. This has given impetus to rapid strides made worldwide in the area of CACE software development and usage and has specifically motivated the GE MEAD Project.

17.1.2 Trends in CACE

The development of CACE software started several decades ago with the development of routines to perform specific functions that had previously been done manually (e.g., root locus, Bode analysis). In the 1970s, the emphasis shifted to the "packaging" of routines to integrate them, share common data structures, and broaden their scope. In this second phase, considerable attention was also given to the development of numerically robust algorithms; libraries such as LINPACK and EISPACK began to supplant the "home-brew" algorithms that had been used before. More recently, the focus has moved to further broadening of functionality (e.g., block diagram interfaces for model building, autocode generation) and to improving the overall environment.

The latter issue—improving the environment—is the specific concern of this presentation. Standard phrases in this regard include "enhanced user friendliness" and "more supportive CAD tools." The areas we chose for improvement are the user interface and support facilities for database management and expert aiding. We discuss the architecture of such an improved CACE environment and specific needs, design considerations, and implementation issues. Much of the description of needs and considerations that follows is based on the **GE MEAD CACE Environment** (MEAD = Multidisciplinary Expert-Aided Analysis and Design[†]) and the experience gained in the course of the GE MEAD Project.

17.1.3 GE MEAD CACE Environment Overview

The GE MEAD CACE Environment (GMCE) has been designed to address the environmental issues outlined earlier while taking maximum advantage of existing software modules. implementing the GMCE thus entailed the integration of commercial CACE packages under a *Supervisor* which coordinates the execution of these packages with an *advanced user interface*, a *database manager*, and an *expert system shell*. The resulting software architecture is depicted in Figure 1. The CACE tools ("core packages") include the PRO-MATLAB®

[†]The origin of the acronym MEAD is the US Air Force MEAD Project (cf. Ref. 15), which is a parallel/synergistic effort to that described here. The USAF MEAD effort was sponsored in part by the Flight Dynamics Laboratory, Wright Research and Development Center, Aeronautical Systems Division (AFSC), United States Air Force, Wright-Patterson AFB, Ohio 45433-6523, under contract F33615-85-C-3611.

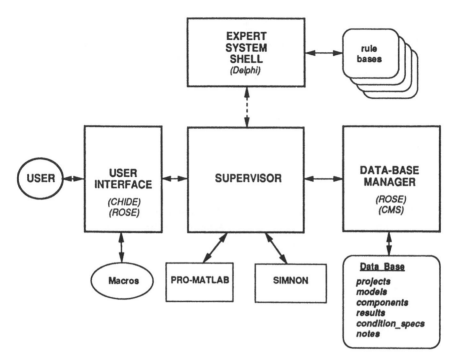

Figure 1 Software architecture for GE MEAD CACE environment.

package[‡] for linear analysis and design and the SIMNON® package for nonlinear simulation, equilibrium determination, and linearization. Other modules are also based on preexisting software: The user interface was built using the GE Computer/Human Interface Development Environment (CHIDE) which rests on the ROSE® database manager; the GMCE database manager uses ROSE and the DEC® Code Management System (CMS®) software for version control; and the expert system uses the GE Delphi® shell which rests on VAX® Lisp. The supervisor and the front end of the DBM are coded in Ada®.

17.1.4 Outline

CACE environments are discussed within the following framework: Section 17.2 itemizes basic functional requirements; Section 17.3 overviews the integration of CACE packages; and Section 17.4 treats concepts for a modern user interface.

[‡]PRO-MATLAB® is a registered trademark of The MathWorks, South Natick, MA; SIMNON is a trademark of Lund University, Lund, Sweden; VAX, DEC, and CMS are registered trademarks of Digital Equipment Corp., Maynard, MA; ROSE is a trademark of Martin Hardwick, RPI, Troy, NY; Delphi is a trademark of GE; and Ada is a registered trademark of the U.S. Government, Ada Joint Program Office.

Section 17.5 describes the CACE database management problem; Section 17.6 surveys incorporating expert systems in CACE environments; Section 17.7 outlines an extended example that illustrates many of the points in Sections 17.3–17.6; and, finally, Section 17.8 provides concluding remarks.

17.2 OVERALL CACE FUNCTIONALITY

The following list captures the *basic* (minimal) functions that are required of a modern environment for CACE:

1. Modeling: linear and nonlinear systems, in continuous- and discrete-time; building arbitrarily structured models from components
2. Simulation: initializing system state variables, setting system parameters, defining input signals, designating simulation variables for storage, running a simulation
3. Steady-state (equilibrium) determination
4. Linearization (see Ref. 12 for the linearization algorithms used in GE MEAD)
5. Linear Analysis: eigenvalues/eigenvectors, zeros, controllability and observability, model reduction, model transformations, root locus, frequency response
6. Linear Control System Design: frequency-domain methods (by manually adding lead/lag or PID compensation, LQR/LTR, H^∞ methods), pole placement, time-domain methods (LQG, LQR)
7. Control System Validation: frequency-domain analysis of linear models, simulation of linear and nonlinear systems

This list is *not* all-inclusive from a control-theoretic point of view or from the standpoint of practical functionality. For example, obvious areas that can be extended are stochastic control (e.g., Monte Carlo analysis capability); nonlinear systems analysis and design (e.g., describing function methods, bifurcation analysis); optimization; discrete-event systems modeling, analysis and design; system model identification; and automatic controller code generation. Furthermore, new approaches and theories are being developed on a continuing basis, making closure impossible.

One objective of any realistic CACE software development project must be to set goals as to functionality. The first phase of the GE MEAD Project was limited to the basic CACE functionality itemized above. Recognizing the need for future extension, the GMCE has been designed to be ''open,'' in the sense of being extensible either by adding built-in functionality or by use of MEAD macros or the ''Package Mode'' access to any CACE functionality of the core software (see Section 17.4.3).

17.3 CACE ARCHITECTURES

Powerful CACE packages exist that cover large parts of the overall functionality outlined in Section 17.2. This enables decoupling numerical functionality from the environmental issues of support and user friendliness that are the main focus of this presentation. The implementation of a modern CACE environment can then be accomplished by providing a "shell" for existing software rather than starting from the foundations of numerical analysis and algorithms.

There are other advantages to this strategy beyond considerations of development time and cost. Most CACE packages are implemented as monolithic software programs with internally defined data structures and algorithms that are immutable. Thus, even extendable packages, such as the MATLAB-derived packages, are limited by the embedded data structures and core algorithms. Because there will probably never exist a single, "frozen" program capable of accomplishing all tasks performed by a control engineer, it seems preferable to shell the best existing CACE packages in an open software environment rather than committing to a specific set of data structures and algorithms.

The approach taken in the GE MEAD Project was based on these considerations. Implementing the GMCE thus required the integration of a database manager (DBM), an expert system shell, and several CACE packages, such as PRO-MATLAB for linear analysis and design and SIMNON for nonlinear analysis and simulation. Access to the software is via a common user interface. The resulting architecture is shown in Figure 1. This system has been carefully designed to integrate different packages into a single, uniform environment despite diverse package interfaces.

At the heart of such an architecture there must be a module that combines and coordinates the various CACE packages and support modules. The GMCE Supervisor serves as the package integrator for the GE MEAD environment (Figure 1; see Rimvall and Taylor [8]). Multiple packages are run under the Supervisor, and data is reformatted or converted, when necessary, to ensure compatibility among packages. The GMCE Supervisor accepts "MEAD commands" and translates these into "package commands"; therefore, the user does not have to learn all of the intricacies involved in using each package unless advanced functionality is to be accessed via Package Mode (using a core package under the GMCE via its own interface—see Section 17.4.3).

The Supervisor contains most of the "intelligence" of the GMCE. It accesses data files that define how to use the underlying packages (e.g., how to convert MEAD commands into package commands), directs the activity of the DBM, and manages the command and data flow between itself and the user interface. It tracks the high-level activity of the user, including knowing what model(s) have been configured. Finally, the supervisor controls the invocation and use of the expert system by activating it, telling it what rule base to load and execute, serving as the conduit for communications between the user and expert system, and handling the expert system's results when it is finished.

The User Interface (UI) of MEAD communicates with all different components of the MEAD system, including the DBM and the expert system shell, through the Supervisor. The Supervisor provides the UI and the expert user with a unified and well-structured command-language interface. This same interface is used by the Expert System when it is invoked to carry out a high-level function. In fact, the GMCE supervisor may be used by itself as an "old-style" command-driven interface to CACE functionality, although this is rarely necessary. The clean separation of the "real" user interface from the Supervisor has proven to be very beneficial in terms of development, test, and refinement and has made it possible to implement the MEAD UI and Expert System independent of the quite diverse collection of underlying modules.

17.4   USER INTERFACES FOR CACE

A primary goal in designing a "user-friendly" CACE environment is to create a user interface that supports control engineers with widely different levels of expertise in using CACE tools. The design requirements necessary to satisfy inexperienced users are very different from those needed for expert users, and this often results in a fundamental conflict that causes engineers from one or the other end of the experience spectrum to be very dissatisfied with a given CACE environment. This conflict may be resolved in two ways: by making the user-friendly features of the interface fast and nonpatronizing and by allowing the user to work flexibly in a variety of modes.

Despite the importance of the UI to the acceptance and success of a CACE package, user-interface considerations have often played a secondary role in their design. Three periods of UI design philosophy can be distinguished in this field:

• The interactive control packages developed during the 1970s had quite crude user interfaces. Most packages, such as KEDDC [10], used rigid question-and-answer or low-level menu interactions. This kind of a UI can be made almost self-explanatory for the novice; however, it becomes very tedious for an experienced user.
• Command-driven interfaces came into prominence with the advent of MATLAB [7]. Primarily, these were designed to "open up" the underlying programs so that users can extend the functionality of a program by adding interactively defined macros and algorithms. However, flexibility often comes at the expense of complexity, forcing the user to recall and accurately enter many rigidly defined low-level commands to get a high-level task executed. This cost is quite high for a novice user or for a person who does not

expect to be a regular user of the package; this factor inhibits many computer-cautious control engineers from using such programs.

• The third generation of UIs, as represented by the GMCE, combines the simplicity of modern graphical interfaces, using drop-down menus, forms, and "point-and-click" techniques, with extendable command- and macro-interfaces as found in the MATLAB family. This provides both the novice and the expert user an adequately powerful and yet fully manageable access to the program.

17.4.1 Graphical Operating Environments

During the last half decade, the workstation and personal-computing arenas have been revolutionized by the advent of completely graphics-based systems. This is best illustrated by the Apple® Macintosh® computer family,[†] which features a graphical operating system relying on icons, menus, and a mouse, enabling the user to perform virtually all necessary actions with "point-and-click" operations. Computers from other vendors feature similar, albeit less dominant, operating environments. These machines have proven to be particularly well suited for inexperienced and casual computer users. Using the same general paradigm, the main GMCE operating environment is menu and form based. This is in distinct contrast to the command-oriented operating environments prevalent in modern MATLAB-based CACE packages and many other programs.

Although "graphics" always played an important role in CACE, it was hitherto mainly used for plotting curves (e.g., frequency and time responses) or for enabling graphical input of models in block diagram form. The control engineer was thought to "need the power" of a command-driven interface, just as software engineers were long thought to need the cryptic details of the UNIX® operating system. In both cases there is a strong risk that users will divide into two distinct groups: expert users and unhappy or nonusers. Incorporating a graphical interface for CACE should substantially increase the number of effective users without penalizing the expert, as long as the interface is sufficiently flexible and well-designed. Such a UI can be illustrated by overviewing the GMCE interface, where the emphasis is on design considerations and the user's perspective.

The GMCE graphical operating environment allows the user to perform all basic controls-related operations in a very consistent manner over mouse-operated menus and forms. The user does not need to know any commands or syntax, and the menu-tree hierarchy is designed so that there is a natural and easy-to-remember path to each desired functionality. In most cases, the menu-tree hierarchy is limited to two or three levels for quick access to all domains. At

---

[†]Apple® and Macintosh® are trademarks of Apple Computer Corp., Cupertino, CA; UNIX® is a trademark of AT&T Bell Laboratories, Holmdel NJ.

the last level of the menu-tree, selection and action forms are used to permit the highly interactive execution of most operations. The GMCE menu-tree is depicted in Figure 2. (This is not a screen image.)

Figure 3 shows a screen-image ensemble illustrating the GMCE graphical operating environment. The top half of this is an actual screen rendering; the bottom half contains forms that are obtained by clicking the buttons as indicated by the arrows. The top-level horizontal menu or Resource Bar is continually displayed across the upper edge of the screen. When a field in a menu is clicked upon, the corresponding submenu or action form appears. Activated fields are displayed in reverse video, and each submenu appears in decreasing hierarchical order to the right of its parent menu.

Action forms vary in size and content, but they are always aligned with the right edge of the screen as shown in Figure 3. The setup buttons enable actions that the user may want or need to perform before executing the function (such as choosing an integration algorithm or defining the end time for the simulation); the "Execute" button triggers the actual operation (e.g., simulate). The bottom row of command buttons is always arranged as follows: the "Display" button to

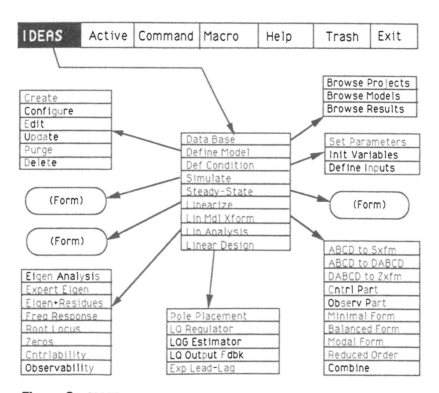

Figure 2 GMCE menu-tree.

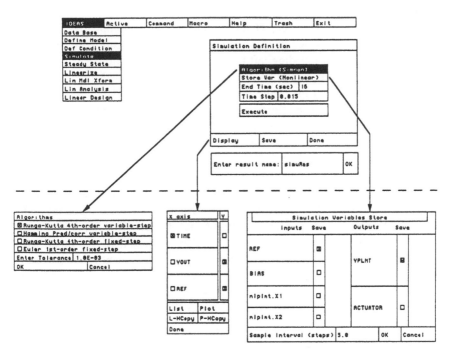

Figure 3 Screen-image ensemble illustrating the GMCE graphical operating environments.

the far left allows the user to view the result and produce hard copies. The "Save" button lets the user save the result in the database. A "Modelize" button is available on forms where the result may be interpreted as a model (e.g., linearization), in which case the result is reformatted into a model and stored as such in the database. Finally, the "Done" button moves back down the menu-tree. Whenever the user is requested to enter any alphanumeric information, such as the name of a result or model, an additional input form will open up below the action form, as shown for "Save."

The GMCE UI is "object-oriented," in the sense that the user selects items and options by point-and-click operations; each object is then presented in a context (e.g., form) that is appropriate to that object class. Various "Browsing Forms" are used to provide the user with the available choices, together with any pertinent information about the selectable items. The information within these forms changes dynamically as objects are created, modified or deleted. Items presented to the user for selection correspond to the database hierarchy, namely, Projects, Models, Components, and Results [16]. Each element may be displayed appropriately (text files loaded in an editor, plottable data rendered graphically).

Three Browsing Forms are depicted in Figure 4. In this example, the projects in a user's database are listed in Figure 4a, and Project "Tallinn" is designated by clicking on the corresponding left button. Clicking the "Models" button on the bottom row brings up Figure 4b, which portrays the Browse Model Form for Project "Tallinn" with a list of the models in that project's database and their key attributes. The user then selects model "NLPPFBS" and chooses an action to be taken with that model; in this case, clicking "Descript" reveals the description and the components of model "NLPPFBS," as shown in Figure 4c. Clicking on "Results" would bring up a similar Results Browsing Form for the same model.

Several additional aspects of the GMCE are revealed on these Browse screens. All entities in the database may be assigned individual notes through the UI by clicking on the appropriate "Edit Note" button. This feature permits on-line documentation at every level. The UI also provides access to other data associations described in Section 17.5; for example, component references and links may be displayed via the "Disp Ref" and "Disp Link" buttons in Figure 4c. These and other details shown in Figure 4 are clarified in Section 17.5 and in the Section 17.7 example that corresponds to Figures 4b and 4c.

17.4.2 Implementation of User Interfaces

Building a modern UI without tools and utilities is a huge task. A faster and more economical approach is to employ the tools and methods of a suitable User Interface Management System (UIMS). The need for a UIMS arises especially when constructing a user interface that requires advanced features found on engineering workstations. The use of a UIMS can greatly reduce the effort required to produce a user interface and ensure a consistent and reliable design. In addition, a UI founded on a UIMS can be refined and extended much more readily than a UI built monolithically without a UIMS. Note, however, that the "wrong" UIMS may not be an asset. It is important to specify the UI requirements and to be sure that the UIMS meets them.

The use of a UIMS permits the separation of the design of the user interface from the application program and the display device. The UI can be maintained as a separate component, thus easing the maintenance of the overall environment. Changes can be made relatively independently in the various modules.

A secondary purpose for a UIMS is to provide advanced UI capabilities to the end user which would not be easily gained otherwise. Thus, a UIMS should provide a comprehensive and extensible set of interface tools, for example, graphical display editor, and support UI capabilities such as user profiles; interactive help facilities; session logging; display graphics; and definition, editing, and execution of command scripts.

IDEAS	Active	Command	Macro	Help	Trash	Exit

HEAD Project Browsing

Name		Created	Updated	Models	Notes
☐ bmark		9-NOV-1989 11:23	22-NOV-1989 17:52	Y	N
☐ demo		14-SEP-1989 12:14	14-SEP-1989 12:47	Y	N
☐ library		14-SEP-1989 13:56	14-SEP-1989 16:13	Y	N
☐ oct26		22-OCT-1989 17:23	22-OCT-1989 17:58	N	N
☐ orlando		14-SEP-1989 16:21	14-SEP-1989 16:51	Y	Y
☒ tallinn		21-NOV-1989 16:58	27-NOV-1989 20:28	Y	N
☐ tampa		14-SEP-1989 16:19	14-SEP-1989 16:56	Y	Y

Models	Edit Note	Dele Note	Select Proj	Create Proj	Dele Proj	Done

Figure 4   (a) Project browsing form.

The GMCE UI has been designed and implemented using a GE-developed, experimental UIMS called CHIDE (Computer/Human Interface Development Environment; see Ref. 5). CHIDE supports the UI design paradigm outlined above and has proven to be very effective.

17.4.3   User-Interface Modes

As mentioned previously, a UI should be designed to facilitate access to CACE package capabilities by engineers with widely different levels of familiarity with the environment. This goal may be achieved by permitting the user to work in a variety of modes. These modes should be available in a flexible interactive framework, so one can always work effectively at the most comfortable level. In the GMCE, there are four such modes:

- *IDEAS Mode.* (IDEAS = Integrated Design Environment for All Systems), using the graphical menu/forms-style UI for basic CACE functionality, as presented in Section 17.4.2
- *M_Command Mode*, using GMCE supervisor commands when this expedites CACE work compared with the more user-friendly menu/forms interface
- *Package Command Mode*, using a core package's native commands when a needed result is not available via M_Commands
- *GMCE Macro Mode*, which includes both direct execution of macro files and a flexible macro-edit facility; macros contain M_Commands, Package Commands, or a combination thereof

IDEAS	Active	Command	Macro	Help	Trash	Exit

MEAD Model Browsing - Project = tallinn

Name	Classes	Type	Created	Updated	Notes	Results
☐ linpint	1	ABCD	24-NOV-1989	24-NOV-1989 09:34	Y	Y
☐ linpp105	1	ABCD	1-DEC-1989	1-DEC-1989 22:11	N	Y
☐ linpp64	1	ABCD	25-NOV-1989	25-NOV-1989 16:43	Y	Y
☐ nlpint	1,2,3	SIMNON	21-NOV-1989	26-NOV-1989 18:26	Y	Y,Y,Y
☒ nlppfbs	1	SIMNON	26-NOV-1989	27-NOV-1989 19:02	N	Y

ACTIONS	Descript	Results	Edit Note	Dele Note	Dele Class	Dele Mod	Done

Context: SIMNON tallinn nlppfbs 1

BROWSE	Edit Model	Update Class	Config Model	Create Model	Done

(b)

IDEAS	Active	Command	Macro	Help	Trash	Exit

MEAD Description Browsing - Model = nlppfbs 1

Name	Version	Designation	Created	Association	Refs	Notes
☐ mcpconn	2	CONNECT	27-NOV-1989 19:01	--none--	N	N
☐ nlpint	2	BLOCK1	26-NOV-1989 18:25	Linked	N	N
☒ statefb	2	BLOCK2	27-NOV-1989 18:39	--none--	N	N
☐ vardef	2	VARSDEF	27-NOV-1989 19:02	--none--	N	N

Disp Compt	Edit Compt	Edit Note	Dele Note	Disp Ref	Disp Link	Done

Context: SIMNON tallinn nlppfbs 1

(c)

Figure 4   (b) Model browsing form; (c) description browsing form.

The availability of a variety of interaction modes supports the inexperienced user as conveniently as possible (primarily via IDEAS) while providing the more experienced GMCE user with an extensible and effective environment for CACE.

The two GMCE command modes are accessed by clicking "Commands" in the Resource Bar depicted in Figure 3. The operating system may also be accessed via this button. These three options are included in the menu under "Commands"; clicking one of these selections opens a command-line area for work to proceed. GMCE supervisor commands are described in Ref. 8, and package and operating-system command languages are commercially documented.

Package Mode is currently only implemented for PRO-MATLAB because this package is more open than SIMNON, that is, it supports the generation of an arbitrary number of result types. This is generally true in comparing nonlinear simulation packages which produce a closed set of results (typically time-history files, equilibrium points, and linearizations) with linear CACE packages like PRO-MATLAB which can produce a wide variety of linear analysis results.

Basic DBM support is provided while using PRO-MATLAB in Package Mode by "layering" two M_Commands over the package commands:

- Data objects may be saved in the Result database for the Model by entering the command "MEADSAVE" that is intercepted and executed by the supervisor; for example, if the result comprises the arrays Q and K in the PRO-MATLAB workspace, then one may enter *MEADSAVE (Q, K, result => thisdata)* and the result will be stored in the DB with the user-supplied name *thisdata*.

- A data object may be saved in the Project DB as a *new model* via the "MEADMDL" command; for example, if the new system is represented in state-space packed form by *Snew (Snew = [Anew, Bnew; Cnew, Dnew])*, then the command *MEADMDL (ABCD, Snew, model => thismd)* accomplishes the goal, where *ABCD* is the type of model (type must be ABCD for continuous-time state-space models or DABCD for discrete time), and *thismdl* will be the name assigned to the model in the database.

Finally, the Macro Facility allows the user to streamline CACE by using custom macro procedures. Macros may be set up to initialize a GMCE session (e.g., to select a project and configure a key model), to perform a procedure defined by a sequence of M_Commands, to execute a task that may require the use of Package Mode, or to carry out a combination of these activities. The user interface facilitates macro invocation by providing a "Macro" button on the Resource Bar (again, see Figure 3) which when clicked produces a listing of all files with the extension "MMAC" for the user to designate and use. Macros may be invoked directly from storage, or they may be loaded into the editor, modified for the task at hand, and then executed. The following simple examples are GMCE macros for startup, for nonlinear simulation, and for evaluating the singular-value decomposition of a linear model:

Initialization Macro:
    Select_project (Tallinn)
    Configure (LINPLNT)
    Configure (NLPLNT#2)
("Tallinn" is the project name; the highest class of "LINPLNT" and class 2 of nonlinear model "NLPLNT" will be configured.)

Simulation Macro:
    Equilibrium (result => steady_state)
    Input (Ref, Step, {15.0, 1.0})
    Simulate (12., 0.01, result => REF_15)
(Find equilibrium and save; make input "Ref" a step of amplitude 15 units start-ing at $t = 1$; simulate for 12 time units with $dt = 0.01$ and save.)

SVD Macro:
    Set_mode (package, promatlab)
    a = unpack_ss(S);
    [u, s, v] = svd(a);
    Meadsave (u, s, v, result => svd)
    Set_mode (UI)
(Enter Package Mode using PRO-MATLAB; split out the *A* matrix; obtain the SVD; save result in the database as "svd"; reset mode to IDEAS.)

17.4.4 Package Unification

Another GMCE UI goal was to provide a totally unified interface for all core packages, with as little as possible left to the user's memory. This is also illus-trated in Figure 3, which shows the GMCE framework for simulation which is identical whether using PRO-MATLAB or SIMNON, except for unavoidable details such as the selection of a suitable integration algorithm when using a nonlinear simulation package such as SIMNON. This is facilitated by the fact that the M_Commands are the same wherever possible, despite the very differ-ent package interfaces.

17.4.5 Database Access

A UI should be designed to provide access to database management function-ality with minimal user overhead. If the user has to do a lot of extra work to use the DBM, then it is unlikely that it will be used. In fact, it proved to be possible to design the UI so that the DBM is an asset with respect to overhead, rather than a liability. This was due in part to devising a natural hierarchical database system organization, and in part to using the "object-oriented design" features in the User Interface as outlined in Section 17.4.1.

The first pivotal decision was that a query language would be excessively difficult to implement and use for DB access. In addition, the nature of the CACE database did not seem to require the typical capabilities of such systems. For example, "find all males over 2 meters tall" has few natural analogs in CACE databases. These factors led us to display data-element information via a hierarchical set of "Browsing Screens" as illustrated in Figure 4.

The second realization that streamlined the UI in relation to the DBM was that the Browsing Screens can be used for CACE functionality in addition to display. Thus, one may browse the models in a given project and immediately designate a model for use. This is called "configuring a model" and is done by hitting the "Config Model" button in Figure 4b. One may also create new models; edit models; purge models; and add, modify, or delete model notes from the same screen.

17.4.6 GMCE UI Limitations

The present platform for the GMCE is a VAX® computer[†] running the VMS® operating system, with the user interface displayed via a Tektronix® 4107 terminal (or higher model number) or an IBM® PC or PC clone running a Tektronix 4107 (or higher) emulator.[¶] This platform is adequate for the type of UI and functionality needed for the GMCE, although it is recognized that the UI could be faster and more flexible if a true workstation environment with high-resolution graphics and window management was used. We hope to port the GMCE to such a platform in the future.

17.5 DATABASE MANAGEMENT

The specific issues of database management that seem to be the most pressing in CACE are related to maintaining the *integrity* of the data base. Primarily, this involves being sure that the model used to generate a result can be identified with certainty and used again if necessary, being sure that the conditions used to generate each result are documented and knowing how models were obtained if they were generated numerically, for example, by linearization. This is a much larger task than simply knowing what is in file refinput_step2.dat in subdirectory [user.tallinn.nlplnt]! In addition, there are support functions such as on-line documentation that can add substantially to the value of the DBM.

Earlier generations of CACE packages provided little or no database management support. It was left to the engineer to decide how to organize data and track the relations among them. Often organization is based on storage, for

---

[†]VAX® and VMS® are registered trademarks of Digital Equipment Corp., Maynard, MA; IBM is a registered trademark of International Business Machines, Armonk, NY; Tektronix is a trademark of Tektronix, Inc., Beaverton, OR.

example, data for a project may be kept in a subdirectory or on a tape separate from data for other projects. Data files for a project may be distinguished by assigning "meaningful" file names. Some packages helped by "tagging" data elements of different types by using different extensions, such as the ".m" convention of PRO-MATLAB. One early package (CLADP) generated filenames by appending characters to a user-supplied "Run name" [2]. All such support is very rudimentary and leaves it entirely up to the user to shoulder the real burdens associated with maintaining the integrity of the database. The first effort to track models and results and to integrate DBM functionality with a CACE environment appears to be Ref. 1. However, the ideas of the complete hierarchy of projects, models, components, and results and of version control were not discussed.

Rigorous database management requirements for CACE were presented in Ref. 16. Database elements were cataloged and categorized, and the relations among them were established. Then an organization for these elements was devised. Two approaches for database access were considered: query language and browsing, as mentioned above. In each case, the CACE software user was the main consideration; this involved determining how the data elements are created and used, how the user perceives their relations, and features that are necessary for "doing the job right." Some of the features in the last area include version control for models that change over the course of a project, recording the conditions (parameter values, etc.) setup before a result is generated, and properly maintaining model components that are used in more than one model.

This line of thinking resulted in the design of the GMCE DBM, which was developed from first principles. Note that we made the pragmatic decision not to be concerned with the exact representation of each type of data element; instead, we used the data elements created or used by the core CACE packages as a de facto standard and only worried about content and format when required for purposes of interpackage compatibility. Work on this aspect of the CACE database definition may be found in Ref. 6.

17.5.1 CACE Data Elements

In terms of data-element categories, the controls engineer works with *models* that are comprised of *components* and a *description* (type, connection, etc.). Associated with each model there are *results*, for example, files containing frequency response data or time-history data. Models and results are often organized within *Projects* (e.g., project = "Tallinn" for the analysis and design example to be presented in Section 17.7). These considerations led to the DB hierarchy portrayed in Figure 5; this reflects the belief that control engineers naturally think of projects and models as being of paramount importance; all data elements produced during CACE activity are "children" of these entities.

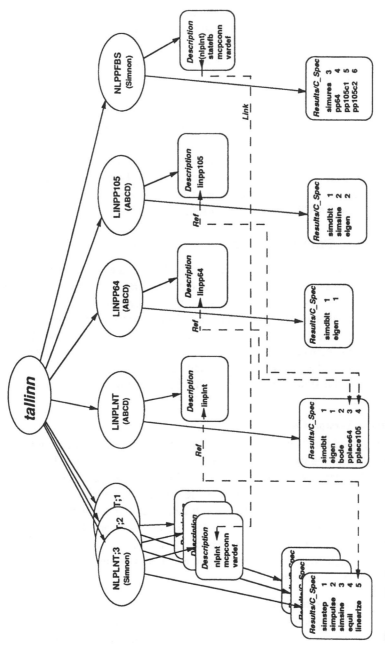

Figure 5 Data-element hierarchy for CACE.

The *Condition Specification* (C_Spec), also depicted in Figure 5, is an important secondary data type directly related to each result. This element contains information regarding operations performed on a model before a specific result is obtained; these include such actions as changing a parameter value, specifying an initial condition and/or input signal before performing a simulation, defining a frequency list used in obtaining Bode plot data, and so on. Condition Specs also record algorithmic conditions, such as setting a tolerance for controllability or observability, selecting an integration algorithm for simulation, and so on. Capturing this data is critical because it is the combination of model instance *and* Condition Spec that determines the result and thereby allows the engineer to document or repeat the result. Hence C_Specs must be kept in the CACE database so they can be recovered for any result that has been saved; this is an integral part of the GMCE database definition.

17.5.2   CACE Data Relations

Although the CACE database categories outlined above are few in number and simple, there are several factors that complicate the DBM task: Models tend to change over the lifetime of the project, some results are also models (e.g., linearizations of nonlinear models or transformed linear models) and components tend to be used in several models, yet they should be stored in one location to simplify their maintenance. In terms of full support for the controls engineer, mechanisms to handle all of these situations must be incorporated, and this should be done so that the corresponding DBM activity imposes little or no burden on the user.

CACE Model Version Control

The primary need for "version control" in the conventional software engineering sense exists in the model level of the hierarchy. The DBM must be able to keep track of system models that evolve over time, for example, as better modeling information becomes available or as preliminary modeling errors are corrected, so that each analysis or design result can be associated with the correct model instance. This requirement motivated the use of a tool that tracks each *version* of a model *component* so that version = 1, 2, 3, . . . refers to the original and subsequent variants of this component, and each *class* of a *model* that incorporates the component.

CACE Component Maintenance

The CACE DBM requirement for tracking models also gives rise to the need for *nonredundant* model management because maintaining the integrity of the Model level of the data base is nearly impossible if copies of various components are separately stored and maintained. The GMCE DBM supports this need via *linking,* which allows the engineer to maintain each component in one model (the "home" model) and use it elsewhere by bringing it out of the home model DB and incorporating it in other models.

CACE Model-to-Result Relations

One remaining relation that complicates the hierarchical DB organization is that which associates a linearization as a *result* obtained using a nonlinear model with a linearization used as a model *component*. The same situation exists with regard to linear model transforms. For example, one may create a reduced-order version of a linear model and desire to save this as both a result and a distinct model for further study. These associations are tracked in the DBM using a mechanism called the *reference*. The user may inspect a linearization result's reference to see if it exists as a component in any model; from the other perspective, a linear model component may be checked to determine if it was obtained as a result generated with a particular nonlinear model so the user can trace that result back to determine how it was obtained (e.g., at what operating point). The value of a linear model is greatly reduced if component traceability in this sense cannot be assured.

In summary, there is a natural hierarchical organization of projects, models, components, and results + condition specs in the GMCE database. Models are tracked over time via class number and version control. In addition, there are relations called links and references to completely maintain the integrity of the database.

17.5.3 Database Software

Several important decisions were made in selecting the implementation of the GMCE DBM:

- It was decided not to use a major commercial DBM package. Such a module would be too costly to license widely, most of the functionality would not be useful or suitable in filling the somewhat unique requirements for CACE, and some needed functionality might be difficult to obtain or unavailable.
- We did not, however, develop our own software for version control. One major concern is the efficient storage of multiple versions of models; without a scheme like that in DEC CMS that rebuilds each discrete version based on stored differences, large amounts of storage would be required.
- The ROSE package was used for low-level data-element relation management because this package was already a part of the UIMS CHIDE (see Section 17.4.2) and considerable software development was thus saved.

Based on these considerations, the GMCE DBM consists of an Ada-coded "front end" that calls CMS and ROSE to execute lower-level storage and version control activity.

17.6 EXPERT SYSTEM SUPPORT

There have been several major studies of augmenting conventional CACE software with AI-based support facilities. For reasons of space, we simply refer to the survey paper by James [3] for an overview of this activity. It is also important to note that there are a number of widely different styles or paradigms of expert-aided CACE, and that the selection is critically dependent on the needs of the user community. An inappropriate paradigm has the potential to detract from the effectiveness of the entire environment. These considerations are discussed in Ref. 11.

Adding an expert system shell (ESS) provides the basis for expert-aiding clear-cut but complicated and/or heuristic procedures that involve unnecessary low-level detail. One concept of expert-aided CACE was originally defined in Ref. 13; the primary difference in MEAD involved adopting a less ambitious model for expert aiding that makes the expert system the *user's assistant* [11] rather than putting it in charge of the CACE effort being performed. This change in perspective was motivated by the specific goal of providing support without getting in the control engineer's way and was the basis for the implementation in MEAD [15].

A second noteworthy feature of the integration of the expert system with MEAD is that the ESS interfaces with the supervisor in exactly the same fashion as the user working through the UI. The ESS outputs MEAD commands and/or package commands to the supervisor and gets the same return as the UI, that is, a result (if it is brief), a filename (for larger results), or messages (errors or information). This aids knowledge capture because this is exactly the output and input of the Supervisor if a user were to perform a given task.

The first CACE rule base to be built in MEAD was a reimplementation of the lead/lag compensator design expert system developed by James, Frederick, and Taylor [4]. The function of this rule-based system is to accept specifications for steady-state error coefficient, bandwidth, and gain margin and to design a lead/lag compensator to meet these specifications if possible. At the end of this task, the expert system performs a simulation of the step response of the designed control system. The user may inspect the result and accept the design if the response is satisfactory or iterate on the input specifications to obtain another trial design. The final compensated linear control system may be modelized if the user wants to use it for further study.

17.7 A CACE EXAMPLE

The following example illustrates the use of the GMCE on a small sample problem. Most of the salient features of the GE MEAD environment are illustrated in the course of this scenario. Note in particular that the database hierarchy por-

trayed in Figure 5 corresponds exactly to this example. The following outlines
the scenario and relates the steps to the various figures:

- Project "Tallinn" is created; see Figure 4a for the corresponding entry in
the Project's Browsing Screen. Work begins in that project by clicking "Select Proj."
- The nonlinear plant model "NLPLNT" is then created, through a process
guided by the user interface that involves defining its components and their
connection (see Ref. 14 for an example of this task). After this is done, the
model can be accessed via clicking on the corresponding entry in the Model's
Browsing Screen for project "Tallinn," as depicted in Figure 4b. (This screen
was captured after the whole scenario was done; hence the additional entries.) The single component "NLPLNT" is changed twice, to refine it suitably; this creates a total of three classes (see Figure 5). All work below is
done with class 3.
- The nonlinear plant "NLPLNT" is simulated thrice, equilibriated, and
linearized; the results and condition specs are listed in Figure 5 and the time-
history plot of the input "REF" and output "YPLNT" versus time is depicted in Figure 6. The linearization is "modelized" to create the linear
plant model "LINPLNT"—see the entries in Figures 5 and 4b; note that the

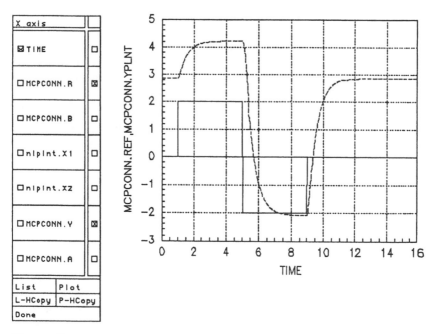

Figure 6 Time-history plot of the input "REF" and output "YPLNT" versus time.

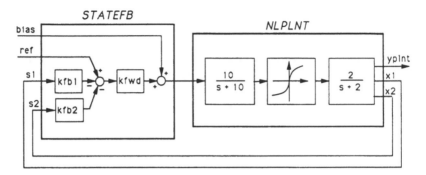

Figure 7 Nonlinear control system NLPPFBS.

"Reference" relation ("*Ref*") diagrammed in Figure 5 will permit this model to be traced back to the linearization result for the model "NLPLNT," and the condition spec will detail the exact conditions of linearization.

- Two linear control systems are synthesized from "LINPLNT" using pole placement with poles at $(-6 \pm 4)j$ and $(-10 \pm 5)j$, respectively. These are "modelized" to install the linear closed-loop models "LINPP64" and "LINPP105" in the database; see also the entries in Figures 5 and 4b. Note that two "Refs" diagrammed in Figure 5 tie these models to the results of the model "LINPLNT."
- The performances of "LINPP64" and "LINPP105" are checked via simulation (time histories not shown); simulation and eigenvalue analysis results are cataloged for these models in Figure 5. The model "LINPP105" was selected as the final linear design based on these results.
- The nonlinear control system "NLPPFBS" is created with gains corresponding to "LINPP105," as depicted in Figure 7. Note that this model uses the same representation of the nonlinear plant component by linking back to model "NLPLNT," as indicated by the "*Link*" relation diagrammed in Figure 5 and by the "Linked" designation in Figure 4c. The performance of this nonlinear control system is validated via simulation and the result is shown in Figure 8.

17.8 CONCLUSION

There has been a growing realization over the last five years that existing CACE packages may be reaching a good state in term of functionality and numerical power, but there are other areas of support that are required to achieve effective CACE environments for less-than-expert users and for use on large projects. Three pressing needs are for a more user-friendly user interface (to expand the

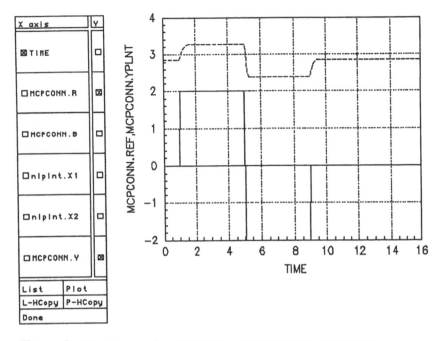

Figure 8 Time-history plot of "REF" and "YPLNT" for NLPPFBS

usefulness of CACE software to less expert users), database management (to rigorously track the many disparate data elements that are generated during the life of a major project), and expert-aiding (to alleviate some of the tedium associated with tasks that presently require a lot of low-level detail and a little heuristic logic). We have assessed and discussed requirements and implementation issues in these areas.

In terms of these features, the MEAD CACE Environments typify such advanced, more supportive environments for computer-aided control engineering. The most important novel features are their flexible user interfaces including a "point-and-click" interactive mode, two command modes (MEAD and Package), and a Macro Facility; their integrated database managers to rigorously maintain all data elements and relations; and the built-in expert system shell to serve as the user's assistant.

In terms of CACE functionality, the present GMCE is a basic CACE package for control system analysis and design. The higher-level functionality of PRO-MATLAB is available through the most user-friendly access mode; all lower-level primitives may be used via Package Mode. The major areas of non-linear simulation and analysis are also accessible using the GMCE graphical interface. A number of extensions and refinements are possible, including improved UI features, more user-friendly handling of linear models, additional expert-aiding, and porting onto a workstation environment. These are described in Ref. 17.

ACKNOWLEDGMENT

The work described above has taken advantage of many of the results of the USAF MEAD Project. Invaluable support has been provided by Aule-Tek Inc. personnel; the contributions of David Kassover, James Trojan, Michael Charbonneau, and Alfred Antoniotti are most gratefully recognized. Professor Dean Frederick of Rensselaer Polytechnic Institute provided extensive consulting and testing services for this project.

REFERENCES

1. Bunz, D., and Gutschow, K., CATPAC—an interactive software package for control system design, *Automatica, 21*, 209–213 (1985).
2. Edmunds, J. M., Cambridge linear analysis and design program, *IFAC Symposium on Computer Aided Design of Control Systems*, Zurich, Switzerland, 1979.
3. James, J. R., Expert system shells for combining symbolic and numeric processing in CADCS, *Proc. 4th IFAC Symp. CAD in Ctrl Systems '88*, Beijing, China, 1988.
4. James, J. R., Frederick, D. K., and Taylor, J. H., On the application of expert systems programing techniques to the design of lead/lag precompensators, *Proc. Control 85*, Cambridge, UK, 1985; also in *IEE Proc. D: Control Theory Appl., 134*, 137–144 (1987).
5. Lohr, P. J., CHIDE: a usable UIMS for the engineering environment. Tech. Report, GE Corp. R & D, Schenectady, NY, 1989.
6. Maciejowski, J., Data structures and software tools for the computer-aided design of control systems: a survey. (Plenary lecture), *Proc. 4th IFAC Symp. CAD in Ctrl Systems '88*, Beijing, China, 1988.
7. Moler, C., *MATLAB User's Guide*. Department of Computer Science, University of New Mexico, Albuquerque, NM, 1980.
8. Rimvall, C. M., and Taylor, J. H., Supervisor design for CACE package integration, *Proc. IFAC Symposium CADCS '91*, Swansea, Wales, UK, 1991.
9. Rimvall, C. M., Sutherland, H. A., and Taylor, J. H., GE's MEAD user interface—a flexible menu- and forms-driven interface for engineering applications, *Proc. CACSD '89*, Tampa, Florida, 1989.
10. Schmid, C., KEDDC—a computer-aided analysis and design package for control systems. in *Advances in Computer-Aided Control Systems Engineering*, (M. Jamshidi and C. J. Herget, eds.), Elsevier, North-Holland, 1985, pp. 159–180.
11. Taylor, J. H., Expert-aided environments for CAE of control systems (Plenary lecture), *Proc. 4th IFAC Symp. CAD in Ctrl Systems '88*, Beijing, China, 1988.
12. Taylor, J. H., and Antoniotti, A. J., Linearization algorithms and heuristics for computer-aided control engineering. *Proc. CACSD '92*, Napa, California, 1992; also in *IEEE Control Systems Magazine, 13*, 2 (1993).

13. Taylor, J. H., and Frederick, D. K., An expert system architecture for computer-aided control engineering, *IEEE Proc.*, *72*, 1795–1805 (1984).
14. Taylor, J. H., Frederick, D. K., Rimvall, C. M., and Sutherland, H. A., The GE MEAD computer-aided control engineering environment. *Proc. CACSD '89*, Tampa, Florida, 1989.
15. Taylor, J. H., and McKeehen, P. D., A computer-aided control engineering environment for multi-disciplinary expert-aided analysis and design (MEAD), *Proc. National Aerospace and Electronics Conference (NAECON)*, Dayton, Ohio, 1989.
16. Taylor, J. H., Nieh, K.-H., and Mroz, P. A., A data-base management scheme for computer-aided control engineering, *Proc. American Control Conference*, Atlanta, Georgia, 1988.
17. Taylor, J. H., Rimvall, C. M., and Sutherland, H. A., Future developments in modern environments for CADCS, *Proc. IFAC Symposium CADCS '91*, Swansea, Wales, UK, 1991.

18

ECSTASY: A Control System CAD Environment

N. Munro

University of Manchester Institute of Science and Technology
Manchester, England

C. P. Jobling

University of Wales at Swansea
Swansea, Wales

18.1 INTRODUCTION

Academic control groups have been developing CAD packages for control system analysis, design, and simulation since the late 1960s, when interactive computing became a practical reality. An interactive suite of design programs to implement Rosenbrock's Inverse Nyquist Array design method for multivariable control system design, known as CONCENTRIC, was implemented at the University of Manchester Institute of Science and Technology (UMIST) in 1971. This was rapidly followed by the Cambridge University package, known as CLADP, to implement MacFarlane's Characteristic Locus design method for multivariable control systems in 1973. Both packages also provided the well-established classical control design methodologies due to Bode, Nyquist, Nichols, and Evans (Root-Locus) for single-loop control system design. In the following decade, many other control groups also developed their own in-house packages, examples being the identification, simulation, and design facilities developed by Åström's group (at Lund, Sweden) in 1976, integrated by their interactive dialogue module INTRAC, and the federated design suite developed by Spang at GE in 1981, until by 1982 the DELIGHT system emerged to provide user-friendly on-line optimization facilities developed jointly by Polak (at Berkeley, California) and Mayne (at Imperial College, London). This latter facility is now widely recognized in the control academic world as an initial attempt to produce a software development support environment and led to

the EAGLES project at the Lawrence Livermore Laboratory in the United States
by C. Herget.

By 1983 various commercially developed CAD facilities had emerged, such
as MATRIX$_X$ (by Integrated Systems Inc., California; see Chapter 16) and
Control-C (by Systems Control Technology Inc., California). Both of these
packages lean heavily on the facilities contained in MATLAB (MAThematics
LABoratory; see Chapter 11), which was available in an early form in the public
domain about 1982 and then commercially in 1984 (developed by The Math
Workshop Inc., California). However, one major feature of control system design was either missing or poorly provided in these packages; namely, the system build and simulation requirement, which depends heavily on good
interactive graphical input facilities. Here, we are referring to the need to be able
to enter a high-level description of a system in a block diagram or schematic
diagram form and then automatically generate the linear or nonlinear equation
sets necessary for initial simulation studies.

In 1985, the Control and Instrumentation (C&I) Sub-Committee of the SERC
commissioned a survey of the then commercially available (and some well-
developed academic) CAD facilities, with a view to determining whether or not
any of these might satisfy the immediate future needs of the academic/industrial
control groups in the United Kingdom. It concluded, at this time, that no such
facility existed, partly for the reason mentioned above and partly because of the
poor integration of major components (i.e., the design, graphics, simulation, and
user-dialogue aspects) of many of the available packages. In 1986, a bold decision was taken by the C&I Sub-Committee to initiate and fund the development
of a new control system CAD environment, which became known as ECSTASY
(Environment for Control System Theory And SYnthesis). This would be developed using modern software engineering tools and would create a fully integrated set of facilities providing data entry and manipulation in a variety of
formats, system analysis tools, a wide range of design and synthesis methods, a
linear and nonlinear system simulation facility, powerful two-dimensional and
three-dimensional graphics facilities, a variety of user-dialogue modes, and a
structured database, all operating in a windowing environment on a range of
modern graphics workstations running under the UNIX operating system.

ECSTASY was intended to service both the present and future needs of
United Kingdom academic control groups and industry. The primary motivations for this software infrastructure were as follows:

1. To provide a common software base to assist the ready transfer of CAD
   tools between the various academic groups
2. To act as a means of transferring the control system design tools already developed in the academic world into industrial use in a consistent framework
3. To provide a common software base for the development and testing of new
   design algorithms and facilities.

18.2   THE ECSTASY SOFTWARE INFRASTRUCTURE

The proposed embryo infrastructure contained six major components, as shown in Figure 1, namely, a Man–Machine Interface (MMI), a Database and Data Management Facility, a set of Matrix Manipulation Tools, a facility to Manipulate System Descriptions in several mathematical forms, a Documentation/Report Preparation Facility, and Simulation Tools.

The Man–Machine Interface facilities provided consisted of both dialogue tools, allowing various dialogue styles such as "Question and Answer" mode, "Command" mode, and "Expression" mode, and various dialogue mechanisms such as keyboard, menu, forms, and button input. In addition, the MMI provided both graphical output of data, in the form of time and frequency response plots, and so on, and the input and output of system schematic representations. The user interface provided a windowing environment allowing several windows such as scrolling dialogue windows, graphics windows, and editing windows to be active simultaneously even on dumb terminals.

The graphics facilities implemented provided both two-dimensional and three-dimensional plotting capabilities with scaling, rotation, translation, panning, and zooming of images. This enabled the currently known graphics requirements for control system time-domain and frequency-domain plots to be readily achieved and also the generation and inspection of various three-dimensional surfaces. Graphics metafiles were provided to allow the archiving of displayed images and the subsequent recall of these images for incorporation into reports prepared on-line.

The Database needs of the control engineering community differ radically from those of the commercial world which are already catered for by various existing facilities of this type. Commercial data tends to consist of very large amounts of relatively simple data types, whereas control engineering data

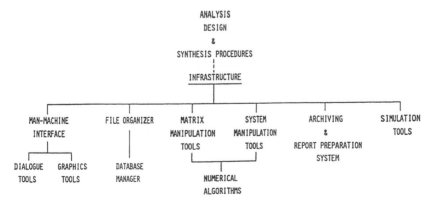

Figure 1   The embryo ECSTASY infrastructure.

consists of modest to large amounts of complex data objects with complex relationships between these objects. The Database facilities provided in ECSTASY
allowed a variety of primitive data types to be defined, and a range of operations
such as creation, deletion, renaming, and copying of data entities were provided. All data entities within the database had date-stamps, access-protection,
and relational pointers associated with them. The database acted as a consistent
central store of all data required by the other major facilities in the infrastructure and also of the data required by the design, synthesis, and identification
tools embedded in the infrastructure. This concept was a major aspect of the
ECSTASY facility.

The necessary Matrix Manipulation Tools, which form the kernel of the
current control system design and synthesis algorithms were provided by
PRO-MATLAB. This component provided a very wide range of easy-to-use matrix manipulations and also has a powerful macro language which allows new
functions to be readily created on-line. It has a user-friendly interface and can
easily be combined with various toolboxes, providing a wide range of well-
established single-variable and multivariable control system design, synthesis,
and identification tools. The PRO-MATLAB package, which uses the robust numerical subroutine libraries in EISPACK and LINPACK, was also interfaced in
ECSTASY to the SLICE/NAG subroutine libraries, making available various
optimization procedures and other algorithms, thus enhancing the functional capability of this component.

The System Description Manipulation facility implemented provided a means
of entering a system description graphically in block diagram form. The various
system components or subsystem blocks may be entered as transfer-function descriptions, state-space equations, or nonlinear differential equations and algebraic relationships, and may then be interconnected to create the desired system
structure by graphical editing actions. For control system simulation purposes,
this facility passed the necessary system information, in either linear or nonlinear equation form, to the simulation facility for time-response studies. This feature of ECSTASY was developed by the Control Group at Swansea University
and is described more fully in sections 18.5 to 18.9.

The Simulation Facilities implemented were provided by the simulation language ACSL, which is already widely used by both the academic world and
practicing engineers. This allows the easy interactive simulation of both linear
and nonlinear system models. It also provides steady-state finders and linearization procedures so that locally linearized models can readily be generated at
various system operating points. This facility was supplemented by another simulation tool called TSIM, which is widely used by various MoD contractors in
the United Kingdom aerospace industries. It was also intended that SIMNON,
the nonlinear/linear system simulation tool developed at Lund, Sweden, would
be integrated.

The Documentation facilities of the infrastructure provided a mechanism for the production of both on-line and off-line documentation. On-line documentation consisted of both simple "help," accessed by means of a hierarchically structured menu facility, as well as on-line access to the full user documentation which described the available facilities and how to use them. Off-line documentation was provided in two forms, namely, a printed version of the latter (i.e., for users of the facilities provided) and a separate printed manual giving internal details of the infrastructure facilities for researchers developing new algorithms and new facilities. The commercially available text-formatting and production system TeX with Preview and Postscript were used to provide this facility, although the FrameMaker equivalent was also available in the development version of ECSTASY.

18.3 DEVELOPMENT PHILOSOPHY

The initial implementation of the ECSTASY infrastructure was carried out on SUN-3 graphics workstations under the UNIX operating system, with further immediate implementations targeted for VAX machines running under UNIX and VMS and μVAX machines. Although some of the major components of the infrastructure were written in FORTRAN IV and some in the C language, all new code developed was written in C. The resulting C-code was verified using the UNIX LINT verifier to ensure transportability into other machines.

Considerable effort was devoted to the creation of a user-friendly Man–Machine Interface. This was intended to be consistent and easy to use with respect to each of the other major infrastructure components. The user interface also detected any relevant event in the user input which required the transfer of data from the ECSTASY database into the local datastructure of the facility being used, and vice versa. User output in graphical form was also intercepted, in particular from MATLAB and ACSL, and was passed for display by the graphics tool component built into the infrastructure. It was considered important that a high degree of flexibility was also provided in the ECSTASY infrastructure. Major components such as the database, graphics facilities, and simulation tools must be capable of being replaced or augmented by other such facilities as future needs demanded. The addition of one other simulation tool has already been indicated above. However, it was thought that it may be necessary at some point to have separate database facilities: one to manage very large datasets and one to manage small datasets in an efficient manner.

The usability aspects of such an infrastructure must be high; that is, it must provide for the implementation of existing control system analysis, design, and synthesis tools and also for the prototyping and development of new tools. It was also considered essential that a clearly defined mechanism was provided for the connection of foreign CAD packages, that is, other CAD facilities developed elsewhere by academic, industrial, or commercial concerns. It was equally important that the initial infrastructure was capable of evolution to take account of new developments in software engineering, computer science, and artificial intelligence. A significant effort is currently being devoted by the engineering community to the exploitation of expert systems design aids and the use of object-oriented programming environments. It is now more than likely that the next generation of control system CAD tools will be written in languages such as POPLOG or COMMON LISP, although even at the present time this remains a somewhat volatile area. Equally, the potential of object-oriented programming environments such as SMALLTALK and C++ will need to be considered, and the combined use of symbolic algebra facilities (such as MACSYMA or Mathematica) along with robust numerical procedures offers interesting possibilities for improved algorithm development.

18.4 THE DEVELOPMENT ENVIRONMENT

The initial project to develop an embryo version of the ECSTASY infrastructure was carried out at UMIST on a very short time scale, with delivery to the SERC of a facility suitable for β-test evaluation in early 1989. This could only be done by making use of a commercially available software engineering environment, and for this purpose, the Software Engineering Tools (SET) developed by the Computer Aided Design Division of PA Management Consultants were used. This environment is based on the concept of software building blocks, as shown in Figure 2, where the user is considered to interact with a desired set of applications programs primarily through a graphics display terminal.

The display interface is designed to handle a variety of both dumb and intelligent graphics display terminals of the DEC, Pericom, Tektronix, SUN, and

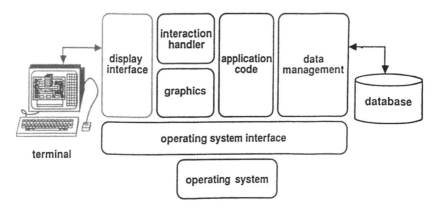

Figure 2 The SET software building blocks.

Apollo varieties. It is also responsible for providing the window management facilities on these devices, allowing a limited but useful windowing facility on dumb terminals. The interaction handler (InSET) looks after the decoding of the user input in a variety of forms. A wide range of interaction mechanisms such as on-screen menus, pop-up menus, forms, graphics tablet, mouse, function keys, and keyboard, as well as input from macros and command files, can be handled. The graphics facilities provided by the graphics module (GraphicSET) are fairly comprehensive, as mentioned earlier, and include two-dimensional and three-dimensional drawing capability together with lines, surfaces, text, and color. The database management facilities (DataSET) provided by the SET environment consisted of a data structuring system, which allowed the definition of various schemas to be used with different datasets, and a relational database management system. The latter is a comprehensive implementation of IBM's Structured Query Language which provides high performance while supporting large, complex datastructures.

An important feature of the SET tools was the operating system interface module OnSET, which buffers the implementation from the host operating systems. This feature was intended to provide the portability of the ECSTASY infrastructure onto a variety of other machines with different operating systems such as UNIX and DEC's VMS. The software engineering tools used provided a development environment which allowed both the rapid prototyping and testing of the user-interface and the database aspects of the infrastructure. Skeleton code-generators were provided which created both C-code or FORTRAN-code based on the interaction model defined by the user. This code could be readily compiled and executed to ensure that the desired behavior would be achieved before specific application code was added, thus accelerating the development, debugging, and documentation stages.

The embryo infrastructure created at UMIST using the software engineering tools described briefly earlier was interfaced to various control system design and identification tools to provide a more complete facility, as indicated in Figure 3, before being released for evaluation at selected academic and industrial sites in early 1989. However, the graphical input facilities of ECSTASY, developed at Swansea, were further subsequently enhanced to provide additional graphical editing features and the determination of plant model representations in symbolic form, using the symbolic manipulation language known as MACSYMA.

18.5 THE DEVELOPMENT OF THE GRAPHICAL INPUT TOOLS

Initially, it was intended that the user-interface components used in ECSTASY would be straight ports of existing from GKS into GraphicSET. However, this proved to be difficult, not least because of the significant differences between

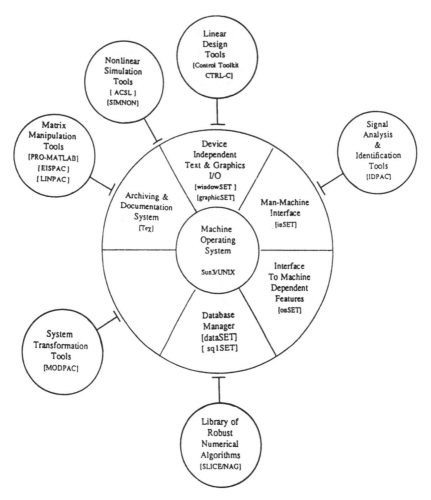

Figure 3   The ECSTASY environment.

the two graphics systems. In the end, a compromise was reached by completely rewriting the graphical interaction and display routines of CES (a Control Engineering workStation facility being developed at Swansea), while maintaining as much as possible of the datastructures of the original applications.

The main differences between the CES and the ECSTASY tools were as follows:

- The use of pop-up menus provided with the GraphicSET system. CES, because of limitations in GKS, makes use of hierarchial menu bars similar to those used in some text-based word processing packages.

- Icons are vector-based images in ECSTASY rather than pixel-based images. This is due to a restriction in PA's GraphicSET module, but it is also an advantage because it allows block diagrams to be easily scaled if this is required.
- A MacroEditor facility was not implemented in CES, although it had been specified. The subsequent and separate development of the MacroEdit facility helped to refine ideas later used in eXCeS (extended X-based CES).
- Dialogue boxes, used for parameter initialization, and so on, were not implemented in ECSTASY due to time constraints. The question-and-answer method of information input actually used was only intended to be a temporary measure.
- OperationEdit, a tool used in CES for inputting transfer-function and state-space models and for initializing certain types of nonlinearity, was not ported to ECSTASY. The standard ECSTASY tools for data input were used instead.
- There were no links to external packages required as this was provided by ECSTASY itself. However, this also meant that dialogue box and form-driven communication with simulation packages, which includes point-and-click selection of signals for plotting, was not provided from BlockEdit.

18.6 DIALOGUE MANAGEMENT

The dialogue definition and management facilities provided by PA's InSET module were not used in the development of the graphical input tools for ECSTASY. This was due to difficulties in combining the formal dialogue with graphical interaction. Instead, all graphics tools provided their own separate dialogues facilities.

18.7 INTERFACE TO ECSTASY

Because the initial versions of the graphical input tools were intended to be ports of CES, many of the original datastructures and file-management facilities were maintained in ECSTASY. In practice, this meant that block diagrams, icons, and icon buffer files were all maintained outside ECSTASY by the tools themselves. (OnSET functions were used to ensure portability of the resulting code.)

The formal links to ECSTASY were by means of ECSTASY extensions to the DataSET facility which ensured that a data object of the correct type was created whenever a block was added to a block diagram in BlockEdit. Initially, the object would only have its name and number of input and number of output fields set, unless it was a predefined block type in which case its default definition would also be recorded. The actual initialization of the block definition, for example, the transfer function or state-space matrices, was done from the ECSTASY transcript window.

To generate a simulation model for a given block diagram, the topology of the (flattened) block diagram was written out into SIMulation statements. These were presorted into execution order to ensure that simulation languages such as TSIM, which have no equation sorter, could be guaranteed to work properly.

Once the blocks have been defined, calling the SIM tool will ensure that the topology and functional data are combined to generate simulation code. The model can then be simulated and if necessary linearized to provide links into the analysis tool of ECSTASY. To provide for closer integration with the ECSTASY database, additional records were defined and used by BlockEdit. These were a system definition record (which was used to record the name of the block diagram file associated with a current model), a macro record (which recorded the name of any block diagram files which were subsystems of the current model), and a connection record used to record the connections between blocks in a block diagram. It was intended that these records could be used by a subsequent replacement to the SIM facility to generate simulation codes directly from the database and, hence, eliminate one of the sources of inconsistency between block diagrams and the ECSTASY database.

Another source of inconsistency could be eliminated by providing a form-based means of defining block functionality from BlockEdit itself. In the current version, blocks are most conveniently defined by means of an ECSTASY macro which is executed just after loading a block diagram. However, there is no guarantee that the definitions in the macro file will match the blocks in the diagram, particularly during definition and refinement of a model. It would be far better and safer to record the functions associated with a particular diagram with the diagram itself.

The BlockEdit Tool implemented consists of two areas: a design area, which is used for constructing block diagrams, and an icon buffer, which is a repository for dynamic blocks. The contents of this latter buffer may be changed by use of the "buffer" menu which loads any of 12 built-in sets of icons. Alternatively, the user may load any icon into any of the icon boxes and so customize BlockEdit to his own requirements or preferences. Customized icon sets can be saved and reloaded from the "buffer" menu. A menu title bar was provided to house the pull-down menus that control BlockEdit. A mouse-prompt window was also provided to indicate the effect of pressing any of the mouse buttons.

The IconEdit tool (shown in Figure 4) has a layout similar to BlockEdit. The design area is used to construct a new icon. An actual size representation of the new icon is shown in the box at the bottom left of the buffer area. In IconEdit, the buffer area is used to store partially complete icons or icons which have been loaded for modification. Again, the mouse-prompt window is used to indicate the state of the tool and the rotate-port box is used to control the orientation of ports on the icon. The "attributes" menu is used to assign the default functionality to a given icon. Icons can be used to represent most of the built-in system

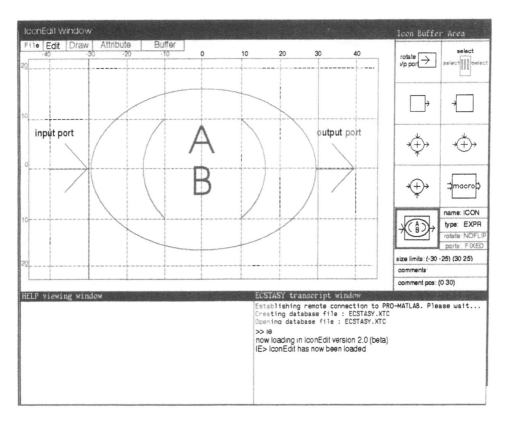

Figure 4 The IconEdit tool.

data types or they can be used as EXPR types, in which case they would be used for user-defined functions.

Figure 5 shows a hierarchial block diagram which represents a position control system. The plant (M6) contains a subsystem consisting of an amplifier, DC motor (itself contained in a further macro M4), gearbox, and integrator. The DC motor (M4) contains a model of an armature-voltage-controlled DC motor with load.

18.8 MacroEdit

MacroEdit can be used to create hierarchial block diagrams, either by top-down refinement or bottom-up reuse of existing models (or a combination). In top-down mode, the user creates a block diagram in BlockEdit as usual, but at some point he adds a special macro block which stands for a subsystem. At some point later in the design, the user opens this macro block which creates a new drawing area into which the subsystem block diagram can be created. To ensure

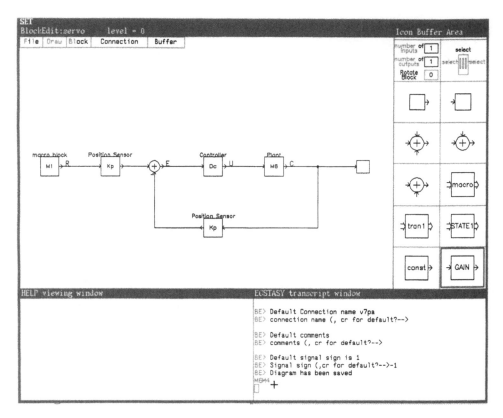

Figure 5   System hierarchical block diagram.

consistency, this new diagram will have the same number of input-output terminals as the original macro block's ports.

For the bottom-up definition, an already existing block diagram is imported into a macro block and its input-output terminals are linked to those of the new parent by means of a split screen as shown (Figure 6) with the supersystem on the left and subsystem on the right. This display is used to enable the user to assign connections in the subsystem to the corresponding ports in the macro block.

Having completed a diagram, "write Sim" is selected from the "File" menu. This creates the Sim code shown in Figure 7. BlockEdit is then closed down; the blocks are defined and then the Sim facility is used to create a simulation-code version of the model.

There is theoretically no limit to the number of levels that are possible in a hierarchial block diagram; the only practical limit is one imposed on OnSET

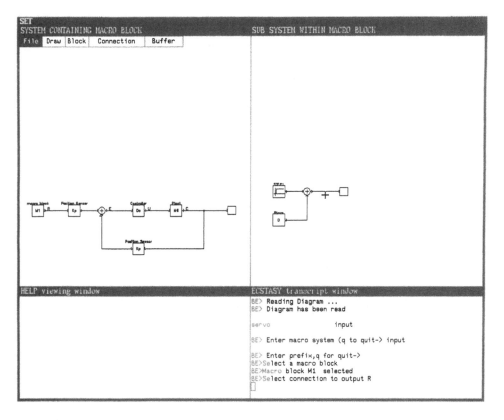

Figure 6 A special "two-up" display.

which limits file names to eight characters plus extensions. As macro block files were named by concatenating as in

M1 M5 M6 . . .

the nesting is effectively limited to four levels. This limit could be removed by use of directories or some other structuring scheme.

When SIM code is generated from a block diagram, the hierarchy is flattened by recursion.

18.9 IconEdit

A facility called IconEdit was provided to enable the user to extend the standard facilities of ECSTASY. Currently, any picture may be assigned to any EXPRession data type, which can then be used in a block diagram. At present, only

```
%      servo.x
%      A sim file generated by BlockEdit
%      for me on 21/3/1990 at 16:16:33
sim acsl serv
STEP1>{ v3 }
{ Rtrim = 0 } >[1]> {v4pa}
{ v1pa=U
{ C=thl }
{ v1pa }>Ka>{ Va }
{ v1=Va }
{ wm=v6pa }
{ Ia }>Kma>{ Qe }
{ Qe }>Ye>{ v6pa }
{ v6pa }>Kma>{ v2 }
{ C }>Kp>{ v7pa }
{ v4 =   + v3 + v4pa }
{ wm }>Ir>{ wl }
{ wl }>I1>{ thl }
{ v3pa =   + v1 - v2 }>Ya>{ Ia }
{ R=v4 }
{ R }>Kp>{ v2pa }
{ E =   + v2pa - v7pa }>Dc>{ U }
end
```

Figure 7   Sim code for BlockEdit example.

EXPR types, transfer-function models, or state-space models can be assigned to a user-defined icon. This was due to a limitation in the SIM translation facility.

The standard set of ICONS provided consisted of some 57 icons arranged into 10 sets. These standard sets can be user-customized to reflect usage or preferences.

18.10   FUTURE DEVELOPMENTS

The graphical user-interface facilities are far from perfect, but they are preferred by most users to the text-based facilities such as SIM, which they were intended to replace.

There are, however, a number of features which need further work. The main ones are as follows:

- Provide a safer link to the database. BlockEdit was added to ECSTASY some time into its development. As such, its development was somewhat constrained by earlier design decisions. The database and code-generation facilities of ECSTASY would be different had BlockEdit been designed-in at the start.

- Ensure consistency. The most dangerous operation in BlockEdit is to load a new diagram into the database. Because of the way the ECSTASY database works, this could have the effect of deleting records due to name clashes or forcing functions to be reinitialized. The need to keep functions in step with block diagrams needs to be addressed.
- All names must be unique, even in a hierarchial model. Variables in a subsystem ought to be local not global, particularly if models are to be reused effectively. Some form of name-space for block diagram data, separate from the main ECSTASY database, would be useful.
- It should be possible to generate a simulation file directly from the database. The SIM language implemented in ECSTASY provides an extra source of problems in this respect.
- Forms and dialogue boxes should have been used to define functions in ECSTASY. They would also be good for controlling simulation runs in the context of the block diagram.
- It would have been nice to have linear block diagrams generating equivalent SYStems records without the need to go into the simulation tool in ECSTASY.

The control group at Lund University, under Professor Karl Åström, were at this time also developing a high-level graphical schematic-input facility, known as Hibliz, in C language and Common Lisp and had indicated a willingness to connect this facility to the ECSTASY infrastructure. It was equally considered important during this development that the emerging tools for "robust control system" design such as Postlethwaite's "Stable-H" package should be connected to the infrastructure. Various other control system design methodologies might well have also been most readily connected to ECSTASY as foreign packages in the first instance. However, there would then inevitably be redundant aspects in this approach because there was bound to be duplication of the man–machine interface, graphics, and numerical facilities. A second phase of development was therefore intended to provide the full integration of these tools into the infrastructure. This would have required significant development effort by the groups concerned.

Recent research in the United States, in particular that being done at the Systems Research Centre, at the University of Maryland had indicated that known difficulties with certain design and synthesis algorithms could be significantly improved if the operations concerned could be reliably performed in symbolic form. Well-developed symbolic algebra packages, such as MACSYMA and Mathematica, provide such tools. The use of MACSYMA, as indicted earlier, would provide symbolic model construction and simplification facilities. It would also allow the use of more compact polynomial matrix algebra tools in the development of analysis, design, and simulation facilities.

Research into database facilities for control engineering was, at this time, being carried out by Maciejowski at Cambridge under the SERC Control & Instrumentation Committee's CDTCS (Computing and Design Techniques for Control Engineering) Programme, and, in particular, the possible advantages of using object-oriented programming environments, such as SMALLTALK or KEE were being explored. However, a workshop organized by the SERC's Computing Facilities Technical Advisory Group indicated that there was then no clear view of whether the object-oriented approach offered distinct advantages for engineering applications of databases over the more established Sequential Query Language approach. Nevertheless, if the ECSTASY infrastructure was to take advantage of current and future developments in the area of expert systems, there was a clear need to pursue further research into Knowledge-Based Systems and recent developments such as the so called "Blackboard Architecture." To explore these areas, it may have been necessary to introduce further major components such as the POPLOG environment developed at Sussex University, SMALLTALK, KEE, and possibly ART (the Automated Reasoning Tool). All of these later facilities would, in addition, allow immediate access to modern powerful man–machine interface tools and facilities which would enable appropriate expert systems "aids" to be explored and developed.

Several other enhancements were being considered with respect to the simulation tools in ECSTASY, which would meet the needs of the electrical power generation and chemical engineering industries. The then CEGB had indicated a willingness to collaborate in the connection of their powerful Plant Modelling and Simulation Program (PMSP) to ECSTASY, and ICI likewise with respect to their Flowsheet Program. It was also important that the simulation facilities to be implemented in the ECSTASY infrastructure should address the needs for discrete-event simulation, which is now becoming extremely important with respect to "safety critical" applications.

There would equally have been great value in implementing the optimization procedures developed jointly by Mayne and Polak at Imperial College and Berkeley University, respectively, and the multiobjective function approach developed by Fleming, then at Bangor University, in the ECSTASY environment. The nonlinear system analysis and design facilities developed by Atherton at Sussex were also considered to be a valuable component of the control engineer's toolbox, and the self-tuning regulator design facilities, created by Åström at Lund, Clarke at Oxford, and Wellstead at UMIST, were then becoming well-established in industry and formed further essential system control techniques.

18.11 CONCLUSIONS

The ECSTASY development started in late 1986 and was completed by the UMIST/SWANSEA development team to specification in December 1990.

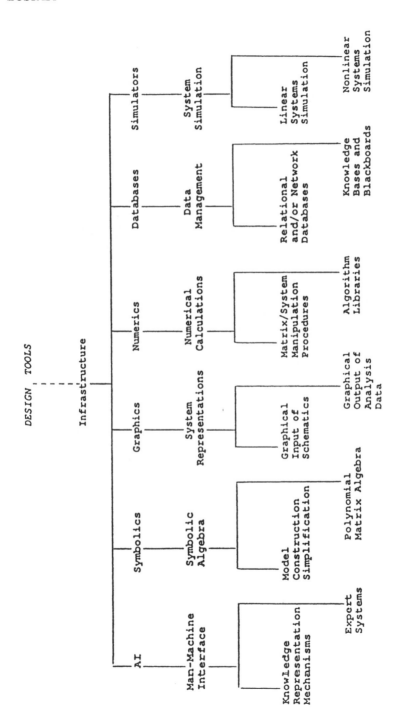

Figure 8 A future form of infrastructure.

During the latter stages of this project, various α- and β-releases were made available to United Kingdom control and industrial groups for testing by its application to real industrial problems. Also, during the latter stages of this project, the essential graphical block diagram input facility for system build (described earlier) was separately developed by Jobling and colleagues at Swansea University and then integrated into the ECSTASY environment at UMIST. The Swansea Control Group had already been developing such a facility using symbolic processing techniques as part of their own CAD activities in the area of system modeling and simulation using block diagrams and signal-flow graphs.

However, in early 1991 the Control and Instrumentation Sub-Committee in the light of recent developments in commercially available software (mainly MATRIX$_X$ which had been progressively significantly improved, particularly in terms of its system build facility, and also, by then, PRO-MATLAB had its new Multivariable Frequency Domain Toolbox and had become more widely available in an optimized PC form) decided to terminate the ECSTASY development. On reflection, it would seem that this was the correct decision because these events and other developments in computer science were beginning to indicate new possibilities for the development of control system CAD environments. These were already becoming very large and complex with consequent serious problems in the maintenance and update area for developers, and in the usability areas for users. The essential help and guidance facilities for the unfamiliar or infrequent user had not always kept pace with the implementation of more and more tools. Expert systems concepts had begun to mature, and there seemed to be a strong need to bring these to bear on the user-friendliness and user-guidance aspects of control system CAD facilities. Object-oriented programming, and the more modular and independent code construction ensuing from the use of this approach, now offered significant advantages for the developers of CAD facilities. Smalltalk-80 was also widely available and well-developed for use on graphics workstations and C++, with its object-oriented preprocessor, was available on workstations and PCs.

Figure 8, developed by one of the authors during a period spent at the University of Maryland, indicates a possible future form of this type of infrastructure. The introduction of the more modern computing components indicated into ECSTASY would have had to be carefully considered and properly phased so that a "bumpless transfer" was achieved.

REFERENCES

1. Rosenbrock, H. H., *Computer Aided Control System Design*, Academic Press, New York, 1974.
2. Patel, R. V., and Munro, N., *Multivariable System Theory and Design*, Pergamon Press, Oxford, 1982.

3. Pang, G. H., and MacFarlane, A. G. J., *Expert Systems Approach to Computer-Aided Design of Multivariable Systems*, Springer-Verlag, New York, 1987.
4. Munro, N., and Bowland, B. J., *The UMIST Computer Aided Design Suite Control System Centre*, University of Manchester Institute of Science and Technology, Manchester, 1981.
5. Edmunds, J., *Cambridge Linear Analysis and Design Programs*, Control Engineering Group, University Engineering Department, Cambridge, 1981.
6. Weislander, J., *Interaction in Computer Aided Analysis and Design of Control Systems*, Department of Automatic Control, Lund, Sweden, 1979.
7. Spang, A., *Workshop on Control Design*, GE Research and Development Center, Schenectady, NY, 1981.
8. Polak, E., Interactive software for computer-aided-design of control systems via optimisation, *20th IEEE Conference on Decision & Control*, San Diego, California, 1981.
9. Lawyer, B., and Poggio, P., *EAGLES Requirements*, Engineering Research Division, Lawrence Livermore, Livermore, CA, 1985.
10. Barker, H. A., Chen, M., Townsend, P., and Harvey, I. T., CES—A workstation environment for computer-aided design in control systems, *Proc. 4th IFAC CADCS*, Beijing, China, 1988.
11. Barker, H. A., Chen, M., Grant, P. W., Harvey, I. T., Jobling, C. P., Parkman, A. P., and Townsend, P., The making of eXCeS—A software engineering perspective, *Proc. IFAC Symposium on CADCS*, Swansea, United Kingdom, 1991.
12. Goodwin, A. J., and Jobling, C. P., *BlockEdit Reference Manual*, SERC, 1990.

19

CADACS: An Integrated Design Facility for Computer-Aided Design and Analysis of Control Systems

H. Unbehauen

Ruhr-University Bochum
Bochum, Germany

19.1 INTRODUCTION

The idea to realize complete development cycles for control system design through the use of digital computers as a primary tool during the identification, modeling, and design phase until the final implementation in a direct digital controller (DDC) goes back as far as the early seventies. One of the first examples of a full-cycle computer-aided design control system (CADCS) package was completed in 1975 by the author and his co-workers [5]. This system, called KEDDC (combined developing and direct digital control system), had already been planned in 1973 to combine works and results of different research projects, guided by the author, with the objective to provide an easy access to advanced methods of system identification and modeling, as well as sophisticated controller design methods—for instance, adaptive controllers—for control engineers. The well-adopted concept of this CADCS package has since been extended systematically by many projects in the following years. Since 1983, the university version KEDDC [6] has been modified and elaborated on for the ultimate goal of a commercial version CADACS, which is used today by around 35 industrial enterprises and 40 educational institutions. CADACS is available today for mainframe, minicomputer, PC, and real-time workstation configurations. As an open system, it is still growing in its broad scope.

According to very recent detailed comparative studies [2,4] CADACS is one of the most comprehensive CADCS program packages current available. The kernel system is coded in portable Fortran 77. Interfaces exist for many different computer types and operating systems. Subsystems are running on PCs or dedicated DDC-stations in HVAC process control systems. It uses simple commands in combination with question-and-answer, menu-driven and form-driven modes. It supports window techniques as well as parallel sessions. For industrial and real-time applications, it offers better guidance than most of the other products from Research and Development Centers. Coupling with special developing tools for control algorithms, for example, PC-MATLAB or MATRIX$_X$, is provided.

CADACS has been designed to cover a broad area of control engineering tasks. The modular program structure comprises, besides classical and modern methods for analysis and synthesis of SISO and MIMO control systems, simulation as well as implementation of control algorithms in real time. The comprehensive library of methods, the flexible facilities of graphical display and archiving, and the state-of-art numerical algorithms are but a few striking features of a CADACS workstation. The application system is tailored for the general user; however, it can be extended by special user-defined modules through use of the library of source program options. The real-time toolbox offers, as an extension to the test operation of controllers designed with this system, the development of new controller software.

This chapter gives an overview of the integrated facilities for analysis, modeling, design, and real-time implementation of the CADACS system. The general description of the system shows the ideas and methods standing behind it. Furthermore, it will be stressed that, from a practical point of view, the design and implementation of a control system should be integrated in a user-friendly comprehensive engineering tool. It will be shown that this system is able to support all classes of methods and algorithms during the stages of design and implementation. The flexibility of this system is obtained by a unified software frame which is independent of the hardware or operating systems.

19.2 FUNDAMENTAL IDEA AND STRUCTURE OF CADACS

CADACS is an integrated development package for analysis and synthesis of control systems which has been developed by the Automatic Control Group in the Department of Electrical Engineering at Ruhr-University Bochum. It supports all phases of control engineering, especially:

*   System identification (on-line and off-line)
*   Control system analysis, design, and simulation
*   Real-time implementation of control algorithms, including adaptive control
*   Documentation of design and analysis as well as real-time control operation.

The system is based on a large number of modern and classical approaches for single-input/single-output and multivariable systems in the frequency and time domain, as well as in the state-space representation. It is characterized by user-friendly handling, modern graphical output facilities, and numerics. Based on a unified concept, CADACS can be implemented in various configurations, forms, and sizes, where additional user-specific parts may be integrated in a simple way.

The development of the system, which is not only available as a PC version, was motivated by the idea of generating a tool which renders possible the use of a manifold of methods for the control engineering expert, as well as for the non-expert in industry and at university, in a unified form.

Figure 1 provides the block diagram for the main parts of the CADACS system, and their interfaces with the applications environment to research and development. These parts are a set of interactive non-real-time programs for analysis and design, a database, an extensive program library, and real-time programs interfacing with experimental hardware. Access to a common database will, of course, improve communication between research groups working on a common project. In particular, access by all users to modeling and design procedures can help to improve the interfacing between system analysts and control designers. CADACS also forms a common software base, hence reducing duplication of program development. It is organized as an open system, parts of which may be added, updated, or removed at any time. This unlimited extendability is of particular importance in an applied research and development environment with on-going new methods and programs. Implementation of existing real-time programs will allow immediate implementation of control algorithms on the same computer or on a slave one, coupled to a mainframe computer, minicomputer, personal computer, or real-time workstation. Controller modules may be integrated into DDC substations of industrial control systems.

CADACS is interactively controlled by an efficient command-driven dialogue, featuring "learning" defaults which adjust to the last user input. Menu, help, and on-line manual facilities provide a friendly environment for a novice, without impeding the expert. Versatile graphics complement the numerical capabilities.

A hardcopy of the user dialogue and graphic output provides complete documentation of the work performed with CADACS. Sessions do not require any planning or programming and can be interrupted at any time. All data can be stored on files for future reference and for building a project database. CADACS encourages the user to thoroughly comment all data on permanent files.

In a workstation environment, CADACS supports window techniques, thus assisting an engineer concurrently working on different tasks [1]. Bit-map displays can provide the user with the capability of displaying multiple windows which represent text and graphics simultaneously. Consequently, the user environment looks actually like an engineer's desk.

CADACS supports a wide variety of control engineering methods. It offers tools for systematic design strategies and helps the engineer to combine his

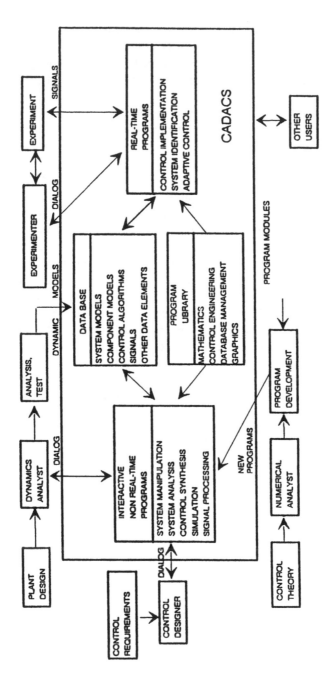

Figure 1 Main components and application environment of CADACS.

intuition with the results of high-accuracy numerical solutions. In addition to modern analysis and design concepts, classical methods are also implemented and may be used in combination with modern control techniques.

19.3 DESCRIPTION OF THE PROGRAMMING SYSTEM

19.3.1 Types of System Representation

Often-needed subtasks for analysis and synthesis are supported in a centralized way. This is valid, for example, for managing numerical system representation forms and transformation from one form to another. In CADACS, 10 different types of representing a system in frequency and time domain are available. Identification, simulation, and, also, approximation are considered as transformation to a different system representation. All paths which are available for transformation, using CADACS, are illustrated in Figure 2, where the numbers define the following characteristics:

1. Sequences of signals of deterministic or stochastic input and/or output signals
2. Discrete values of the autocorrelation and/or cross-correlation functions of two signals
3. Discrete values of the impulse response

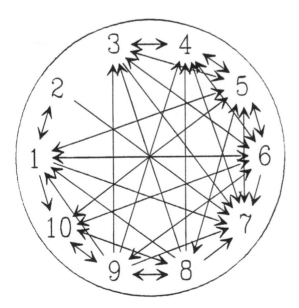

Figure 2 Types of system representation and possible transformations in CADACS.

4. Discrete values of the step response
5. Transfer function (matrix) in s-domain
6. Transfer function (matrix) in z-domain
7. Discrete values of the frequency response (frequency response matrix) or spectra
8. Continuous-time state-space description
9. Discrete-time state-space description
10. Polynomial matrix fractions in s- or z-domain

A large number of alternative ways for modeling, computation, transformation, and simulation offer a high degree of flexibility, which is necessary to handle the broad spectrum of tasks occurring during control engineering projects. For instance, the result of the modeling phase can be transformed to a representation which is better suited for controller design.

19.3.2 Database

According to the different types of system representation and components of dynamic systems, the database offers for each type of control engineering fundamentals, a specific binary type of file. In addition, further supplementary fundamentals are necessary. These exist for signals, transfer functions (matrices), coefficient matrices, state-space systems, frequency tables, frequency response tables, spectra, description of the graphical layout, structure descriptions, and control engineering objects in textual format.

19.3.3 Tools for Basic Functions in Analysis and Synthesis

Frequently used standard tasks of moderate size, like transformations to different representation forms, database definitions or basic calculations are grouped into several central interactive modules, so-called "managers," which together form the core system of CADACS. There are about 300 commands implemented in the core system which play the role of elementary utility functions. The core system is augmented by individual modules for particular tasks and methods. The following "managers" are available.

Signal Manager (SDISP)
Handling of signals and computation of signal characteristics:

* Input and output of signal data, signal editor, graphical representation of signals
* Generation of signals and sequences of signals from measured signals or standard ones such as step, ramp, impulse, PRBS, white noise, and so on
* Analysis of signals, for example, correlation analysis
* Operations with signals ($+$, $-$, $*$, $/$)

System Manager (SMGR)

Handling and analysis of transfer functions in s- and z-domain:

- Input and output of transfer functions: transfer-function editor
- Alphanumerical and graphical pole/zero analysis, cancellation and elimination of poles and zeros, decomposition into partial fractions
- z-transformation and inverse transformation; w-transformation and inverse transformation
- Calculation of z-domain transfer functions for change of sampling time
- Sensitivity analysis with and without feedback connection
- Test on stability
- Calculation of loss functions
- Parallel, series, and feedback connection of two systems described by transfer functions

Frequency Manager (FRMG)

Handling and analysis of frequency response tables, entries of frequency response matrices or spectra:

- Input and output of frequency response tables: frequency and frequency response editor
- Computation of the frequency response from transfer functions, input and output signal, step response, impulse response, and state-space description
- Display of Bode, Nyquist, and Nichols charts with and without automatic step size
- Calculation and display of principal gains
- Inversion of frequency response tables, dead-time correction, Popov's chart
- Determination of amplitude and phase margin
- Calculation of step response from frequency response
- Parallel, series, and feedback connection with other systems

Matrix Manager (MMGR)

Handling of matrices of continuous- and discrete-time systems in state space:

- Input and output of matrices, matrix editor, generation of submatrices and hyper ones
- Computation of matrix properties, such as rank, determinant, singular values, condition number, eigenvalues, eigenvectors, characteristic polynomial, different norms, and definity
- Operations with matrices ($+$, $-$, $*$), inversion, balancing
- Solution of linear underdetermined, determined, or overdetermined systems of equations $AX = B$
- Solution of linear matrix equations of the type $AX + XB + C = 0$

- Alphanumerical and graphical pole-zero analysis (decoupling and invariant zeros)
- Three-dimensional net diagram and structure display of matrices and systems
- Computation of discrete-time from continuous-time systems, and vice versa
- Transformation to (quasi-) canonical forms (17 types) as, for example, observer form, controllability form, Jordan form, and so on
- Computation of the transfer-function matrix from state-space description
- Minimal realization of a state-space description or of a transfer-function matrix
- Solution of pole-placement problems (three methods)
- Test on controllability and observability with computation of Gramians
- Decoupling analysis
- Parallel, series, and feedback connection with other systems

Polynomial Matrix Manager (PMGR)

Handling of polynomial matrices of continuous- and discrete-time systems described by polynomial matrix fractions (PMF):

- Input and output of polynomial matrices, polynomial matrix editor, definition of left and right PMFs
- Operations with polynomial matrices ($+$, $-$, $*$)
- Determination of matrix properties such as rank, determinant, row and column degrees, determinantal degree, properness, and coprimeness
- Inversion of a polynomial rank
- Dual coprime factorizations, left PMF from right PMF and vice versa
- Computation of PMFs for a state-space description or for a transfer-function matrix
- Inversion of transfer-function matrices
- Realization of PMFs in state-space description
- Determination of greatest common divisors
- Alphanumerical and graphical analysis of poles and zeros (invariant zeros)
- Solution of diophantine equations

Graphics Manager (GRMGR)

The Graphics Manager is a centralized graphical representation of control engineering objects in diagrams. All tasks for graphical representation within CADACS are handled by the Graphics Manager. This module can be seen as a graphical device driver, which is able to transform the graphical information generated by CADACS programs into diagrams. The principle of operation of the graphical display is depicted in Figure 3. The information to be represented graphically, including a diagram request, is put into a mailbox. The Graphics Manager takes this information and maps it to a specific diagram. Besides standard diagrams, such as a Bode chart and a Nichols chart, a large number of further diagrams is available. The user defines, by a so-called problem menu,

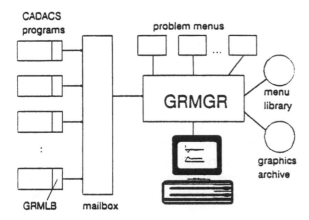

Figure 3 Function principle of the "Graphics Manager."

the form and type of graphical representation for his problem-oriented data. For standard representations, a menu library is available which can be extended for each individual user or a specific application. For that purpose, a diagram definition language is available. This form of centralized organization of graphics display allows different programs to map clearly and in a unified way into a single diagram or to map results on a single graphics working sheet in several different diagrams. Therefore, results of a program for analysis and design may be mapped together with simulation results which were generated by standard simulators or using real-time simulation.

Documentation Manager (DOKMG)

Documentation of signal sequences generated by real-time experiments:

- Storing of experimental data (signals, parameters, etc.) in real time
- Graphical representation and signal monitoring in real time (in combination with GRMGR)

Cross Manager (XMGR)

This manager has the task of changing the binary format of the CADACS internal database into text format, and vice versa, for data exchange between different CADACS systems or for communication with other software systems. Besides the CADACS text format, the generation of binary files for PC-MATLAB™ and MATRIX$_x$™ is fully supported.

19.3.4 Tools for Analysis and Synthesis

For more complex tasks in analysis and synthesis, specialized tools in the form of closed programs are provided. Complete techniques are, therefore, compactly grouped together in a single module, which allows, initiated by the user, one to carry out the computer-aided analysis or synthesis in whole. These programs often offer special facilities for simulation, necessary for verification and judgment of the results. This does not exclude, due to any necessity for extended simulation, the use of standard simulation tools in the course of analysis and synthesis. The special program architecture of CADACS renders this possible.

Programs for Identification

PTAPP   Approximation of step responses by lag elements with user-defined order or with system-order detection: six different methods and models types. Direct and indirect curve-fitting techniques

PRONY   Approximation of step responses by exponential functions with order detection. Laplace transformation and determination of transfer function

UNEK   Approximation of frequency response tables or spectra by rational functions. Model reduction in frequency domain. Direct and indirect curve-fitting techniques.

IMIMO   Parameter estimation for MIMO systems using left-coprime PMDs and different numerical algorithms (recursive and nonrecursive)

MLID   Parameter estimation for MISO systems using least-squares and/or maximum-likelihood technique

BALRE   Model reduction by balancing transformations, using different criteria and algorithms. H_∞ model reduction.

HBAL   Harmonic analysis of nonlinear systems, using standard or user-supplied nonlinearities.

NLID   Parameter estimation for nonlinear systems of Hammerstein, Wiener, or Kolmogorov-Gabor type, including structure selection

Programs for Controller Design

KOMPR   Design of continuous- and discrete-time compensators, deadbeat controllers, simple or extended compensators for stable, unstable, and/or nonminimum phase SISO plants

OBSDN   Design of observers, full- and reduced-order observers, with and without disturbance feedforward, disturbance observers, pole-assignment approach, using three different methods. Synthesis for continuous- and discrete-time observers

RICAT   Design of linear quadratic optimal proportional or proportional-integral state feedback controllers. Design of stationary Kalman filters. Disturbance feedforward and disturbance observation. Different types of reference feedforward. Specification of stability margins by eigenvalue bounds. Support for the choice of weighting matrices, using root locus plots in the s- or z-plane: choice of a numerical method out of 13 different possibilities. Handling of continuous- and

discrete-time systems, including discrete singular systems. Calculation of continuous-time weighting matrices in the discrete-time case

PAROP   Multiobjective optimization of control structures, using performance
index vectors

ROPTI   Parameter optimization for different types of SISO controllers and
parameter determination for PID-controllers, using different design
rules

WOKU   Root-locus generator for computation and graphical representation
of root-locus plots with respect to various parameters, for both
continuous- and discrete-time systems. Also the consideration of
sampling time as root locus parameter. Graphical display of poles
and zeros.

INAM   Controller design, using Inverse Nyquist Array (INA) technique.

POSY   MIMO servo-compensator design, using polynomial matrices, spectral factorization, and diophantine equation-solving techniques.

19.3.5 General Tools for Simulation

The programs discussed in the previous subsection include internal simulation
routines, which are tailored for the specific purpose of analysis or design. On the
one hand, these are verification tasks, where the result is plotted in a TY-diagram;
on the other hand, there are algorithms which include simulation in their course.
For example, it is necessary to get the value of a specific performance index,
which is based on time signals, to simulate the control system several times during an optimization step. Throughly organized, specific simulation tasks are integrated in a fixed manner into the modules for analysis and synthesis and
consist of elements from the control engineering library within CADACS. The
spectrum of simulation is extended by generally applicable simulators. A control
engineering CAD system must cover the complete scope, starting from special
classes of processes and standard configurations up to universal tools for the
handling of control engineering problems.

Standard Simulator DIGSI

For commonly occurring cases of signal generation and verification of models
and controller design, which are based on the results of different CADACS programs, the standard simulator DIGSI is used. This module allows the simulation
of linear control systems in the maximum standard structure, illustrated in Figure 4. For each block, any suitable system representation is usable. Continuous-
and discrete-time blocks may be mixed. All signals may be of multivariable
type. For external signals, the use of a manifold of generated signals, as well as
the use of signals from the database (i.e., measured signals), is allowed. Besides
simulation of the total structure, it is also possible to generate step and impulse responses of single components. The principle of function of the standard

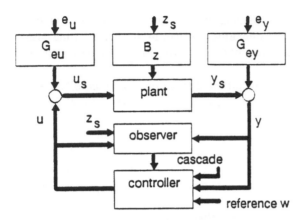

Figure 4 Block diagram of the standard simulator.

simulator may be best explained using the course of a simulation. After activation of DIGSI, the block "plant" is specified by definition of a file name. This file contains a suitable system representation of a dynamic system. Valid types of systems are the forms 5, 6, 8, 9, and 10, according to Figure 2.

After this specification, already simple simulations like generation of step, impulse, and ramp response may be carried out. To build the structure according to Figure 4, a configuration must be performed, where the type of the blocks "controller" and "observer" have to be specified. Besides the internal structure of these blocks, the form of the external signals has to be specified also. The user may, for example, specify for the internal structure of the block "controller" a proportional or proportional-integral state regulator with and without observer, with disturbance observer, with cascade feedforward of external and internal signals, with reference or disturbance feedforward or prefilter for the reference signals. By the configuration of the structure, only the skeleton of the overall system will be prespecified. The data necessary for simulation can be added later, that is, after the design of the specific block. This means that during configuration there is no necessity that all data already exist in the database. All design tools may be called from within DIGSI, due to the structure of CADACS. The results of the design process are then put to file and can be imported into DIGSI for the specific block by specifying the actual file name.

Simulation itself requires specification of simulation time and of documentation step size, which is necessary for the recording of signals on file or for graphical representation. Choice of the type of graphical representation is made—as is the general case in CADACS—by filling in a form of GRMGR, where the user specifies diagram entries for the desired signals according to the identifiers of Figure 4.

Equation-Oriented Simulation Using SIMGR

A small number of powerful function blocks are available to the user of DIGSI. In most applications, these are sufficient for linear systems as depicted in Figure 2 in most cases of application. For more complex nonlinear systems, other universal tools are necessary. Here, two different concepts, the block-oriented and equation-oriented ones, for representation of the simulation problem can be commissioned for CADSD systems. The block-oriented concept offers function blocks which have to be parametrized and connected by the user. This approach parallels the programming of analog computers. For nonlinear problems with many state variables, the compilation into a block diagram is time-consuming and has the danger of getting difficult to survey. An equation-oriented formulation of the problem in form of a vector differential equation

$$\dot{x}(t) = f[\mathbf{x}(t),\mathbf{u}(t),t], \qquad \mathbf{x}(0) = \mathbf{x}_0$$

using a simulation language, is preferable in this case. The simulator SIMGR ("Simulation Manager") combines both types of formulation possibilities in its simple programming language. Here, first of all, the nonlinear system of vectorial differential equations is specified by describing the right-hand side of the above equation using algebraic functions. Besides a set of useful, truly algebraic or transcendental functions, a number of control engineering functions, such as hysteresis, dead zone, and so on, blocks according to the systems representation illustrated in Figure 2 and signals from the database can be used. These functions are considered to be scalar function blocks which have to be assembled to yield the right-hand side of the system of differential equations. Blocks may be used in the algebraic relations as variables. The example of a SIMGR program, given in Figure 5, describes the simulation of a nonlinear control system with dead time and a linear controller. The outputs of blocks are defined by identifiers Bx, where x is an integer number.

19.3.6 Real-Time Toolbox

CADACS supports the real-time implementation of all control schemes, which are designed using CADACS. A multipurpose real-time suite serves as a development tool in basic studies or pilot experiments. This subsystem was designed to provide a very short development cycle of a controller. It is organized on a PC as a toolbox for Turbo Pascal, and consists of several parts. A command-driven monitor serves as an intelligent interface between the user and the related real-time task, which itself runs on the same PC. The user can specify all the data necessary to implement the controller and to run the real-time task parallel to any other CADACS activities. He can change or list parameters and can call components from the database. For monitoring of the real-time response, the "documentation manager" is linked between the real-time task and the

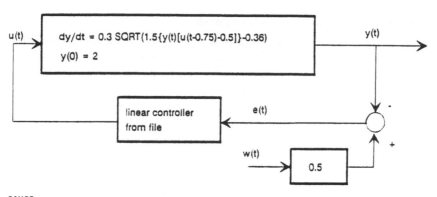

SIMGR

```
B1=X1,2                    [plant state y=X1=controlled value]
B2=B1*(B3-0.5)             [Term in brackets]
B3=OTIME(B6,0.75)          [dead time of manipulated value]
B5=0.3*SQRT(1.5*B2-0.36)   [block for right-hand side]
B5=F1                      [definition of right-hand side F1]
B6=TF(B10,'contr')         [linear controller from file 'contr']
B10=0.5*B11-B1             [control error e(t)]
B11=SIGNAL('Meas',3)       [reference signal w(t) from file 'meas']
```

Figure 5 Example of a SIMGR program.

"graphics manager." Recording of signals and internal values can be performed simultaneously with other CADACS activities.

On a PC, which can be connected directly to the plant to be controlled, the system supports the evaluation of prototype control systems. The implementation of analysis modules, as well as real-time tasks on the same computer, results in a very powerful tool for advanced control engineering. It has become feasible to pack the entire power of CADACS into a transportable system which can be taken to the plant to be analyzed and controlled.

Such an implementation is most suitable for basic controller studies in laboratory-size processes where, during the development phase, new software parts must be generated. This philosophy has proved to be suitable for developing adaptive controllers, running later in DDC substations in automation communication networks [3]. In the first stage of development, the adaptive controller runs locally on the same system. Later on, the control software will be downloaded to the target processor of the DDC substation and runs supervised by the development system in the same way as running on the development system. The real-time toolbox contains examples of different kinds of real-time tasks. They are realized both as source-code programs and as turn-key ready programs.

The following real-time programs are available:

DATST/DATAC	Data acquisition with optional test signal generation
DDCST/DDC	Decentralized MIMO controller with feedback and feedforward dynamics described by transfer functions
ZSTEW/SFEDB	MIMO state-feedback controller with optional observer
APPCS/APPCA	Decentralized adaptive MIMO controller, using pole-assignment or LQ control laws
MRACS/MRACA	Model-reference adaptive control.

19.4 USER INTERFACE

It is the task of the user interface to render possible the interactive usage of the package. It is necessary to offer a unified interface for all tools of a CAD system, to simplify the logical chaining of different steps, which are used to solve a specific problem. Results will get evidence by graphical representation. The graphics interface should be almost independent of the hardware used. The user should be able to use predefined standard diagrams as well as diagrams designed by himself.

The handling of the system will be clarified by the following commented example of a dialogue. During implementation of a real-time state feedback controller, the step response of the plant shall be examined graphically. For that purpose, the simulation problem is interlaced into the dialogue of controller implementation, as indicated in Figure 6.

CADACS allows a free and simple type of dialogue. Using unified and very basic commands from the overall set of main tasks, one task, that is, "Digital Simulation," is chosen. Having done that, all commands of DIGSI are available

Prompt Command	Comment
ZSTEW: aa	monitor call from real-time environment
KEDDC: DG	command 'digital simulation'
DIGSI: RE	local command for reference to data base
FILE NAME FOR PLANT = ? \DATEN\STR.MAT	specification of the file containing the plant
DIGSI: UF	local command to generate step response
SIMULATION TIME = ? 10.5	specification of additional information
DIGSI: EX	end 'digital simulation'
ZSTEW:aa	dialog for implementation

Figure 6 Example of a dialogue for a simulation problem.

to the user. With these commands, consisting of two characters, a subtask is scheduled from within DIGSI, which can be a calculation or a local question-and-answer dialogue. A standard set of commands gives access to status information, command menu, protocol, or graphical information. In addition, a help facility is available which offers the user detailed analysis of errors with explanation and help. Each subtask can be called on at any time in such a way that the running main task is suspended and then continued at a later time. The system is divided into several single programs and the exchange of data is made by files. The central program control and a number of utility commands is executed by the central monitor KEDDC. The number of CADACS programs, which can be used simultaneously on the computer, is only limited by the disk of EMS storage used for program swapping.

Although the combined command and question-and-answer dialogue seems to be complicated, it will be simplified by an intensive program guidance and support. Using ?? a local menu of commands will be displayed. In the case of an error, the "help" command displays an explanation of the last error with an additional error help and support to correct the situation. Further, the user does not need a printed manual because it is available on-line, offering more than 1400 screen pages of information. During dialogue, each page of the manual can be called, giving comments to the methods used, and explaining the principle of operation of a specific program.

19.5 AN APPLICATION EXAMPLE

The following example will show a typical application of CADACS. Based on measurements of a real-world plant, a dynamic model will be developed and then used for the controller design. Figure 7 shows the measured excitation current (input u) and the corresponding generator voltage (output y) of a turbo-generator set. To determine a model of the turbo generator, the input (u) and the output signal (y) are used for parameter estimation, using the maximum-likelihood estimation program package MLID. The print of the corresponding protocol based on MLID is presented in Figure 8. For model verification, the frequency response also was measured and approximated by an s-domain transfer function. Figure 9 shows the measured and approximated frequency response, using the UNEK package. Results obtained using the MLID model are compared, using step response, Nyquist chart, and Bode diagram. The step response was generated using DIGSI.

Based on the above identification results, a state feedback controller for the turbo-generator was designed and simulated. For this controller/observer design, actually a multivariable model, including the above submodel, was used. The simulation was done using the standard simulator DIGSI. Figure 10 shows the simulation comprising the two manipulating variables u_1 and u_2 (excitation cur-

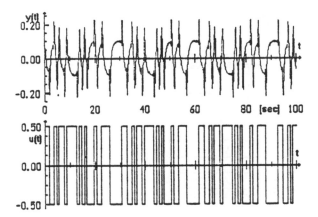

Figure 7 Measured input and output signals from data acquisition for a turbo-generator plant.

```
############################################################
MAXIMUM LIKELIHOOD PARAMETER ESTIMATION
############################################################
SIGNALS FROM FILE................/KED/DATEN/s21swr.SIG
SIGNAL #  OUTPUT SIGNAL Y...... 3
SIGNAL #. INPUT SIGNAL U1...... 1
EVALUATION FROM SIGNAL VALUE... 1
EVALUATION TO SIGNAL VALUE..... 1001
STEPPING BY VALUE.............. 1
# OF SAMPLES USED..............1001
ORDER OF SUBMODEL.............. 3
SAMPLING TIME...................9.999999E-02
VALUE OF LOSS FUNCTION..........1.560120E-02 (6.973582E-04)
STANDARD DEVIATION ERROR SIGNAL 5.583120E-03 (1.247800E-04)

PARAMETERS (STANDARD DEVIATION) PARAMETER (STANDARD DEVIATION)
-------------A------------------  -------------B1*z** 0------
 1.00000                          .000000
-1.78758   (7.249973E-03)        -2.694543E-02 (8.002027E-04)
 1.20544   (1.278333E-02)         .131878      (1.472958E-03)
-.343712   (6.392595E-03)        -.119545      (9.225194E-04)
```

Figure 8 Protocol output from MLID for the turbo-generator model.

rent and turbine valve position, respectively) and the two controlled variables y_1 and y_2 (generator voltage and frequency, respectively) for a change of the reference values w_1 and w_2 at times $t = 0$ sec and $t = 5$ sec.

19.6 CONCLUSIONS

This chapter has described the approved concept of CADACS which has been developed since 1973 and which is today well-accepted by many industrial and

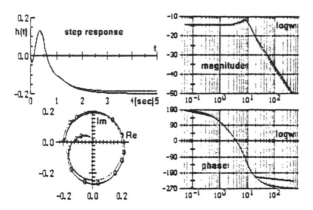

Figure 9 Comparison of the measured and identified system characteristics.

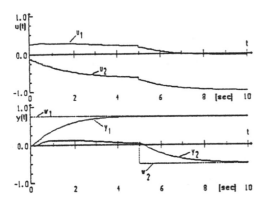

Figure 10 Time–behavior plot of the simulated turbo-generator set.

educational centers. The modular structure of the CADACS system comprises not only a broad spectrum of classical and modern numerically robust methods for identification, modeling, simulation, and design of SISO and MIMO control systems, but also provides the simultaneous implementation of real-time control algorithms for DDC-operation. Thus, CADACS offers an excellent guidance in industrial and real-time applications.

Nevertheless, CADACS provides coupling with PC-MATLAB or MA-TRIX$_X$, especially for the rapid prototype development phase of new algorithms; after a successful test, they will be recoded in the implementation language of CADACS. To save reimplementation time, primitive operations can be translated in a systematic way. Therefore, CADACS represents one of the most powerful integrated CADCS tools today, which as an open system renders it possible to integrate newly developed control engineering methods. It is hoped that CADACS will contribute in future even more to the release of control engineers from routine work and, hence, provide more time for real creativity.

REFERENCES

1. Schmid, C., A workstation concept for computer-aided analysis and design of control systems, *Proc. 7th IFAC/IFIP/IMACS Conference on "Digital Computer Applications to Process Control,"* Vienna, 1985, pp. 691–694.
2. Schmid, C., Techniques and tools of CADACS, *Proc. 4th IFAC-Symposium on "Computer-Aided Design in Control Systems"* (*CADCS '88*), Peking, 1988, pp. 67–75.
3. Schmid, C., Adaptive controllers: Design package, in *System & Control Encyclopedia. Supplementary Volume 1*, (M. G. Singh, ed.), Pergamon Press, Oxford, 1990, pp. 1–4.
4. Schumann, R., CAE von Regelsystemen mit IBM-kompatiblen Personal Computeren, *Automatisierungstechnische Praxis*, *31*, 349–359 (1989).
5. Unbehauen, H., Schmid, C., Böttiger, F., Bauer B., and Göhring, B., KEDDC—Ein kombiniertes Prozeßrechnerprogrammsystem zum Entwurf und Einsatz von DDC-Algorithmen, PDV-Report KFK-PDV 37, Gesellschaft zur Kernforschung mbH, Karlsruhe, Germany.
6. Unbehauen, H., KEDDC—Ein Programmsystem zur rechnergestützten Analyse und Synthese von Regelsystemen, *Z. messen, steuern, regeln*, *30*, 78–82 (1987).

20

CACE Software Implementation Using Ada

Magnus Rimvall

General Electric Corporate R&D
Schenectady, New York

20.1 INTRODUCTION

Since its introduction in 1980, the Ada™[†] computer language [1] has seen increased use in both defense-oriented and civilian projects. A recent survey of Ada-based software activities [2] lists over 300 such projects. Of these projects, a majority fall into one of two categories: those involving embedded, real-time software and those resulting in general-purpose language tools of "self-serving" value (language analysis tools, compiler-compilers, program generators, etc.). Only a few projects are in the area of CAE or general-purpose engineering software.

The present emphasis on Ada projects resulting in real-time software and language tools is not accidental. Ada was selected from several "vendor-bids" by a US DoD committee around 1980; this committee had the specific charter to specify and select one high-level computer language to replace the myriad of computer languages used by DoD. The primary focus was to define a single language for embedded, real-time systems. This indeed succeeded, and Ada is the only widespread computer language with real-time capabilities built into its

---
*This work was sponsored by General Electric Corporate R&D, Schenectady, NY, Dept. of Automatic Control, Swiss Federal Institute of Technology (ETH), Zürich, Switzerland, and the main ETH research commission.
[†]Ada™ is a registered trademark of the U.S. Government, Ada Joint Program Office.

standard language elements (with the exception of Jovial, one of the older languages Ada was to replace). Moreover, the Ada requirements documents also foresaw the availability of a range of standardized Ada language tools, initiating the trend to provide funding for such projects.

Despite its origin, Ada turned out to be more of a general-purpose language than a real-time language, and it is very well-suited for nonembedded, non-real-time scientific and engineering applications such as CACE (Computer-Aided Control Engineering) and general CAE. It is, however, yet to become widely used in these applications; the built-in inertia of the Fortran community and the initial unavailability of good Ada compilers has slowed its proliferation into these areas considerably. Nevertheless, Ada is an excellent tool both for implementing bread-and-butter mathematical libraries hitherto only available in Fortran and for implementing complex interactive engineering applications. This chapter will discuss the potential of using Ada in CAE, based on the experiences gained during the implementation of the CACE projects IMPACT [3] and GE-MEAD [4,5].

The IMPACT program was the first major CACE program to be implemented in Ada. It was developed at the Swiss Federal Institute of Technology (ETH) in the years 1985–1988. IMPACT is a general-purpose CACSD package: It gives the user access to a wide range of control algorithms in an interactive manner. Furthermore, through the use of the flexible IMPACT command language, the user can define new algorithms based on already existing IMPACT primitives. The versatility of the command language is complemented with datastructures needed in automatic control, for example, polynomial matrices, transfer-function matrices, and linear system descriptions (see Refs. 3 and 6 for more information on the IMPACT program itself).

The GE-MEAD project involves the integration of several CACE packages under a supervisor which coordinates the execution of these packages with a database manager and an advanced user interface. The database and the user interface were based on proprietary database languages and user-interface management systems (UIMS), respectively. The GE-MEAD supervisor was implemented in Ada and consists of a full command-language interpreter with procedural/macro capabilities, and a datadriven translator from generic GE-MEAD commands to specific commands for the supported CACE packages (see Refs. 4 and 5 for more information on the GE-MEAD program itself).

By Ada standards, the IMPACT and GE-MEAD projects should be rated a medium-sized software project (and a large-sized project for being totally university based/self-funded). IMPACT involved two to four software engineers and programmers (M.S. and Ph.D. students) during about 5 years and resulted in a program consisting of some 85,000 lines of Ada code (Ref. 2 lists only 39 projects larger than this). The GE-MEAD supervisor, which was in part based on code used in IMPACT, involved one to two software engineers during a little more than 1 year. Thus, the projects were large enough to really utilize the advantages of a language such as Ada (the author would still argue that, despite all advantages of Ada, stand-alone projects having less than, say, 2-4k LOC are implemented faster and more efficiently in a smaller language such as Pascal).

The issues discussed in this chapter are generic in nature and do not depend on any particular Ada compiler or run-time system. However, for the record, it should be stated that during its first year, the IMPACT project relied on a very early version of a Telesoft compiler running under VAX/VMS. The initial experiences with that environment prompted a switch to the DEC VAX/VMS compiler, which was used throughout the rest of the two projects.

20.2 THE Ada PROGRAMMING LANGUAGE

Given the meticulous and lengthy process leading up to the definition and adoption of a language fulfilling DoD's requirements and also given the long delay before the first usable Ada compilers became available, it should not be surprising that this whole extensively published process resulted in giving Ada a certain aura of mysticism. This aura, which is the result of overestimating both the capabilities and the overall complexity of the language, has been used both to advocate the language as a panacea to the whole software crisis and to criticize the language as impractical and prohibitively inefficient. The reality is, as always, somewhere in between.

In the cases where Ada is seen as the panacea to the software crises (as if a better implementation language alone could solve the much more fundamental software *design* and *maintenance* problems), the language is often thought to have many of the capabilities otherwise only found in the so-called fourth-generation or nonprocedural computer languages. This is not the case. Ada does not, and was never designed to, inherently support such things as DBMS query functionality, AI capabilities, or automatic program generation. Instead, Ada is primarily procedural in nature, with some object-oriented capabilities. Thus, Ada belongs to the same group of third-generation languages as Pascal, C, Objective-C, and Modula-2. Ada will, when used correctly, be a better and more efficient tool and it will generate easier-to-maintain and more reliable code than its third-generation cousins, but the programming paradigm and general programming abstraction level is the same within that group of languages.

On the other hand, Ada is often criticized for being too complex and unwieldy for practical use. This, also, is a misconception. It *is* true that, by any metric, Ada is a very comprehensive language, and many potential users are intimidated by its sheer size. For example, the grammar of the language contains over 175 productions (50% more than in Pascal [7]), the reference manual comprises some 250 pages (twice that of the original Pascal Manual), a typical Ada compiler (DEC/VMS) is 1.6 MB large (three times that of the VMS Pascal and five times that of the VMS Fortran compiler). However, sizes like these alone are not necessarily a good measure of complexity:

- Ada was derived mainly from the Pascal language, which, in turn, has its roots in the first structured language, Algol-60. Thus, anybody having experience with abstract data types and structured programming will be immediately familiar with the basic Ada language elements. Moreover, to start programming in Ada, only a basic set of language elements have to be known (corresponding to a Pascal or C subset of Ada).
- Each nonbasic feature of Ada (e.g., exception-handling, generics, tasking) can be picked up and learned by the individual programmer whenever needed.
- Ada is a highly structured and a very strongly typed language. This enables the compilers to find many more errors much earlier than, for example, a C or Modula-2 compiler. In particular, many logical errors that can only be found through run-time errors in other languages are flagged immediately at compilation time in Ada. This makes the language particularly friendly to new users or inexperienced programmers.
- The syntax and semantics of Ada are logical, predictable, and very robust.

20.2.1 The Robustness of Ada Code

The importance of a sturdy language syntax for the implementation of good programs cannot be overemphasized. In languages like Fortran, most bugs do not derive from logical errors in the program design, but rather from type-in errors, forgotten declarations, and so on. In Ada, a large percentage of these bugs will be caught by the compiler. An example will illustrate this 8. Consider the following incorrect Fortran program (a comma is missing between the 1 and the 100 in the DO statement):

```
C*****INCORRECT FORTRAN
      SUBROUTINE CACSD
      DO 999 I = 1 100
      CALL CONTR
999   CONTINUE
      RETURN
      END
```

The lack of rigid lexical and semantical rules in Fortran will leave the missing comma undetected; thus every compiler will produce an executable code equivalent to the following nonsensical Pascal code:

```
(*Equivalent Pascal*)
PROCEDURE cacsd;
   VAR do999i : INTEGER;
BEGIN
   do999i := 100;
   contr;
END;
```

Thus, a one-character error (the comma) translates the Fortran iteration loop into a dummy statement and a single subroutine call. Advocators of structured programming and forced declaration will argue that the same kind of error cannot happen in structured languages. Indeed, Pascal *seems* to have a more robust syntax:

```
(*Correct Pascal*)
PROCEDURE cacsd;
   VAR i : INTEGER;
BEGIN
   FOR i := 1 to 100 DO contr;
END;
```

However, even in Pascal, a small lexical error can entirely change the logic of a program. In fact, the following one-character error (a superfluous semicolon after the DO) *again* transforms a loop to a dummy statement and a single procedure call:

```
(*Equivalent Pascal*)
PROCEDURE cacsd;
   VAR i : INTEGER;
BEGIN
   FOR i := 1 TO 100 DO; contr;
END;
```

Ada, however, has a much more robust syntax. Different constructions have more distinct syntax. For example, the correct version of the above illustration looks as follows:

```
-- Correct Ada
procedure CACSD is
begin
   for I in 1 .. 100
      loop CONTR; end loop;
end CACSD
```

Indeed, in Ada, there is not even a marginal risk that a small typographical error entirely changes the logic of the program, as the incorrect Ada program would have a quite different syntax (due to explicit null statements, loop/structure ends, etc.):

```
-- Incorrect Ada equivalent to
-- erroneous FORTRAN/Pascal program
procedure CACSD is
begin
```

```
        for I in 1 .. 100
           loop null; end loop; CONTR;
     end CACSD
```

This kind of robust and yet logical syntax is characteristic for all language constructions in Ada.

20.3   THE USE OF Ada IN CAE

Despite the availability of modern third- or even fourth-generation languages, Fortran still dominates the field of engineering and scientific software. The reason for this prolonged endurance of a technologically outdated language is best summarized in one word: *reuse*. Software development is both expensive and time-consuming. Thus, substantial savings can be made each time a component is reused rather than reimplemented. Fortran, despite its shortcomings, shows some real advantages in this respect:

*   The size and scope of the existing Fortran software base surpasses that of any other computer language by several orders of magnitude. Libraries of algorithms and methods can be found for virtually every scientific area. For example, control engineers may find many generic mathematical algorithms in public-domain Fortran libraries such as LINPACK and EISPACK. Large number of control-specific algorithms are available in libraries such as SYCOT, SLICE, and RASP.

    The wide use of Fortran has also been sustained by its own inertia. Applications programmers often find it too expensive to use anything but Fortran, as all the base algorithms are available only in that language. Most of the public-domain base software, on the other hand, was once funded through national research programs, where the development of algorithms was the primary research objective. The porting of these libraries to another language has no further research value, again leaving no alternatives. Even the commercial vendors of mathematical software such as IMSL are yet to offer alternatives to their Fortran libraries.

    There is a slight glimmer on the horizon for proponents of Ada. The more software the DoD requires to be coded in Ada, the larger this software base will get. However, this software base has to be made available to the software community at large in a better maintained basis than is presently done with the public-domain code in Ref. 9.

*   From a software engineering view, Fortran is a very weak language. It has weak typing, weak consistency checking, and nonexistent inherent error-handling. Yet, when setting up Fortran libraries, these drawbacks turn into distinct advantages. An example will illustrate this; practically all mathematical Fortran libraries use the same parameter-passing mechanism: All inputs

and outputs to an algorithm are passed as simple one-dimensional arrays (in parameter fields or common blocks). Thus, explicit type conversion is seldom necessary when connecting algorithms from different sources. In languages supporting data abstraction (e.g., Pascal and Modula-2), frequent type conversions become necessary when mixing different libraries. Moreover, Pascal or Modula require that the dimension of each data element be predefined, something the library vendor cannot and should not specify. In Fortran, the lack of dimensional checks make library units accept data of any dimension—at the risk of run-time errors such as array incompatibilities and memory overwrites.

In the rest of this chapter, we will study the suitability of Ada for scientific programs, both for implementing algorithmic libraries to be reused within other programs, and for implementing scientific stand-alone, interactive programs. The use of Ada in real-time systems has been discussed frequently; thus, we will concentrate on off-line software with CACE as an important case.

20.3.1 Data Abstraction

The kernel of scientific programs most often consists of a set of mathematical algorithms operating on problem-specific data. In CACE, this data may represent a linear model in state (matrix) or frequency (polynomial) form, a time or frequency response from that model, and so on. To reuse any such software, the algorithms have to be generic enough to be applicable for a large variety of such datasets. This is hard to attain with most structured languages, but quite possible using Ada. Let us consider some simple examples involving linear systems in the time domain (i.e., sets of correctly dimensioned matrices A, B, C, D). First, we have to define a datastructure to hold the said matrices.

In Fortran, no specific datastructures are available to store away a complete system description. We *could* define four arrays with fixed dimensions. This, however, is inadequate in an interactive program, where it is not known at compile time how big a model will be created at run time. The only general method is to store away all four matrices in a single one-dimensional array of "adequate" size. An extension of this principle leads to the one-dimensional stack used by most commercial, Fortran-based CACE programs [10]. The approach is simple, but the resulting code is neither very readable nor very safe. In our simplistic case, we would declare an array CSYS, together with the number of states IX, inputs IU, and outputs IY:

```
DIMENSION CSYS (1000)
INTEGER IX, IU, IY
```

This declaration is valid for all models with less than 1000 individual elements (models up to order 30 or so). Assuming that the *A* matrix is stored first (elements 1 ... IX*IX) and the *B* matrix thereafter (elements IX*IX + 1 ... IX*IX + IX*IU), the coefficient relating input signal 3 to state 4 would be accessed as

```
I     = IX*IX + (3−1)*IU + 4
COEFF = CSYS(I)
```

Code like this is neither readable nor maintainable. Only extensive testing of any code of this nature will guarantee its correctness, no automatic compile or run-time checks of dimensions, indices, and so on, are possible. The code is, however, very easy to (re-)use. If a controllability algorithm is called with two one-dimensional arrays and two dimensions as input parameters, the parameter list is simple to construct:

```
I     = IX*IX  + 1
CALL CTR11(CSYS(1), CSYS(I),IX,IU)
```

On the surface, structured programming languages supporting data abstraction, such as Pascal, C, or Modula-2, seem much better fit to implement readable and *safe* programs involving complex datastructures. Our linear model could be defined as

```
CONST
     ix = 5; iu = 1; iy = 1;
TYPE
     amatrix = array[1 .. ix,1..ix] of real;
     bmatrix = array[1 .. ix,1..iu] of real;
     ..
     linsys = record
                 a : amatrix;
                 b : bmatrix;
                 ..
             end record;
VAR
     csys : linsys;
```

Now we may access elements in a much more natural fashion. For example, the two Fortran examples become more legible:

```
coeff  := csys.b(3,4)
ctr11(csys.a,csys.b)
```

Also, the safety of this code is much higher. For example, if an index element gets out of range due to some logical programming error, run-time checks will make sure that the erroneous command is aborted. However, the illustrated approach is not generic enough to be of practical use:

- The first-generation structured languages (e.g., Pascal and Modula-2) require all array and record declarations to have fixed, predetermined sizes. However, the implementor of basic control software cannot possibly know the exact order of the models his algorithm will be applied to. The only solution to this problem would be to define "high enough" maximum dimensions, forcing a waste of memory space for each smaller linear system (if we, for example, maximize the order to 30 states, 884 out of 900 elements of the *A* matrix will be wasted each time we use the program on models of fourth order).
- Any mathematical base library has to be based on some fixed datastructure as well. In particular, if we have a base-library working with a particular datastructure, it is not usable for any other structures. Thus, any change to dimensions, precision, component names, and so on, will require reediting and recompilation of the library.

Ada, on the other hand, is the first major language to really address the problems of reusing software while supporting safe and readable datastructures. For example, so-called *discriminants* may be used to obtain full-dimensional flexibility and yet retain a clean and readable structure:

```
type matrix is array(positive range<>,
                     positive range<>) of float;
type linsys(x_dim, u_dim, y_dim : positive) is
        record
            a_matrix : matrix(1..x_dim, 1..x_dim);
            b_matrix : matrix(1..x_dim, 1..u_dim);
            ..
        end record;
my_sys : linsys (5,2,1);
```

This code is as legible as its Pascal counterpart, yet the defined structures have no inherent dimensional limits. Moreover, dynamic versions of structures like these may be created using pointer techniques similar to those available in other structured languages. In fact, IMPACT and the GE-MEAD supervisor both use pointers and dynamic structures to give the interactive user of the package the freedom to create any number of data structures "on the fly," each having arbitrary dimensions. This is done totally portably, as only standard language elements are used. The advantage of working with dynamically discriminated variables is obvious; at no time do we allocate more memory space than required, yet there are no upper dimensional limits.

All routines called with parameters of the above-declared types can utilize these abstract datastructures fully. For example, our routine ctrl1 can be implemented as follows:

```
procedure ctrll (system    : linsys) is
   inputs        : positive   := system.u_dim;
   my_matrix : matrix (1..system.x_dim,
                                 1..system.x_dim);
begin
   ..
end CTRll;
```

Note how the dimensions of system may be accessed and used in an arbitrary fashion within the routine.

20.3.2 Generic Programming

A second fundamental problem with reusing structured programs with abstract data types derives from the strong typing itself. Strong typing does increase both program readability and code reliability. However, code segments defined using a particular data declaration can normally only be reused within other programs using exactly the same data declaration. For example, if a user has access to an algorithm using double-precision reals, and he wants to use the same algorithm on single-precision data, languages such as Pascal and C offer only two alternatives: The user either has to change the declaration in the heading and recompile the code (this assumes that he has access to the source code) or he can convert his data to double precision structures and call the existing algorithm (not very useful if what he had was an object copy of a single precision program, and his data was in double precision). Ada provides mechanisms for implementing one *generic* version of the code. Later, a programmer using that code may *instantiate* the code for any number of special cases. In our example, the following small generic package could be defined:

```
generic
type gen_elem is private;
package gen_linear_opers is
   type matrix is array(positive range <>,
                        positive range <>) of gen_elem;
   type linsys (x_dim, u_dim, y_dim : positive) is
          record
              ..
          end record;
   ..
   procedure ctrll (system : linsys);
   ..
end gen_linear_opers;
```

This package may be used several times, each time operating a different data-structure. For example, single- and double-precision instantiations (variants) are trivially made available as follows:

```
single_lin is new gen_linear_opers (float);
double_lin is new gen_linear_opers (long_float);
```

However, it is also possible to define the same package operating on a completely new data structure, such as complex numbers. Then, however, we first have to redefine all primitive operations performed on the structure within the generic routine. For example, the algebraic routines (e.g., +) have to be extended to the new type (overloaded).

```
type complex is record
                re,
                im : float := 0.0;
              end record;
  ..
function "+"(c1,c2 : complex) return complex;
```

The algebraic routine "+" must, of course, also have an implementational part which relies only on already known (predefined) operations:

```
function "+"(in_1, in_2 : complex)
                return complex is
begin
  return (in_1.re + in_2.re, in_1.im + in_2.im);
end "+";
```

We can now instantiate the generic package

```
complex_lin is new gen_linear_operation (complex);
```

The potential of generics extends much further than shown in these simple examples. Less-unified operations such as list and tree-handling, lexical analysis and parsing, memory management, and complex tasking operations may be implemented once and reused through generics.

20.3.3 Object-Oriented Programming Using Ada

Section 20.3.2 had an example of another very powerful feature in Ada, namely, *overloading*, that is, providing equally named subprograms, each implementing a certain operation on a different structure. A routine may be overloaded any number of times, as long as the rules for strong typing make the choice of routine unambiguous to the compiler. This feature can be used to produce *object-oriented* programs. It allows an implementor to define several disjoint sets of datastructure and procedure combinations, each of which implements a complete object/methods set. The compiler will determine which operation is to be called, based upon the data types. Note that, in Ada, overloading is static and not dynamic. Thus, there may be no binding of objects and methods "on the fly" within interactive programs (e.g., as in Objective-C [11]).

In our discussion on reusability, we concentrated on problems as seen by the programmers reusing code. However, there is another side of the same coin, namely, the problem of maintaining and upgrading an existing piece of code once it has become widely (re-)used. There is always a risk that a "fix" or enhancement will cause the upgraded code to behave erroneously when operating within other programs. In particular, whenever internal details of a library routine is known, such as the structure or size of some internal data, any application program may make explicit use of that knowledge and any subsequent changes to the internal datastructure may cause the application program to behave differently. Thus, whenever internal structures are made known and accessible to applications programmers, these internal structures should never change. This could, however, become a limitation, for example, theoretical development of new algorithms dictate a change. The solution to this problem is obvious: Do not make internal structures accessible to the applications programmer. Ada provides a tool for enforcing this by letting data types be defined as *private* (known only within a particular package). In our example, we could have defined a library package as follows:

```
package linear_operations is
    type linsys is private;
    function new_linsys (a,b,c,d : matrix)
                            return linsys
    procedure ctrl1 (system : linsys);
     ..
private
    type matrix ... ;
    type linsys( ... ) is ... ;
end linear_operations;
```

When *using* package linear_operations, it is not possible to make use of any knowledge on how linear systems are implemented. Thus, later we can make arbitrary changes to the innards of the package without risking that any application packages will behave differently. For example, we could exchange the way system component matrices are stored to sparse_matrix without risk.

20.3.4 Error-Handling in CACE

In the first example of this chapter, we mentioned that strong typing and explicit dimensional checks assure a correct program execution. However, it is often not enough to abort the program whenever some dynamic dimensional check indicates a program error. For example, much time and effort may be lost to a user if an interactive package suddenly aborts, deleting all interactively created and dynamically saved data. Instead, the user should be given a friendly error message and be given the chance to continue normal execution of the program. On more general terms, detecting errors and taking appropriate corrective action is of paramount importance in all software.

The most common way of avoiding catastrophic error situations has been to test for potential error conditions and take preventive actions *before* the error occurs. This, however, requires a very defensive programming with an abundance of tests to be effective, and there is still no assurance that all cases have been covered. In Ada, on the other hand, all errors (both expected and unexpected ones) can be automatically detected through the use of *exceptions*:

```
procedure divide_135;
   -- body where a division by 0 may occur
exception
   when numeric_error =>
      text_io.put ("Division by zero attempted");
end divide_135;
```

If a division by zero is attempted in divide_135, control will automatically be transferred to the correct exception handler at the end of the procedure, where proper action can be taken (e.g., to write an error message). Thereafter, the exception *may be propagated* to the calling procedure for further treatment. Note that the exception mechanisms may be used for *all* kind of errors. Both in IMPACT and the GE-MEAD supervisor, errors caused by the user entering an illegal command sequence as well as errors caused by software bugs are caught, reported, and treated by one general error-handler. Thus, in all cases the continuing execution of the interactive program is guaranteed.

20.4 THE Ada SOFTWARE ENGINEERING ENVIRONMENT

The traditional view of software development had programmers working with only three or four simple tools: an editor, a compiler, a linker/loader, and (maybe) a symbolic debugger. This simplistic view of CASE (Computer-Aided Software Engineering) has definitely changed over the last decade, and auxiliary software tools ranging from pretty printers to complete CASE workstation environments have become irreplaceable to the skilled programmer. However, Ada is unique in that a certain amount of software engineering support is embedded within the language itself.

The DoD language requirements foresaw not only a language definition according to which compilers and linkers could be constructed, but also required the definition and implementation of software engineering environments. These environments are referred to as APSE (Ada Programming Support Environment) and are intended to give the Ada designer/programmer the same working environments regardless of the kind of machine on which he is working. Moreover, a minimal set of software engineering tools, each having a standardized interface, has to be made available. Unfortunately, even 9 years after the selection of Ada *language*, the designs of standardized language environments are still on the committee and prototype levels [12].

Despite the lack of official APSEs, most of the existing (but nonunified) language environments still offer more programmer support than traditional compilers. One example of such support is the consistency assurance between interdependent parts of a program. The compilation environment of Ada will make sure that all dependent modules (packages) are recompiled if any change is made to a lower-level module. This is particularly important in multiprogrammer environments, where lack the synchronization of module updates could otherwise lead to serious problems.

Another level of software design support may be obtained by using software engineering tools such as CADRE/Teamwork, which explicitly supports Ada code development or structured modeling environments such as ISI/Autocode, which will automatically produce real-time code of the entered model/controller.

20.5 QUALITY AND RELIABILITY OF Ada CODE

The quality of any software is primarily dependent on the quality of the individual programmer. Thus, no structured language or software engineering environment will make a bad programmer into a good programmer. Similarly, no implementational guidelines or software review process in the world will guarantee that the code is bug-free. However, this is not to say that such measures are all in vain. In IMPACT, the programming effort was accomplished by a very heterogeneous set of persons, namely, students working on short-term projects. Each student was obligated to use standard Ada constructions only; he/she was also given a set of implementational rules to be followed. Comparing the quality of the software written by these students with code implemented in other languages, two self-evident conclusions could be drawn:

- Excellent students produce good code in any language.
- The converse is true for poor students.

More importantly, however, is the conclusion:

- There was a noticeable improvement in the quality of the code produced by average to good students when using Ada. Of the programs produced by the two dozen or so students involved in the IMPACT/Ada project, three-fourths could be used without major redesign. In control software projects using primarily Fortran and Pascal, less than 50% of the individual contributors were able to produce code that could be used afterward to any practical extent.

Why did the students of the IMPACT project produce better than average code? The distinct difference between the two groups of students can only partially be attributed to such factors as project suitability and project leadership. Another major factor must lie in the combination of stringent implementational guidelines and the strong enforcement of typing and modularization within Ada. The IMPACT guidelines required students to produce well-tested, stand-alone and yet integrated modules. Furthermore, the stability of these modules was enhanced by a required use of private types and standardized memory management and error (exception) schemes. However, all these restrictions would not have been enforceable in less strict languages such as the different members of the C family. The lessons learned during the IMPACT project heavily influenced the decision to use Ada for the GE-MEAD supervisor, and the good experiences with Ada in IMPACT were confirmed in the GE-MEAD supervisor.

Another reason to expect a better maintainability when using Ada, albeit not a major factor in the VAX-only IMPACT project, results from the very strict attitude of the Ada committees not to allow any supersets or subsets of Ada. This ensures a high portability of the code between different compilers and between different operating systems. In fact, if IMPACT were to be ported to a new machine, less than 1000 lines of code (out of 85,000) have to be reviewed.

The increased quality of the Ada code, however, takes it toll in code volume. More than in any other structured languages, the code of Ada programs gets voluminous due to declarations, error-handlers, module specification, and body separation. Although this code explosion can to some extent be offset by consistent standardization and "generification" of portions of the code, programs such as IMPACT and GE-MEAD supervisor are probably implementable within 30–50% less lines of code using C or any other less-stringent language.

20.6 PERFORMANCE OF Ada PROGRAM(MER)S

The size of the Ada language makes the construction of Ada compilers a formidable task [13]. Indeed, it took 3–4 years for the first good-quality compilers to hit the market. Moreover, due to the complexity of the language, Ada compilations are still relatively slow. However, the separate compilation feature of Ada cuts down on the size of each incremental compile (compilation of a single module), which dominates during the edit–compile–test cycle of software development. This makes the overall efficiency very competitive with that of other language environments.

The learning time of the Ada programming language is surprisingly short for anybody familiar with structured programming, abstract data types, and (if needed) object-oriented programming. Students at the ETH in general had a good background of Pascal and/or Module-2, and they needed only about a week before they could write simple Ada programs. They usually became proficient in the language within another one to two weeks. The GE-MEAD implementors had solid knowledge of Fortran, C, and Ada eliminating any learning curve. Obviously, programmers with a lesser background will need more time, but even then the stringency of the language gives the compiler a chance to teach the user through his mistakes—and usually this can be accomplished in a fairly user-friendly manner.

20.7   RATING Ada

Table 1 summarizes the suitability of Ada versus alternative languages for implementing an interactive CACE package such as IMPACT. The items in the table have been discussed in some detail in the previous chapters.

20.8   CONCLUSION

The Ada programming language, although originally designed for real-time embedded systems, is very well-suited for general-purpose use in scientific and engineering software. It uniquely supports several language features much needed

Table 1   Suitability of Different Languages for CACE Software.

	Fortran	Modula-2	C	Objective-C C++	Ada	Logic Progr. Languages
Programming issues						
Structured programming	0	2	2	2	2	1
Control Data Structures	0	1	1	1	2	0
Robust code	0	1	1	1	2	0
Error-handling	0	0	0	1	2	1
Modularity	1	2	1	2	2	0
Object-oriented Progr.	0	0	0	2	1	1
Suitability for						
Numerical algorithms	2	1	1	1	2	0
Symbolic operations	0	1	1	1	1	2
Practical issues						
Basic libraries available	2	0	1	0	1	0
Wide spread use	2	0	2	1	1	1
Software tools	1	1	2	2	2	1
Portability	2	1	2	1	2	1
Summary	10	10	14	15	20	8

both for implementing off-line support libraries (data abstraction, generics) and for increasing the quality of on-line interactive CAE software (object-oriented programming, error-handling). Moreover, the Ada language environments are often richer from a CASE point of view than those of other languages.

REFERENCES

1. *Reference Manual for the Ada programming language*, U.S. DoD Report ANSI/ MIL-STD-1815A, January 1983.
2. Ada usage database, *Ada Information Clearinghouse Newsletter*, 6(4), 26–37 (1988).
3. Rimvall, M., and Cellier, F. E., IMPACT—Interactive mathematical program for automatic control theory, *Proc. 6th International Conference on Analysis and Optimization (INRIA)*, Springer-Verlag, Berlin, 1984.
4. Taylor, J. H., Frederick, D. K., Rimvall, Cm M., and Hunt, A. S., The GE-MEAD computer-aided control engineering environment, *Proc. 10th IFAC World Congress*, Tallinn, Estonia, 1990.
5. Rimvall, C. M., and Taylor, J. H., Data-driven supervisor design for CACE package integration, *Proc. IFAC Symposium CADCS 91*, Swansea, Wales, 1991.
6. Rimvall, M., Man-Machine Interfaces and Implementational Issues in Computer-Aided Control System Design, Dissertation 8200, Swiss Federal Institute of Technology (ETH), Zürich, Switzerland.
7. Jensen, K., and Wirth, N., *Pascal User Manual and Report*, Springer-Verlag, Berlin, 1975.
8. Bucher, W., private communication, 1984.
9. Ada Software Repository Newsletter, Management Assistance Corp. of America, White Sands, NM, Issue 108, February 1989.
10. Moler, C., *The MATLAB user's guide*, University of New Mexico, Albuquerque, 1980.
11. Cox, B., *Object Oriented Programming—An Evolutionary Approach*, Addison-Wesley, Reading, MA, 1987.
12. Boyd, S., (Chairman), APSE Builders Working Group Overview, *ACM Ada Lett.* 7(1), 29 (1989).
13. Ganapathi, M., and Mendel, G., Issues in Ada compiler technology, *IEEE Computer*, 52–60 (February 1989).

21

DB-Prolog Environments for Computer-Aided Control Engineering

J. M. Maciejowski and C. Y. Tan

Cambridge University
Cambridge, England

21.1 INTRODUCTION

Consider the way a bank clerk or travel agent interacts with a computer. The interactions are with a database and are phrased entirely in terms which are at the forefront of the clerk's or travel agent's mind: "What is the cheapest non-smoking seat available to Rome tomorrow?," "Credit this account with £1000," and so on. Examples of corresponding interactions for a control engineer are "Is this system stable?" (a database query) and "Connect these two systems in series" (a database update), with the systems in question being identified by pointing or clicking with a mouse. Contrast this with what the control engineer actually has to do today. Working with MATLAB, for instance (which is a great advance on what we had before), the control engineer has to perform the following gymnastics. First, she* must recall *how* to answer the query, then what the name of the appropriate data item is. For instance, it might be called des_cl_A—the name being cleverly chosen to indicate that it is the *A* matrix of a state-space representation of a designed closed-loop system—or perhaps des_cl_den if it is the denominator polynomial of a transfer-function

---

*In the interests of brevity, in this article all control engineers will be assumed to be female, and bank clerks and travel agents will be assumed to be male.

representation. Having done all this, the query "Is this system stable?" can easily be formulated as

all(real(eig(des_cl_A))<0)

or, perhaps,

all(real(roots(des_cl_den))<0).

The update example is equally good fun. Suppose that one of the systems we wish to connect in series represents a gyro in transfer-function form (held in MATLAB variables gyron and gyrod), and the other is a state-space model conveniently stored in a1, b1, c1, d1. Then all you have to do (using the *Control Toolbox*) is to convert the gyro to state-space form:

[a2,b2,c2,d2]=tf2ss(gyron,gyrod)

then think of a name for the new system and perform the series connection:

[anew,bnew,cnew,dnew]=series(a1,b1,c1,d1,a2,b2,c2,d2)

Let us focus on what it is that makes the bank clerk's life so much easier than the control engineer's. First of all, the style of interaction is declarative rather than procedural. This means that the user specifies *what* is wanted or *what* is to be done, rather than *how* to do it. Having to think about the algorithm required is a great distraction and, of course, is fraught with the possibility of error. The bank clerk's system knows how to credit an account, and it knows how to look at what kind of account it is, in order to debit the correct transaction fee. A good control engineering environment should know how to check the stability of a system, and it should know how to check the available representations of that system to decide which of the possible algorithms it should use. Second, the ability to identify the object being considered by pointing, or some other simple method of selection, is a great improvement on having to specify it by name or, even worse, a collection of names. Third, the syntax of languages traditionally used by engineers is a distraction—would readers familiar with MATLAB care to consider whether all the parentheses in the examples given above are in the right places? Fourth, the bank clerk's system does not force him to specify irrelevant information nor to invoke unnecessary computation. At the time of saying "These two systems are connected in series" the control engineer should not need to specify a name for the resulting combination—a name may never be needed, and if it is, it should be specified in a separate operation. She should also not be forced to invoke any computation other than the minimum required to record the fact that "These two systems are connected in series." Further computation is only required when some *query* is expressed which involves the series connection. Only then will it become clear what the appropriate computation is. The connection may even be "undone" before any computation is required.

The key to providing "fourth-generation" languages in the commercial world has been the availability of good database management systems, which allow the kinds of data which occur in commercial applications to be modeled accurately. Indeed, one definition of these languages states that "fourth-generation" languages are languages which operate on databases rather than on smaller entities such as memory locations. In addition to the obvious role of a database as a repository of the current "state" of an enterprise, a powerful role has emerged for database management systems as integrating platforms, around which various software services can be provided, such as user-friendly interfaces. We believe that the key to providing similar facilities for engineering users lies also in the provision of appropriate database management facilities, and this view is not particularly novel [8]. However, the provision of such facilities is considerably more challenging than for commercial applications.

Perhaps the biggest difference is in the complexity of the data which must be represented. In the commercial world, entities such as Accounts, Employees, Seats, Flights, and so on are relatively simple to model and are represented very effectively by building up tables, or "relations," in modern relational database managements systems. For control engineering, we need entities such as Block diagrams, Specifications, Simulations, Multirate systems, and so on, which are very complex and are not naturally represented by tables. Another difference is that the operations which need to be performed in commercial databases are mostly simple arithmetic or logical ones: Sum the numbers in each column, form the union of two tables, and so on. For control engineering, we need arbitrary numerical (and symbolic?) operations. Finding eigenvalues or solving Riccati equations are "elementary" operations for control engineers.

Much work in database research is concerned with the problems of handling enormous amounts of data. If a database contains gigabytes of information, then retrieving particular chunks of it can take a very long time, and careful design of retrieval strategies is called for. On the other hand, in engineering design, this problem is much less severe. If a database is associated with one control engineering project, then it is unlikely to hold more than a few megabytes of information and will often hold much less than that; relatively crude storage and retrieval strategies are adequate.

Another factor which makes provision of commercial database facilities relatively easy is the expected behavior of the user. As far as a travel agent or bank clerk is concerned, the data types and algorithms which are available in the database are fixed. They are created by a "database designer," who is distinct from the end users, and any modification to them requires shutting down the system and starting up what is effectively a new system. (Relatively minor additions of algorithms may be handled by a "database administrator," who is usually distinct from both end users and the designer.) This is very different from the control engineer. Although she certainly wants a comprehensive set of data types and algorithms built into the system, she will also wish to refine the data types and add algorithms in response to new kinds of design problems that may arise, or to reflect the working practices of her organization, or to capture a bright idea she has had. There are occasions, in other words, when the engineering user needs to act as a database designer as well as a database user.

One final difference to which we wish to draw attention is that transactions in commercial databases last seconds or minutes, whereas those in an engineering database may last hours, days, or even longer. If a database is shared by more than one user—a project design team, for instance—then the usual strategy of simply "locking" the database to other users while a transaction occurs is not acceptable.

Having highlighted the differences between commercial and engineering environments in which a database management system may need to operate, what about the similarities? Why should it be promising to pursue the analogy with commercial systems if there are so many differences? The obvious first similarity is that in both cases it is desirable to have a well-ordered record of the latest state of an enterprise, whether it is a set of bank accounts or an engineering design project. This record can be made permanent if desired, by archiving the database at intervals, so that an "audit trail" is available; this is particularly valuable for design reviews, for example. Second, the management system should ensure that the database is internally consistent; so that if one makes a minor change in a subsystem—an amplifier gain, perhaps—then the consequences of that change are accounted for when answering any future queries. (This is a facility taken for granted in "spreadsheets," which are particular kinds of "fourth-generation" interfaces to database systems.) Third, the existence of a database management system allows the provision of user-friendly interfaces which support *ad hoc* queries and updates, of the kind given in the earlier examples. The reason for this is that a well-designed database contains entities at the same level of abstraction as the concepts at the forefront of the user's mind. Finally, as already mentioned, a database management system can provide a very effective integration platform, largely because it provides a convenient channel for exchanging data between various pieces of application software at precisely the level of abstraction required by that software—for example, if one algorithm needs a whole system, including various representations, comments, experimental conditions, and so on, this can be passed to it as a single entity; if another algorithm requires a single matrix extracted from that system, it can be passed just that matrix.

In the next section, we shall demonstrate the capabilities of an advanced software environment for control engineering, which provides a "fourth-generation" interface to the user. Having convinced the reader of the utility of such a facility (we hope), we shall then go on to discuss the functionality which is required, and how we have provided it. We then discuss the utility of the facility, considering various tradeoffs which occur.

21.2   AN EXAMPLE SOLUTION: DB-Prolog

We have developed a system which we call DB-Prolog. This has advanced database management capabilities together with general programming capabilities, a single uniform language being provided for both kinds of activities. It has access to external application software for performing certain computations—in our case MATLAB for general "scientific computing" and control-theoretic calculations, and TSIM for dynamic system simulation—and provides a graphical interface for the management of block diagrams and similar entities [15,16]. It also provides "forms" for querying and updating the database in cases for which the graphical interface is inappropriate.

We have specialized DB-Prolog to Computer-Aided Control Engineering (CACE) by defining three "environments" within it. The first is for "Control system analysis and design," by which we mean largely linear analysis and design of the kind supported by control theory. The second is for model building, and the third is for simulation.

21.2.1   Control System Analysis

The first example shows how DB-Prolog can be used to work through the F-14 IEEE Benchmark example [12]. The problem is specified by the block diagram shown in Figure 1, which shows the pitch control loop of the F-14 aircraft. Various properties of this loop are to be computed. The first thing to do is to enter the system description into the DB-Prolog database. Because the problem is specified by a block diagram, it is very appropriate to use the graphical interface for this task, and Figure 2 shows the result. In this figure, it can be seen that the problem has been structured more hierarchically than in Figure 1, with the blocks labeled dynamics, filters, and comp being aggregations of some of the blocks visible in Figure 1. The graphical interface allows the user to "zoom" into any of these aggregated blocks and obtain the block diagram of its constituents. The graphical interface and the underlying database allow arbitrary nesting of such structures. The database also contains, or is capable of finding, aggregated attributes of these structured blocks, such as their time responses or their state-space representations. Figure 2 shows some queries being made by means of pulling down menus on the graphical interface.

Now, suppose that the user wishes to issue the query "Is the dynamics block (shown in Figure 2) stable?" She can do this as follows. First, she points to the dynamics block with the mouse. This causes a menu to appear with

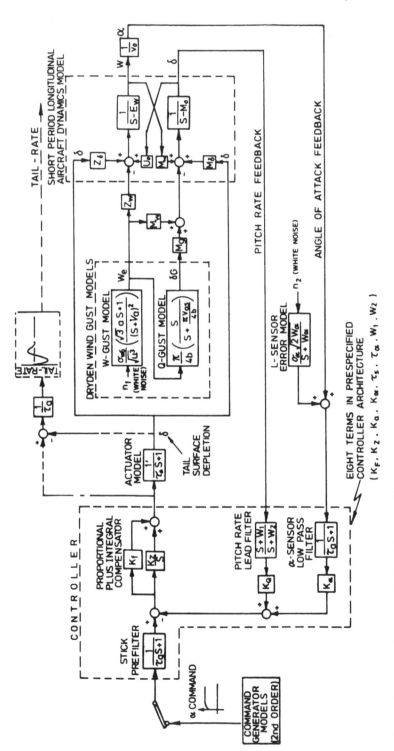

Figure 1 Block diagram of F-14 pitch control loop.

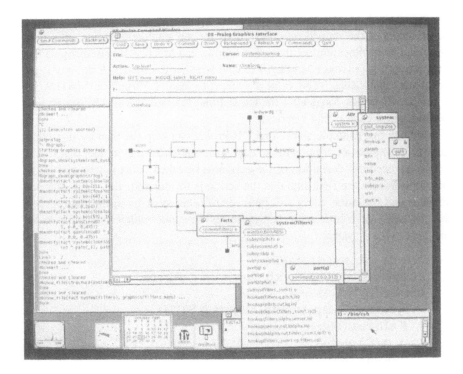

Figure 2 F-14 loop in DB-Prolog's graphical interface.

options "Operations," "Attributes," and "Facts," and she selects "Attributes." Another menu then opens; this contains a number of attribute names for *dynamics*, and the user selects "stsp" (the "state-space" attribute). When this selection is made, the string

system(dynamics)^stsp(Inputs,Outputs,A,B,C,D)

is written in the DB-Prolog Command Window, which is visible in the top left-hand corner of Figure 2. This string requires a little editing to formulate the desired query, and in this sense the Command Window resembles a "form" in that it provides a "skeleton" which the user has to complete. In this case, the user has to append the string

,#all(real(eig(A))<0==1

to complete the query, then click on the Send Commands button to execute the query. The explanation of this is the following: The comma is the standard Prolog way of expressing the conjunction of two statements, and Prolog "unifies" the argument A which appears in both statements. That is, Prolog first determines the "A" attribute of the system *dynamics*, then uses that attribute as the argument of the second statement. This second statement begins with a #, which in DB-Prolog signifies that what follows is a MATLAB statement; so the second statement invokes MATLAB to find whether the matrix A has any right half-plane eigenvalues. (The value 1 is used to signify "true" in MATLAB.) Note that DB-Prolog looks after the translation of A into MATLAB format, and the translation back of the response.

Although this example shows that DB-Prolog can help the user by supplying her with the names of entities in the database, the reader may object that it still involves procedural definition of the action to be taken. This is true and is, in fact, unavoidable the first time that the action is formulated. But if the user anticipates that the same query will be made of other systems in the future, then she can easily do a little more editing to define a *rule* which will apply to all entities of type "system":

```
attribute system(S)^stable.
rule system(S)^stable with (key = [S]) where
  (system(S)^stable :-
      system(S)^stsp(Inputs,Outputs,A,B,C,D),
      # all(real(eig(A))<0)==1.
    system(S)^stable :-
      < other rules >
  )
```

(Note that we use the entity type "system" to represent linear time-invariant systems.) The particular system called dynamics has been changed (by editing) to the variable name S. (It is a Prolog convention that variable names begin with uppercase letters.) Clicking on the Update Database button will add the rule to the database. If the "Attributes" menu is now pulled down for a "system" entity the "stable" attribute will appear on it. Selecting this by clicking will cause something like system(dynamics)^stable to appear in the Command Window, and clicking the Send Commands button will cause the query to be answered using the new rule. So from this point on, stability can be queried in a declarative manner. The example shows that several rules can be defined for determining stability. <other rules> can be alternative rules for establishing the stability of systems, such as the following rule for determining stability from a transfer function:

```
system(S)^trfn(Inputs,Outputs,Numerator,Denominator),
#(length(Numerator)<=length(Denominator)&all(real(roots
(Denominator))<0))==1;
```

Note however that if DB-Prolog contains rules for converting transfer-function representations to state space, then this rule would only be invoked for systems which could not be converted to state-space form, such as improper or irrational transfer functions, because DB-Prolog will invoke conversion to state-space form when trying to satisfy the first rule. Note also that the stability of hierarchical systems which contain subsystems will also be determined by these rules, providing that DB-Prolog contains rules which tell it how to obtain state-space or transfer-function representations of such systems from their structural descriptions. (DB-Prolog currently has such rules for linear systems.)

The example also has the phrase with (key = [S]). This is not essential, but it tells DB-Prolog that a "system" has only one "stable" attribute, so that it will never waste time trying to find more than one "stable" attribute for a given "system." (In database parlance, S is a "primary key" for entities of type "system.") This shows that the DB-Prolog user/programmer must have some knowledge of the way DB-Prolog works to use it effectively.

It is also worth noting that the addition of rules of this kind cumulatively builds up a high-level knowledge base in the domain of control engineering.

21.2.2 Model Building

The next example shows how DB-Prolog is used for "model building." This is perhaps the most time-consuming stage of control engineering. We shall consider a simple problem: modeling an electric circuit with just a few components. A circuit is also defined most easily by means of a graphical interface, though in this case a "block diagram" is not quite appropriate. The problem is that, although the individual components can be represented by "blocks," their input-output causality is not defined until all the circuit connections have been made, and the semantics of junctions is different from that implicit in block diagrams [5].

Figure 3 shows an electric circuit (called network) consisting of a voltage source (labeled e), a capacitor (labeled c), and three resistors. The circuit has been defined by mouse operations in the graphical interface. The fact that r2 exists and is of entity type resistor has been stored in the database. This can be seen on the facts menu. By "pulling down" a further menu from this fact, the set of facts specific to this resistor has been made visible. The fact that r*ali = v is an equation describing the behavior of this resistor (this is Ohm's Law) has been highlighted, with the consequence that this fact has been written to (the third line of) the Command Window, with the full database reference being given there. The first two lines in the Command Window are the results of two earlier mouse–menu interactions. In this way, all the equations describing the behavior of r2 have been positioned in the Command Window, perhaps as a prelude to editing them to formulate some complex query. (Further facts about r2, such as the value of the parameter r, will become visible in the resistor(r2) menu as they are added.)

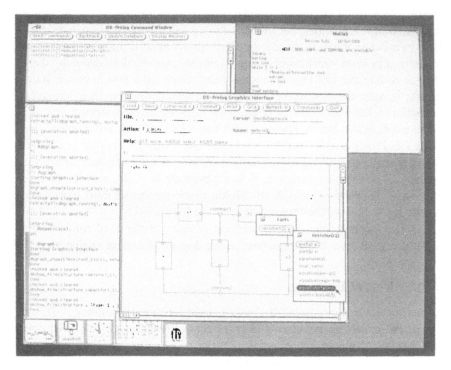

Figure 3   Electric circuit modeled by DB-Prolog.

The two junctions shown in Figure 3 (labeled common1 and common2) have semantics which are inferred from the nature of the "ports" which they connect. Each port of all the components in an electric circuit is defined to contain two "channels"; one of these is an "across" variable (the voltage) and the other is a "through" variable (the current). DB-Prolog contains rules which define the behavior of these two kinds of variables when they are connected together and is able to infer the consequences of these rules. For example, Kirchhoff's current law is inferred for common1 and the corresponding fact about the circuit is stored in the database:

$$r1\hat{}b\hat{}i + r2\hat{}a\hat{}i + c\hat{}a\hat{}i = 0$$

The models allowed in the model-building environment are again allowed to be hierarchical, so that the whole circuit model could become a submodel of a larger model. The environment which we have provided can be specialized to various applications by defining suitable subtypes of "model." In some cases, it may be worthwhile to specialize the graphical interface too. For instance, to support process modeling it would probably be desirable to provide a richer set of "icons" than just "blocks." One could have icons for "separators," "cooling towers," "compressors," and so on, each of these being a subtype of "model." (This could be done for electric circuits too, of course.)

21.2.3 Simulation

The third component of our CACE environment is an environment for simulation. It is advantageous if a simulation environment is to some extent independent of the simulation languages which are employed. This aids migration from one language to another and the exchange of information between organizations using different simulation languages. There are also tasks, such as "flattening" of hierarchical models, or equation sorting, which are logically independent of any particular simulation language and which are more naturally programmed in an environment such as DB-Prolog than in the languages in which simulation languages are usually written. Furthermore, writing rules for such tasks in DB-Prolog makes the whole environment more open: The user can check exactly what the rules are and modify them if necessary.

The next example shows the use of DB-Prolog for simulating a dynamic system. The actual simulation (i.e., numerical integration of differential equations) should not be, and is not, performed in DB-Prolog. That is a job for special-purpose simulation software, and DB-Prolog sends this task out to TSIM. But there is much work which a control engineer has to do before that can be done, and here DB-Prolog can provide a lot of help.

One of the preliminary tasks is the identification of any ill-conditioned algebraic loops, which may indicate a modeling error and certainly indicates potential numerical difficulties. One may wish to find even well-conditioned algebraic loops because they usually slow a simulation down. One can envisage automatic ways of dealing with such loops, or ways which require intervention by the modeler, for instance to decide how to remodel a system to remove a difficulty. In either case, the first step must be the determination of whether any algebraic loops are present. Figure 4 shows a block diagram with some blocks having more than one input or output port, and some of these ports containing several channels. It also shows that the user has queried the database for the existence of any algebraic loops in this system, and the second (larger) menu is the response to that query—it shows that there are several algebraic loops and the length of each one. Figure 5 shows a particular loop being queried (on the menu) and, in response, the loop being displayed on the block diagram. This example is incomplete because the next step should involve DB-Prolog either offering advice to the user on what to do next or taking some action itself. However, we have not yet entered rules for either of these possibilities into DB-Prolog.

This is a typical "presimulation" activity which needs to be performed. Others are setting up particular simulation "experiments" and the translation of models into simulation languages, possibly including the assembly of submodels which may have been written in simulation languages other than the target language. (But we have not implemented such functionality yet.)

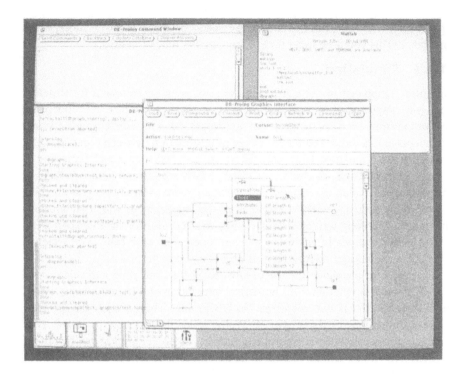

Figure 4   Querying for algebraic loops.

Our final example shows the use of DB-Prolog in use as an integrating tool. Figure 6 shows a block diagram of a position control system (called position) in the DB-Prolog Graphics Interface. Menu interactions (not shown) have been used, together with some editing, to construct the line of code.

dbcall tsim(position)ˆplot(2, Subsystem, Port).

in the DB-Prolog Command Window. This single line has the effect of invoking the TSIM simulator to simulate the system and passing the results to MATLAB, which displays them in the MATLAB Graphics Window. DB-Prolog handles the translations required among itself, TSIM, and MATLAB and can be instructed to store the results in a structured way if desired.

21.3   FUNCTIONAL REQUIREMENTS

21.3.1   Requirements for Control Engineering

We can infer some of the required functionality from the examples given above. It should be possible for the user to define various kinds of entities which cor-

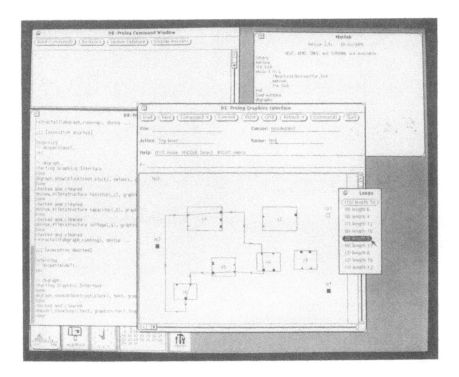

Figure 5 Querying an algebraic loop of length 8.

respond to either actual or conceptual entities that occur in control engineering: dynamic models, linear models, frequency responses, and so on. The need is not just for the definition of the *data structures* but also of their *behaviors*—for example, how to compute the RMS value of an output variable of a linear system or how to find the algebraic loops in a simulation model.

In addition to defining entities and their behaviors, the user must have convenient means of querying and updating instances of those entities. She must also be able to change her mind, so that entity definitions and/or behaviors can be changed in the light of experience. These two requirements must be met simultaneously, which is quite challenging.

The user must be able to access external application software, such as MATLAB or TSIM, as naturally as possible, and should be able to formulate queries which involve calls to such software relatively easily.

21.3.2 Requirements for the Software Environment

Generalizing from the requirements listed above, we see that, first, we need the capability to define data types (structures and behaviors) within some *semantic*

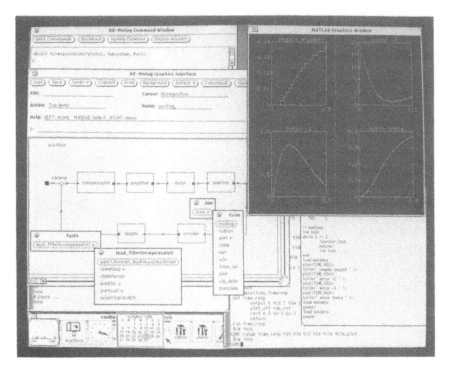

Figure 6   Using DB-Prolog to invoke TSIM and MATLAB.

data model [7]. This term is used to denote data models which are more powerful than the relational model used in commercial applications, and hence open the possibility of defining entities in the database which are in close correspondence to entities in the application domain. Because both structures and behaviors must be defined, this is close to requiring an object-oriented data model. However, there is no particular need for the behavior definitions to be bundled together with the structure definitions (although it is advisable to do this) or for access to these definitions to be restricted by dividing them into "public" and "private" parts. In fact, control engineering at least seems to require a rather random pattern of accesses to parts of entities, and this indicates that strict object orientation is probably not desirable.

An important requirement is that the data schema (namely, the *kinds* of entities representable in the database, roughly equivalent to data *types* in programming languages and to *classes* in object-oriented terminology) should be modifiable dynamically while the environment is running. It should not be necessary to shut the environment down to make such modifications. Inheritance of entity types should be provided to enable efficient type definitions by reusing existing ones.

The long-term storage capability of a database management system is not directly indicated as necessary by our examples. However, the organizing power of such a system strongly indicates that it should form part of the solution, and the long-term persistence is a very welcome and significant bonus resulting from that.

In addition to formulating queries and updates to a database, there is a requirement for general programming capability in order to formulate *rules*, namely, algorithms for deriving data when answering queries, and *constraints*, namely, algorithms for checking the consistency of data in the database. The two capabilities together, if provided in a "seamless" way so that the user does not have to switch between two radically different language styles, constitute a *database programming environment* [2,3]. If existing applications software such as MATLAB is integrated with the environment, it seems unavoidable that the user be exposed to the existing language style, which may be rather different from the style of the parent environment, at least if the full capabilities of the applications software are made available to the user. This is evident in one of our examples. This should be less of a problem with newly written software if its use in an environment such as DB-Prolog is anticipated when it is developed. It is also possible to present the user with a more uniform language by providing a "filter" between the user and the application software, as was done in the GE-MEAD project [13], although this entails losing some of the functionality of the external software.

For a more detailed analysis of the functional requirements of a system for Computer-Aided Control Engineering (CACE), see Ref. 17. The overall requirement is for a system which can certainly be considered an object-oriented database system, as judged by the criteria used in Ref. 4, for example.

21.4 THE DB-Prolog IMPLEMENTATION

21.4.1 The Structure of DB-Prolog

DB-Prolog is best thought of as being layered. At the lowest layer, it consists of the logic programming language Prolog, which is coupled to a database management system (DBMS). This DBMS is not at all sophisticated and its purpose is to provide a persistent (disk-resident) version of the usual memory-resident Prolog database. It is, in fact, implemented using the UNIX "dbm" tool. Also at this lowest level, loosely coupled interfaces are provided to external software. Currently, we have interfaces to MATLAB and to TSIM, which is a simulation language [19]. External software is coupled through "mail box" files, but an interface based on pipes could also be developed. This kind of interface is quite versatile, with the particular merit of being able to handle external software which is designed to be used interactively, rather than through subroutine calls. Adding an additional external software product requires a little tailoring to match that specific product, but the same generic mechanism works in all cases.

The next layer defines and supports a "semantic data model" [7]. This is analogous to the "relational data model" supported by relational DBMSs such as Dbase or Ingres, but it is much more suitable to the needs of engineering design. It is "semantic" in the sense that it can be tailored to match closely the semantics of the application area. This data model allows entity types to be defined, and these types can be specialized to entity subtypes, which automatically inherit properties of the parent type; for example, the entity-type "model" may have a subtype "tank." Entity types can be structured by being endowed with attributes; a "model" may have attributes such as "name," "port," "equations," and these attributes may themselves be entities of considerable complexity. A "tank" may have additional attributes such as "level," "capacity," and so on.

A very important mechanism for capturing semantics is provided by *constraints*, which can be imposed on and between attributes. These constraints can be of arbitrary complexity, both with regard to expressing them and checking their satisfaction, which may require an arbitrary amount of computation. The expression of complex constraints is made possible by the general power of the Prolog language; their computation is made possible by the availability of external software, such as MATLAB, which can be called on as necessary, together with the deductive capabilities of Prolog. A relatively complex constraint which may be imposed on an entity of type "controller" may be that "it should be open-loop stable," for example. If this is imposed, then entities which are not open-loop stable will not be allowed as entities of type "controller" in the database.

Individual entities (objects) are instantiated by stating *facts*, such as Aircraft(F14) which declares that "F14" is an entity of type "aircraft." Some of the attributes of an entity may, however, be computed automatically by *rules* which are associated with entity types—these are similar to "methods" in object-oriented systems. For instance, there may be a rule for computing the "transfer-function" attribute of an entity of type "system" if its "state-space" attribute is already known.

Typically, design entities evolve through their lifetime. Many different versions of the same entity may be produced. These entities can be related using the built-in version control mechanism based on the *ivo* relation (read as: *is-a-version-of*). For example, a modified version of "F14" may be defined as

```
fact Aircraft(F14-mark2) ivo Aircraft(F14) ^ [
        except( ... ) /* unwanted old attributes */
].
```

Queries on "F14-mark2" will then be evaluated first against the "F14-mark2" entity. If the required data is not found, the query will be subsequently evaluated on the "F14" entity except for the omitted attributes. This process is transitive in the sense that data may be inherited from any older version recursively.

The third and highest layer of DB-Prolog is application-specific. In this layer, the facilities provided by the middle layer are used to tailor the environment to a particular application area. The tailoring process consists of

1. Modeling the concepts and data in the application area in terms of DB-Prolog entity types.
2. Providing an appropriate user interface for the manipulation of these entities. The interface may be as simple as a set of alpha-numeric commands, or as complicated as a graphics interface coupled to an expert system.
3. Interfacing DB-Prolog to any necessary external software packages. For example, MATLAB is useful as a "back end" processor for CACSD applications, whereas TSIM is good for simulating dynamical systems.

Multiple applications can be run concurrently in DB-Prolog. Due to their common execution model and syntax, it is very easy to cross application boundaries and make use of facilities in other applications. For example, a user running a simulation in TSIM can generate a set of test signals in MATLAB, feed them into the simulation, and display the results back in MATLAB, all in a single command, as shown in Figure 6. Furthermore, the entity types developed for individual applications can themselves be cross-related, resulting in a single homogeneous computer-aided engineering environment.

DB-Prolog is implemented in Poplog, which is a software development environment in which several programming languages are available [14]. The two languages which have been used to implement DB-Prolog are Prolog and Pop-11.

21.4.2 The CACE Environment

For the Control Engineering application area, we have provided the entity-type *block*. This has attributes of types *port*, *subblock*, and *hookup*. *Blocks* are supposed to correspond to blocks such as those found in block diagrams; however, all their attributes are structural: *ports* are interfaces to the rest of the world, *subblocks* are lower-level constituent blocks, and *hookups* describe how the subblocks are connected together (in fact, which ports are connected together). Our *blocks* correspond to Cellier's "stylized block diagrams" [5]. This entity type is the only one which is common to the whole of the CACE environment, as shown in Figure 7.

The entity names used here do not correspond exactly to those used in Ref. 18; we now prefer this set of names. One of the important features of DB-Prolog is that it is very easy to make such changes. We enforce the constraint that each *subblock* is itself a *block*:

```
constraint block(B)^subblock(BB) with (key=[B,BB]) where
        block(B)^subblock(BB) :-(BB).
```

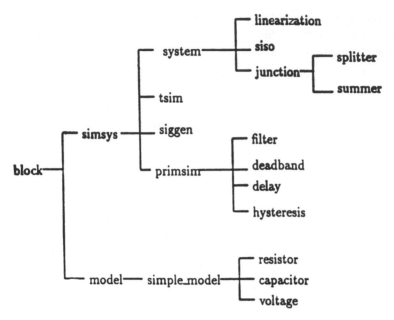

Figure 7   The type hierarchy provided so far for CACE.

The graphical interface referred to in previous examples has been developed
to deal with entities of type *block*, and it is tied to the structure of *blocks*: It can
be used to represent connections between "ports," to zoom in on a block and
display the structure of "subblocks" within it, and to interrogate the attributes
of a particular block by clicking on it with a mouse. This whole interface is (X)
window-based and is in the modern "WIMP" style. The graphical interface is,
in fact, a separate program, loosely coupled to DB-Prolog in a similar manner to
that used for coupling to application software. It works by assembling sequences
of DB-Prolog statements, which are executed by DB-Prolog. In other words,
one could get exactly the same DB-Prolog functionality by typing in DB-Prolog
statements oneself at the keyboard.

For linear control system analysis, we specialize *blocks* to represent (linear)
systems by defining new entities, using DB-Prolog statements such as

```
attribute system(Name)^
        stsp(Inports,Outports,A,B,C,D).
system(S) isa block(S).
```

For model building a suitable paradigm of what is required is provided by the
modeling language Dymola [6]. (An alternative would be the more recent
Omola [1]). We have introduced the entity-type *model* for this application,
which is based mainly on the Dymola language, but with some modifications, to

be more compatible with the existing *block* entity type. *model* is defined as a subtype of *block*.

To allow several signals to be bundled together in one *port*, each *port* has an arbitrary number of *channel* attributes, as proposed in Ref. 10. A channel may carry one of two types of signal: *through* or *across*. The significance is that "through" variables meeting at a point must sum to zero, whereas "across" variables must be equal. Moreover, the units of the connecting channels must be the same and, in a port with multiple channels, the ordering of the channels within the ports much be correct also. A constraint has been written to check for inconsistencies. The resulting definition of the "port" attribute is as follows:

```
attribute model(Model_id) ^
        port(Port_id) ^ channel(Order, Channel_id, Across_or_
        through, Unit).
```

Note that this is the definition of a *port* entity when it is an attribute of a *model*. A *port* which is an attribute of a *system* has a different definition because its semantics are different: It has directionality (*input* or *output*), and the meaning of a connection, and of a junction involving several ports, is more rigidly defined. The *port* attribute of a *block* exists, but has no definition.

Following Dymola, the dynamic behavior of a *model* can be completely defined by a set of differential algebraic equations.

For simulation, we have defined two sets of entity types. The first set is a language-independent set, which supports the description of simulations at a language-independent level. The second set is language-specific. So far, we have only defined entities in this set for the TSIM simulation language, but we envisage providing similar entity types for Simulink, and perhaps ACSL. The language-specific entity types are isomorphic to entities which appear in the corresponding language and, in effect, provide an interface to that language.

The entity type which represents a language-independent simulation is *simsys*. This is a subtype of *block*, as defined previously, but with directed *ports*. This means that entities of type *simsys* can again be manipulated by the graphical interface. A *simsys* entity has only one attribute, *defn*, which is its definition. Some of the information contained in the *defn* attribute is redundant in the sense that it is already implicit in its *hookup* attribute. However, in *defn*, such information is in a nonhierarchical "flattened" form, which has been extracted from the hierarchical description available in *hookup*. In particular, it may contain the equations correctly ordered for simulation. The type definition of *simsys* is

```
attribute simsys (name) ^
        defn(Ports, Vars, Params, States, Cons, Eqns).
simsys(S) isa block(S).
```

The *system* entity type (which represents LTI state-space systems, remember) has, in fact, been defined as a subtype of *simsys*, as shown in Figure 7. This allows any *system* to be simulated, and it does no harm because there is nothing wrong with regarding a linear state-space system as a kind of simulation. The fact that it has additional attributes and properties, not shared by other kinds of *simsys* entities, allows us to do more with *systems* than just simulate them.

It is debatable whether entities such as *filter* and *deadband* should be part of the model building or of the simulation environment. It seems to us that these entities are considerably abstracted from the phenomena which they represent and cannot be introduced until questions of causality have been thoroughly understood for a particular model. They are, therefore, more appropriate in the simulation environment, and so are defined to be subtypes of *simsys*, as shown in Figure 7. On the other hand, standard components which are not abstracted to the same degree, such as resistors, capacitors, valves, boilers, or amplifiers, are more appropriately represented by *model* rather than *simsys* entities.

TSIM has been integrated with DB-Prolog at two levels. The first, lower level, is between Poplog and TSIM via a modified mailbox protocol for sending text commands from DB-Prolog to TSIM and receiving data from TSIM. The second, higher level, is at the entity-type level, where the TSIM-specific *tsim* entity type has been designed to represent abstract models of simulation processes running in TSIM. *tsim* is a subtype of *simsys*, which contains additional attributes required by TSIM; these hold information on which integration algorithm is to be used, details of the particular integration run required, and so on. The user can simulate a *simsys* entity in TSIM just by making it into an attribute of a *tsim* entity. Additional TSIM-specific attributes will be obtained by built-in rules and by means of a dialogue with the user. Commands issued to the *tsim* entity will be translated into TSIM commands and sent to the concurrently running TSIM process.

At the entity-type level, the *tsim* entity type is a specialization of *block* and contains attributes and rules which capture the concepts and data in TSIM, so that the user can interact with TSIM by manipulating and querying *tsim* entities only. Possible interactions range from the translation of *simsys* entities into TSIM processes, assigning/modifying initial values of states and parameters, setting up time-response runs, feeding input data from external sources, executing the runs and recording the outputs, perturbing the system to get a linearized model, and so on.

More details of the entity types provided in the CACE environment are available in Ref. 11.

21.5 EVALUATION

Is it all worthwhile? Does the use of an environment such as DB-Prolog help the engineering user in the ways that we outlined in the Introduction? Our experience so far leads us to think that an environment of this kind is potentially very helpful, although there are, no doubt, features of DB-Prolog's functionality, language, and interface style which could be improved.

A key question is whether the "overhead" incurred by the user in having to face the increased complexity of an integrated environment outweighs the benefits which are obtained. After all, software such as MATLAB is popular largely because it is extremely easy to use, at least in the sense that one can easily develop algorithms with it. The answer depends, first, on the details of the language and interaction style; whereas these can undoubtedly be improved, one imagines that there is a limit to the improvement which can be obtained and that any really flexible and general-purpose environment will have to incur a certain penalty of complexity. The answer also depends very much on the pattern of use. If the user spends all her time doing once-off computations, for example, computing the RMS value of the output of a linear system excited by white noise, and never repeating those computations, then it is probably faster and easier to work directly with software such as MATLAB. The payoff comes if such computations are expected to be repeated; in that case, they are made into rules which are attached to specified entity types, and they can subsequently be invoked not only simply and reliably by the user when required but also automatically by the environment when it is in the process of answering more complex queries.

For most users, this is the more typical pattern of use. It can be expected that initially a relatively large proportion of effort is devoted to defining rules for the environment, but that this rapidly reduces as the most useful rules become "known." When the relatively stable period is entered in which the problem of specifying data (supplying the correct arguments, in conventional programming terms) becomes greater than the problem of defining rules (algorithms), then the use of an environment such as DB-Prolog becomes more convenient than the conventional applications software to which we are currently accustomed.

Of course, the amount of reuse of entity-type definitions and rules which is possible depends on having well-designed types and rules. It is largely for this reason that the ability to redefine these is so important: The user must be able to *redesign* the environment in the light of experience. Also, one anticipates that a good set of type definitions and rules for each application area would be discovered relatively quickly and that, thereafter, each user could be supplied with this set to provide at least an initial environment which would be immediately efficient.

This assessment is probably too pessimistic because it assumes that data is easily specified when using conventional software. When defining the rules for the detection of algebraic loops, for instance, this was not true. The only information which is required in this case is very high-level information about subsystems, and the pattern of their interconnections—essentially whether a particular subsystem has a direct (nondynamic) connection between its inputs and outputs, and how it is connected to other subsystems which have such direct connections. If this information is available when using MATLAB, for example, it is buried in a very obscure way in some matrix components of subsystems, or possibly held in variables specially created and named for that purpose. The ability to construct a DB-Prolog predicate which specifies unambiguously and clearly the information required is a great help in such cases, even during the initial development of rules. This is not to say that the development of such rules necessarily becomes trivial. We found, for example, that it took about 2 weeks to write the rules for the detection of algebraic loops. But, in retrospect, it is clear that all that time was spent on "high-level" tasks such as researching the required algorithms and deciding the required functionality. The "coding" itself was then very straightforward, and very little time went on low-level "overheads."

There is clearly a tradeoff between the ease of use of the environment and its flexibility. This is apparent in our Command Window, in which we attempt to provide a "forms" type of interface. This is clearly quite a long way from "forms," as they are typically provided in commercial applications, because the user may have to do a substantial amount of editing of the strings which DB-Prolog writes in the window rather than just typing in simple responses in specified areas. But we think that it serves essentially the same purpose as a "form," namely, it restricts the attention of the user to those matters which DB-Prolog cannot decide on its own. An earlier version of the Command Window was, in fact, closer to a conventional "form," but we abandoned that because we found it to be too restrictive: It made certain tasks easier, but others impossible. Of course, one could think of providing alternative styles of "forms," with more restrictive ones provided for some users, possibly in conjunction with an "expert advisor" on how to use the environment.

One problem which we have encountered is that changes to the environment sometimes require changes to the graphical interface. When implementing the facility for detection of algebraic loops, for example, we added the capability to display lines showing direct connections between the inputs and outputs of blocks, as shown in Figure 5. Until then, lines could only be drawn externally to blocks, not internally. Although this change was straightforward to implement, it required some familiarity with X-Window programming [15]. This is in contrast with the addition of the rules which actually find the loops, which were coded in the DB-Prolog language itself. Ideally, such modifications to the graphical interface should be done by a user working at a higher programming level, perhaps by "dragging" and otherwise manipulating icons. This is not possible at present, but will probably become possible when better graphical interface building tools become available.

On the whole, however, we have found the flexibility inherent in DB-Prolog to be extremely powerful. We have modified the entity-type definitions for

CACE several times, and we are by no means sure that we have found the optimal ones yet. In any case, the set is surely not yet complete, and probably never will be. Also, the environments for modeling and simulation were created at a very late stage in the development of DB-Prolog, which involved undoing some of the previous entity-type and rule definitions and defining new ones. The fact that we were able to do this provided a very convincing justification of our approach. (For more details of what this involved, see Ref. 11.)

The use of Prolog has proved very beneficial to us. Not only is it a good language for rapid prototyping, it is also a natural language for database queries and updates [9]. At present, it executes rather slowly on many machines, but much improved performance is anticipated on alternative architectures.

21.6 CONCLUSION

We have described a software environment for Computer-Aided Control Engineering which is radically different from those proposed previously. It goes some way toward providing engineering users with a style of interaction which is as comfortable and productive as that enjoyed by commercial users of computers while respecting engineers' more challenging requirements. We have gone a considerable distance toward providing a declarative style of interaction and "fourth-generation" interfaces.

We emphasize that DB-Prolog is a vehicle for proving ideas. We do not make any claims about it being the ideal environment for CACE, but we believe that future software environments for much engineering analysis and design will exhibit features present in DB-Prolog:

* User interaction by means of querying and updating a semantic database, assisted by convenient interfaces
* Easy integration of an infinite range of external application software
* A single language for both data management and data definition (i.e., for querying, updating, and "programming")
* Openness of most aspects of the environment, allowing user modification of structured data types and algorithms
* Permanent (disk) storage of the results of analysis and of design decisions

REFERENCES

1. Andersson, M., Omola—An Object-Oriented Language for Model Representation, Report CODEN: LUTFD2/(TFRT-3208), Dept. of Automatic Control, Lund Institute of Technology, Sweden (1990).
2. Atkinson, M. P., and Buneman, P., Types and persistence in database programming, *ACM Comput. Surveys*, *19* (1987).

3. Atkinson, M. P., Buneman, P., and Morrison, R., Binding and type checking in database programming languages, *The Computer Journal, 31* (1988).
4. Cattell, R. G. G., *Object Data Management: Object-Oriented and Extended Relational Database Systems*, Addison-Wesley, Reading, MA, 1991.
5. Cellier, F. E., *Continuous System Modeling*, Springer-Verlag, New York, 1991.
6. Elmqvist, H., A Structured Model Language for Large Continuous Systems, Report CODEN: LUTFD2/(TFRT-1015), Dept. of Automatic Control, Lund Institute of Technology, Sweden (1978).
7. Hull, R., and King, R., Semantic database modelling: survey, applications and research issues, *ACM Computing Surveys, 19* (1987).
8. Eastman, C. M. and Lafue, G. M. E., Semantic integrity transactions in design databases, in J. Encarncacao and F-L. Krause, eds., *File Structures and Data Bases for CAD*, North-Holland, Amsterdam, 1981.
9. Gallaire, H., and Minker, J., (eds.), *Logic and Databases*, Plenum Press, Oxford, 1978.
10. Maciejowski, J. M., A Core Data Model for Computer-Aided Control Engineering, Report CUED/F-CAMS/TR.257, Cambridge University Engineering Dept., Cambridge, 1985.
11. Maciejowski, J. M., and Tan, C. Y., Control engineering environments in DB-Prolog, *Proc. 1992 IEEE Symposium on Computer Aided Control System Design*, Napa, March 1992, pp. 55–61.
12. Rimer, M., and Frederick, D. K., Solutions of the Grumman F-14 benchmark control problem, *IEEE Control Syst. Mag., 7*, (1987).
13. Rimvall, M., and Taylor, J. H., Data-driven supervisor design for CACE package integration, *Proc. IFAC Symposium on Computer Aided Design in Control Systems*, Swansea, July 1991, pp. 33–38.
14. Sloman, A., and Hardy, S., *The POPLOG Environment—User's Guide*, University of Sussex, Sussex, United Kingdom, 1982.
15. Tan, C. Y., A Graphical Interface to DB-Prolog for Control Systems Design, Technical Report CUED/F-INFENG/TR43, Cambridge University Engineering Dept., Cambridge, 1990.
16. Tan, C. Y., 1990, *Database Graphics Interface User's Manual*, Technical Report CUED/F-INFENG/TR44, Cambridge University Engineering Dept., Cambridge, 1990.
17. Tan, C. Y., DB-Prolog: A Database Programming Environment for Computer Aided Control System Deisgn, Ph.D Thesis, University of Cambridge, Cambridge, 1990.
18. Tan, C. Y., and Maciejowski, J. M., Control systems design in a database programming environment, Preprints, *IEEE CDC Conference*, Brighton, United Kingdom, December 1991.
19. *TSIM User's Guide*, Cambridge Control Ltd., Cambridge, 1988.

22

Graphical System Input for Design

P. H. M. Li and J. O. Gray

University of Salford
Salford, England

22.1 INTRODUCTION

Control system design is a complex process in which the designer has to manage the various constraints imposed by physical laws, user requirements, and economic factors as well as to visualize the form, characteristic, and performance of the dynamical system in question. In large systems, interconnectivity complexity, the multiplicity of detail, and proliferation of data result in significant cognitive loading which strains the capacity of even experienced designers. Human beings are, in general, relatively slow information processors and are poor at performing repetitive and tedious tasks [1,2]. However, humans are good at handling abstract concepts and symbols, and monitoring an abnormal situation as well as the normal progress of system operation. Therefore, one of the goals of the study in Computer-Aided Control System Design (CACSD) has been to devise human computer interfaces whereby computers can be effectively harnessed to augment the natural capability of human designers in terms of information processing and mental capability.

This chapter commences by reviewing some of the interface formats employed by existing simulation and control system design packages and highlights the information mismatch between a designer's mental model (which could be in the form of block diagram representation) and the description formats normally used by these packages. An ''impedance'' mismatch generally exists between the two forms of description, and a graphical interface approach is proposed to

531

address the problem. The general requirements for a user interface in the context of CACSD environments is initially outlined which emphasizes the user friendliness, adaptability, flexibility, and extensibility of the software structure that should characterize such an environment. A description of the three functional requirements fundamental to the implementation of a graphical input system is then given which allows the direct diagrammatic description of dynamical systems for implementation by the appropriate CAD algorithms. These three functional requirements, namely, the graphical editing, graphical database management, and graphical database interpretation, are then elaborated further by using a specific example which describes the implementation of a graphical front end for the TSIM simulation language [3]. However, the philosophy adopted here for the development of a diagrammatic input facility for a CACSD environment is generic. Examples of the operation of the graphical environment are given and the results evaluated. Concluding comments on the graphical environment are given and the future directions of CACSD development are discussed with particular emphasis placed on the information management requirements of CACSD tools at an organizational level.

The simulation software package TSIM was originally authored by a small group at the Royal Aircraft Establishments [4] but is now supported and marketed by Cambridge Controls Limited. Although written specifically for aircraft control studies, the package is sufficiently general in nature to be applicable to other application areas. Digital as well as analog simulation can be performed allowing computer control procedures to be considered. Moreover, a mixture of linear and nonlinear dynamic equations can be simulated. The whole process of the package is based on the Fortran language and needs a mathematical model to be formulated before any simulation can be performed. Normally, the input format of TSIM consists of sets of linear and nonlinear dynamical equations which must be abstracted from the engineering configuration. Very often, a simulation or control system analysis problem is posed in block diagrammatic form and, thus, TSIM is an excellent target package for the implementation of advanced graphical input procedures but the approach adopted can be applied to a wide range of similar packages and algorithmic procedures.

22.2 GENERAL INTERFACE REQUIREMENTS

The general requirements for a human computer interface are that the interface should be easy to use and fit for its intended purpose [5]. Software development without due consideration for its potential users will be of little value to the community regardless of its functionality. On the other hand, user-orientated software renders its facilities accessible even to the tyro. In the context of CACSD, the function of the interface is to facilitate the mapping of the symbolic conceptual model of the human designer onto the numeric procedural format employed by the control or simulation algorithms. In engineering design, such conceptual models are often expressed in terms of graphical notations. A software environment which embodies such an interface should be capable of assisting the user in managing the complexity of design by providing a mechanism for the representation of hierarchical and composite design models [6]. Provision for future extensions should also be made to allow for the addition of new simulation or control algorithms. A CACSD software system which is designed without any consideration for its extensibility will soon be obsolescent as the requirements on the software continues to change with the progress in control technologies. As any control system design or a simulation package may not contain all the tools that a designer requires in a particular development, the software structure should be such as to allow the addressing of different packages within a single homogeneous environment [7].

The widespread use of Fortran language in the engineering community has influenced the development of simulation packages in that the model description language is very much similar in syntax and format to the Fortran constructs in which the software is written. Engineers usually communicate ideas in the form of concepts and symbols, and the tools used in the design of systems are usually diagrammatic and closely related to the mental models held by human designers in that abstract concepts and concurrent processes can be represented. Such a diagrammatic approach allows a natural expression of parallel activities which is generally easier to comprehend than the sequential disposition of parallel components in conventional textual programming languages. However, this diagrammatic form of representation is not fully supported by conventional control and simulation algorithms which usually require procedural and numerical inputs. For example, as shown in Figure 1, to input a dynamical model to a CAD environment, a control engineer has generally to translate mentally the conceptual model into a suitable format for the appropriate algorithmic processes: an operation which is prone to error and inconsistency. However, if a design environment had the ability to accept and interpret information in a more natural anthropomorphic way, errors generated at the human computer interface would be greatly reduced [8] and the resulting ease in communication would improve designer efficiency. A better approach would be to allow the user to describe his conceptual model graphically to the computer as shown in Figure 2. This graphical input capability means that facilities for the insertion, editing, storage, and interpretation of graphical information are required. The functional requirements of a graphical CACSD system are depicted in Figure 3, where the *graphical editor* is intended for the graphical definition and manipulation of system components and, with the assistance of this tool, the user is able to draw, on the screen, various symbols to represent the different system components and define

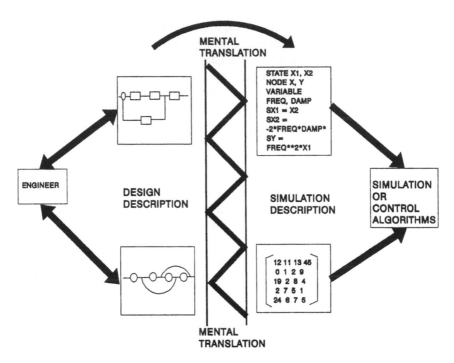

Figure 1   Impedance mismatch between the user's conceptual description and the simulation description.

their interconnection by joining the components using directed lines. Having formulated a system graphically, the graphical information has then to be stored for subsequent use. The *graphical database management system* plays an important role in providing a mechanism for the filing, maintenance, and retrieval of these graphical data. The operations carried out can be compared to a conventional database management system, the difference being that graphical instead of textual data are managed. At first glance, the graphical input requirements are similar in functionality to that of a conventional computer-aided drafting package which assists the user in drawing arbitrary diagrams, but, however, the similarities end here because the graphical objects defined in the graphical CACSD system are not just passive symbols but are able to embed in the pictorial representation certain functions, which when invoked, will perform active operations. To achieve this, a *graphical database interpreter* has to be implemented to extract the passive and active elements of a graphical object. This is a mechanism that enables the system to interpret the information contained in the object as well as the operations that have to be performed. Subsequent to this, the system simulation algorithm performs the numerical processing to determine the

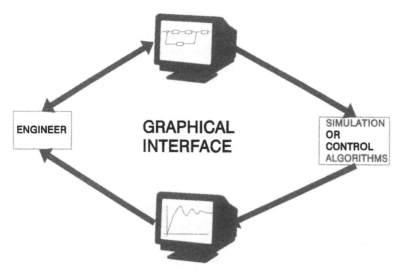

Figure 2 Graphical input of designer's conceptual model.

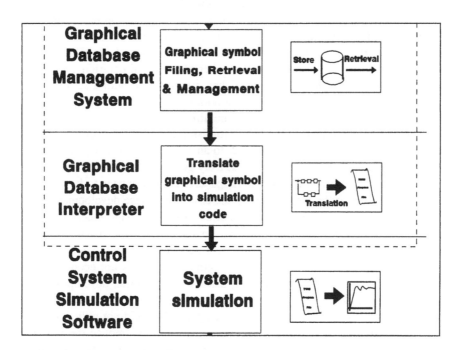

Figure 3 Functional requirements of a CACSD system.

characteristics and performance of the model formulated by the user and the results are then displayed through the results presentation function. In Figure 3, the functions enclosed in the dotted area that define the input requirements of CACSD systems will be described in more detail in the following sections.

22.3   A GRAPHICAL EDITOR

The purpose of a graphical editor is to support the creation, representation, and editing of graphical notations [9]. These may be classified into three categories:

1. General-purpose editors that offer a set of graphical primitives for drawing arbitrary diagrams
2. Notation-specific editors that possess a knowledge of the symbols of a graphical notation and how they may be connected or combined to generate diagrams
3. Generic editors that can be customized to support a wide variety of notations.

The first approach supports only the drawing of graphical symbols, but not the functions embedded within them and is, therefore, considered unsuitable for representing engineering graphical notations. The second approach has been adopted by a few proprietary tools [10,11] available now in the marketplace which support the block diagram definition of dynamical systems. These tools, which usually cover only a very small number of graphical notations, are often inadequate and inflexible, and, because they are equipped with a set of predetermined features built around a particular notation, cannot generally provide all the manipulation functions a user might require.

Generic editors are clearly the most promising candidates to address this issue and yet they are the least researched, to an extent that currently no credible tools are available that can be regarded as truly generic. Considering the current trends in the use of graphical notations, generic editors appear to be the only rational direction for the future. Taking the case of software development as an example, commercial software suppliers often extend or adapt widely used notations to suit their own working practices. In some cases, they even invent entirely new notations to serve the needs of a particular project. Because some notations may be used for only a short length of time, it is not economical to invest in developing tools for them and, usually, the potential development delays would be unacceptable. As a result, paper-and-pencil methods are often adopted for short-term solutions. A generic approach would not only overcome these problems in a simple and practical manner but also reduce the required diversity of such support tools. This implies that a potential user will only have to learn how to use one tool to work with many different types of notation. Furthermore, having a common tool eliminates the problems of integration and mapping between different notations.

22.4 A GRAPHICAL DATABASE MANAGEMENT SYSTEM

The function of a graphical database management system is to maintain a permanent self-descriptive repository of data which is called a database. It provides facilities that control the organization, storage, retrieval, security, and integrity of a database. Central to a database management system is the underlying database model which governs the operations and behavior of the system [12]. As the graphical database model is meant to support the representation of graphical notations such as the block diagram notation, the inherent data structure of the model should thus also capture the three levels of abstraction common to all graphical notations as follows:

- **Lexical** which denotes the set of symbols of a graphical notation, also known as the *alphabet* of the graphical notation.
- **Syntatic** which denotes the set of rules governing the combination of symbols in producing a diagram (or diagram segment) without regard for its meaning. A rule, in general, is understood to be predicate over the alphabet of the graphical notation. The set of all such rules is called the *syntax* of the notation.
- **Semantic** which denotes the meaning attributed to each symbol in the notation.

Taking the block diagram notation as an example, if the basic unit of a block diagram is a symbol, the *semantics* of the notation is then the dynamical or computational function attributed to each individual symbol. A library of predefined symbols such as shown in Figure 4 captures the *alphabet* of the notation. The *syntax* consists of procedural rules for combining symbols and would, for example, determine that a signal flow-graph element cannot be bidirectional. To facilitate the representation and manipulation of these three levels of definition, the database model should also support the following mechanisms:

1. Information latency
2. Inheritance
3. Aggregation

Information latency allows different aspects of an object to be selectively accessed. This can be achieved by devising a level code mechanism which associates each level of information with an attribute and by enabling or disabling the attribute; a particular level of information can be selected or deselected. If a symbol is considered, for example, to be a picture formed by the superimposition of a set of transparencies, the level determines to which transparency a subsymbol belongs. The concept is illustrated graphically, in Figure 5, by the analogy of having different types of graphical information on a map of the United States placed on different transparencies. By inserting or removing a

MENU 1	MENU 2	MENU 3	MENU 4	MENU 5	
Switch	Delay	Absolute	Carpet	Delay(logic)	QUIT REDR COMP HELP
And_gate	Integrator	Sign	Interpolat	Limit	ALL ADD SUB
Or_gate	Integrator	Lag	LVDT	Function_g	MODE PAGE YMEN XMEN
Xor_gate	Multiplier	Deadband	RVDT	Testdata	MENU SCAL
Inverter	Divider	Hysteresis	Timer	Highest_wi	-16 23 21 15
Sum(2)	Power	Comparator	Timeout	Lowest_win	14 11 10
Subtract	Remainder	Magnitude	Hold(trig)	Input_val	8 7
Sum(3)	Z_transfor	Hysteresis	Hold(samp)	Output_var	
Summer	Square_roo	Quantizati	Hold(logic)	Define_var	
Product	Flow_root	Table	Delay(cont)	Driver	

Figure 4   A library of control system building blocks.

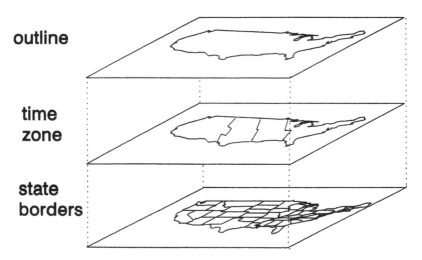

outline

time
zone

state
borders

Figure 5   An example illustrating the idea of information latency.

transparency, different aspects of the map can be highlighted and, in so doing, it is possible to remove unwanted detail as required. In the same way, each component in an object can be associated with a level code, which corresponds to putting different levels of detail on separate transparencies. By selectively enabling and disabling the level codes, information latency and different views of the same object can be achieved. This concept gives rise to some major advantages in terms of flexibility in that a dynamical configuration system can be described using a variety of models with different levels of detail, as opposed to a fixed model. In other words, each dynamical system block can have, at the same time, models for a selection of control system design packages or different simulation languages. A possible implementation of this approach is shown in Figure 6 in which a dynamical system is modeled in TSIM code and also the code of a similar proprietary language ACSL [13]. By selecting the appropriate sets of level codes, TSIM codes embedded in the symbol can be visible to the interpreter while the ACSL codes are not. On the left-hand side of the diagram, only the TSIM model is visible to the interpreter because the currently selected level code is 1. Alternatively, if the level code 2 is selected instead, the ACSL code is generated. In such a manner, a common input format can be used with a range of algorithmic procedures and packages with implications for ease of use and the wide applicability of common input format.

Inheritance allows the functions and properties of a previously defined object to be incorporated within a newly created object. In other words, an instance of the previously defined symbol or a reference pointer to it is included in the new symbol. As an example, when an instance of a two-input OR gate symbol is included in an electronic circuit simulation diagram, the simulated OR gate will perform the same function as that of the previously defined symbol which is to "OR" the two inputs and to give the corresponding response without further definition.

The **aggregation** mechanism permits the composite definition of an object, that is, a complex object can be constructed from a group of basic objects and inherits their functions or properties. The new object can be manipulated as one single entity. To give an example, a second-order dynamic object can be formed by the aggregation (or concatenation) of two first-order objects and the new object can be treated as a single entity with its properties inherited from the two constituent objects which are combined to give the second-order characteristics. The components of a composite object are referenced by a recursive pointer defined in the object instead of being physically incorporated into the newly formed object. This does not just save computer memory space by not duplicating objects but also facilitates the subsequent modifications to any of the constituent components in that any change made to one of the components will be automatically propagated to the parent composite object through the referential structure.

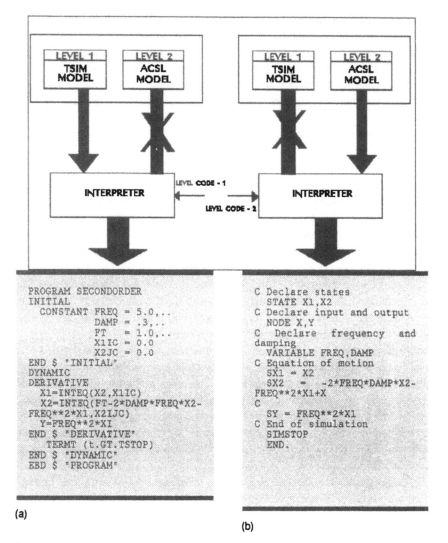

(a)

(b)

Figure 6   Selective interpretation of (a) TSIM or (b) ACSL model by activating the appropriate level codes.

A symbol model which supports the above-mentioned mechanisms is shown in Figure 7. The fundamental unit of the database model is a symbol. It consists of any combination of elements supplied as graphics functions such as lines, text, and other symbols. A symbol is defined as a series of graphical structures of subsymbols which are generally transparent to the user. A subsymbol is, in turn, made up of a level code, a sequence of primitives (see below), and either an empty clause or a pointer to another subsymbol. The level code affects the interpretation of all the primitives called, but has no effect on any of the subsymbols to which it points. The interpretation can, for example, dictate which parts of a symbol are displayed. By selectively enabling and disabling the level codes, information latency and different views of the same object can be achieved.

Primitives are the basic structure of the graphics system which generate output to the screen. They are either actual procedures that generate output or pointers to include instances of previously defined symbols and for these reasons, they can be classified into four different categories, namely, the recursive, intrinsic static, intrinsic variable, and vector set.

Recursive primitives are pointers to earlier defined symbols and when included in a newly created symbol will allow the properties and functions of

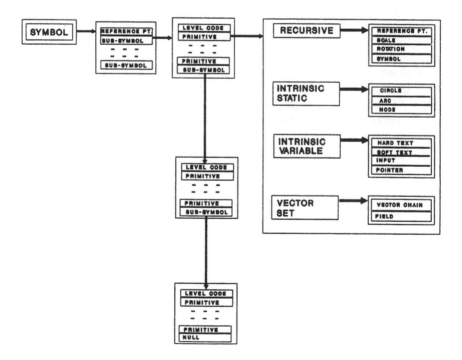

Figure 7 Symbol structure.

previously defined symbols to be inherited. Intrinsic static primitives have a fixed number of variables associated with their call. The procedure "circle," for example, will always require the same number of arguments, that is, center location and radius. On the other hand, the intrinsic variable primitives can take a variable number of arguments such as a text string which can take a variable number of characters. Vector set primitives are functions that are made up of a series of coordinate pairs such as vectors and areas. This symbol structure allows symbols to be combined together to form a composite object which can be manipulated as one single entity.

It can be observed that an engineering database is substantially different from a business database in that interaction with a business database usually deals with a complete database, one that contains essentially all the data and does not undergo significant change. Engineering design generally does not deal with a complete database. A design database starts out nearly empty and is filled up as application programs, specifications, and other constraints generate more data. Only when the design is finished is a complete database achieved. For that reason, a graphical database should be organized in such a way that will allow classification of design objects according to their levels of abstraction, that is, a system primitive, a symbol, or a diagram. The database should also allow a symbol to vary its size and complexity as a result of the incremental development of a design. The ability of a database management system to handle complexity and be greatly enhanced by the implementation of the information latency, aggregation, and inheritance mechanisms. As described above, information latency allows the relevant information of a design to be presented to the engineer while the unnecessary minutiae are hidden from him. Aggregation enables a group of related objects to be treated as a composite object and manipulated as a single entity. Inheritance allows the functions and properties of previously defined objects to be shared by newly created objects.

The development of an engineering graphical database is usually performed in an incremental and hierarchical manner by constructing a complex system from basic building blocks, and the resulting database may consist of many objects. The user should be able to move up and down the hierarchy, make changes to any graphical objects, create new objects, or delete existing ones. Figure 8 shows an abstract hierarchy of this kind.

The ultimate purpose of the database model described above is to facilitate the representation of a graphical notation which can be efficiently translated into a form compatible with a control system design or a simulation package. To demonstrate the effectiveness of this model, an example will now be given which shows the definition of the block diagram notation for the TSIM simulation package based on the database model. Similar results can be defined for other commercially available packages and the approach can, thus, be considered generic.

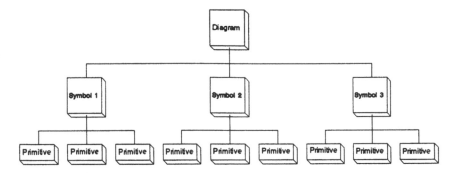

Figure 8   Graphical symbol hierarchy.

There are two approaches to the implementation of a block diagram input interface to TSIM and they are described as follows:

Tight coupling implies that the existing TSIM software is modified to allow a graphical shell to be built around it. The user interface is handled by the graphical shell which links closely to the internal operations of the simulation software.

Loose coupling implies that the existing TSIM software remains unchanged but that a graphical interface is placed between the user and the TSIM software which translates the user's graphical input into commands or simulation code that is compatible with TSIM.

The tight coupling approach requires less effort to implement because the graphical shell, possessing the knowledge of the internal operations of the simulation software, can make direct routine calls to the functions within the simulation software without the need to perform any transformation or translation. However, this approach will require access to the source code of the simulation software which, in many cases, is not always possible for commercial reasons. Unauthorized modification to the software will also invalidate the software reliability guarantee issued by the vendor. When applied to the design of a safety critical system, the validity and reliability of the simulation results will, thus, be doubtful. This approach falls into the second category of the graphical editor classification described previously, that is, the notation-specific editor approach, and as a result, the graphical shell cannot be extended to cater for other notations or other simulation packages. The graphical shell is, in fact, an integral part of the TSIM software and any change to the software due to the addition of new control algorithms will not be so easily accommodated and the flexibility of the process will ultimately be circumscribed.

The loose coupling approach requires detailed analysis of the graphical notation employed by the graphical interface and the simulation language used by

TSIM to define the various levels of mapping between the two notations. A more complex graphical structure will also be needed to accommodate this multilevel mapping. However, the obvious advantages are that the copyright and reliability issues will then become irrelevant and that the graphical interface can be extended to accommodate new changes to the simulation without undue complication. The same interface can also be used to drive other proprietary simulation packages and, thus, should have a broad application base. An example of this was given earlier in Figure 6, where the graphical interface can contain a simulation language definition for both TSIM and ACSL. A comparison between the two approaches is summarized in Figure 9. The loose coupling approach, being flexible and amenable to future extensions, is, therefore, the preferred option.

TSIM as a simulation language has its own semantics, lexicon, and syntax as opposed to that of the block diagram notation. The coupling of the two notations can, however, be achieved by taking a functional view of the TSIM language. Consider, as an example, a TSIM program generated from a block diagram shown in Figure 10, where the functional correspondence between the two notations can be observed. The semantics of a dynamical block in a block diagram corresponds to a group of TSIM statements which define a certain dynamical or computational function such as an integrator or a summer. A repository of TSIM functional blocks can be defined which captures the lexicon of the block diagram notation. The program flow of the TSIM program is analogous to the signal flow in the diagram which defines the syntax of the notation. Having defined the

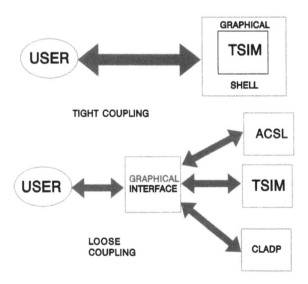

Figure 9   A comparison between the tight coupling and loose coupling approach.

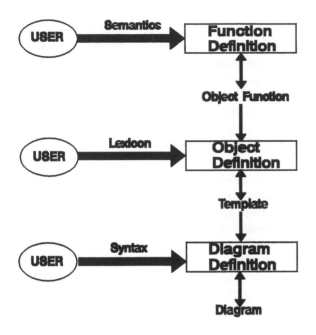

Figure 10 The three-stage block diagram definition model for TSIM simulation language.

functional mapping, a definition mechanism is required. The logical approach to the development of a block diagram database which supports these three-level mappings is to have the development carried out in three stages of definition as follows:

Function definition allows the expression of the function of a block operator or object in terms of procedural TSIM statements, such as dynamical, gain, non-linear relationship, integrator, and so on.

Object definition defines a library of functional objects which capture the lexicon of the TSIM block diagram notation. These objects include logic operators, summers, dynamical blocks, and so on.

Diagram definition specifies the syntax and signal flow of a block diagram.

These three stages of definition are depicted in Figure 10. The development of the block diagram database commences with the definition of the function of each of the basic units of the dynamical or functional elements of a control system which forms the first layer of the database definition. Since the TSIM simulation package is the target application of the graphical CACSD system, the function of each of the basic elements, called objects, should also reflect the form and syntax of the simulation language. Therefore, **function definition** requires an intimate knowledge of the format and syntax of the TSIM simulation language and, for that reason, a detailed analysis of TSIM is essential. Functions are the only means of communication between the function layer and the object layer within the language. This approach has the advantage of separating the semantics of a notation from the overall object definition and, in such a manner, the function of an object can be modified without affecting the whole object. For example, the function of an object can be changed to support the ACSL simulation language instead of TSIM by just modifying the function definition without affecting the rest of the object definition.

The second layer, called the **object definition,** enables the user to specify a family of basic objects interactively through the graphical editor described earlier. Each object of a notation is defined by drawing it in a fully generalized form and, hence, specifying its components, the geometry of the components, and their spatial relationships. The components of a block diagram representation consist of the pictorial representation, the operational functions embedded in the representation, and the input-output nodes. All of these elements are incorporated into an object definition by first drawing the icon of the object, followed by incorporating the **function definition** described above, and finally designating the input and output nodes on the diagram of the object. The user can construct a library of objects of various functions and a specification produced in this form is called a **template,** which captures the lexicon or alphabet of the TSIM block diagram notation. Such a library is shown in the right-hand column of Figure 11. Again templates or object libraries are the only means of communication between the **object definition** and the **diagram definition** layer. The two obvious advantages of the approach are as follows:

1. A notation has to be fully specified before actually being used. Using computer programming as an analogy, it would make little sense to use a programming language before having a specification of its syntax in a finalized and fixed form. This will avoid undesirable references to undefined objects.
2. The template acts as a unique and self-contained source of reference for the notation. Modification to a notation will only concern the template itself; the structure of the graphical editor remains unchanged.

The third layer, the **diagram definition,** is supported by a diagram editor which allows a user to construct block diagrams interactively by defining the topology and signal flow of a group of dynamical or computational blocks selected from a **template** defined in the object definition stage. The editor has no a priori knowledge about any graphical notation. However, when loaded with a template for a graphical notation, it will behave as if it was originally designed for the notation. In other words, it will understand the alphabet and the syntax of the notation and allow the user to create a diagram using the notation. The creation of a block diagram is achieved by selecting from a template the appro-

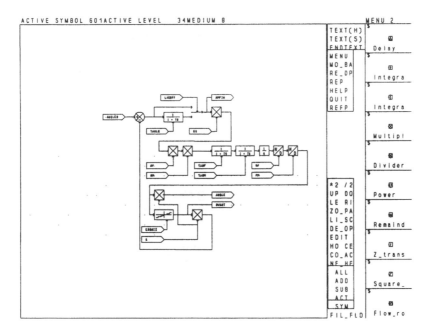

Figure 11 A TSIM model block diagram.

priate objects and defining the syntax of the diagram which governs the interaction and signal flow of the constituent objects as illustrated in Figure 12.
Because the behaviour of the diagram editor is decided by a template, the user
can switch to another diagrammatic notation such as signal flow graph by simply switching the template, without leaving the editor. This feature augments
the notation independence and thus the flexibility and extensibility of the graphical system.

The approach of using a three-stage definition of the block diagram database
facilitates the division of labor in an engineering organization. For example, an
engineering company which uses a set of graphical notations for its activities
may assign the task of creating and maintaining the notation to a group specialized in that task, whereas the notations themselves are used by the rest of the
employees, such as engineers. The latter group will only use the diagram editor
and need not even know how the templates are created. Furthermore, by cen-
tralizing the notation specifications, one can easily control the way a notation is
used within an organization and enforce appropriate standards, thereby ensuring
greater consistency in all procedures.

To accommodate the various levels of information in an object and to allow
these levels to be selectively accessed, a level definition scheme is implemented
which associates each level of information with a level code. The level codes

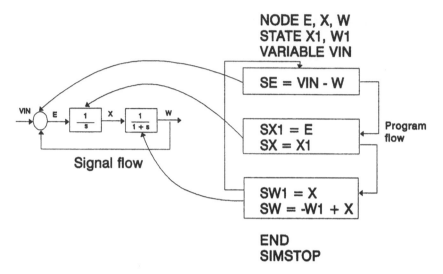

Figure 12 An example illustrating the functional mapping between the block diagram notation and the TSIM simulation language.

allow parts of a picture to be separated, for example, in the case of graphics terminals, by colors. This is a very useful feature when defining a block diagram because sets of functional elements of the diagram can be grouped together for interpretation in terms of display or function extraction, using the level code as a tag describing this functional grouping. In other words, the elements within the diagram used for graphical representation and those elements used only for specifying the function behind the graphical representation can be separated, and unwanted information can be made invisible by disabling the level visibility attribute associated with a particular level code. The level definition for the TSIM database that will be described later is shown in Figure 13.

The level code takes the form of an integer value which is specified by the user prior to defining a new subsymbol. The user has, for example, to select level code 3 before defining the nodes of a symbol. This value will then be stored alongside the subsymbol and will subsequently be used, when the graphical database is searched, to determine the visibility of that subsymbol by comparing with the currently activated level table.

This concept gives rise to some major advantages in terms of flexibility in that a dynamical configuration system can be described using a variety of models with different levels of details, as opposed to a fixed model. In other words, each dynamical system block can have, at the same time, models for a selection of control system design packages or simulation languages. In the implementation of the TSIM block diagram database, four particular submodels exist to specify the principal object function, the object definition, the diagram defini-

Level Code	Descriptive Data Types	Visible to the interpreters				Visible to the User At			Symbol Level Description
		Syntax Check	Lexical Analysis	Semantic Analysis	Run Code	Object Level	System Level	Run Time	
1	Basic Graphics					*****	******		
2	Procedural Code		*******			*****			Function Definition
3	Nodes		*********			*******			
4	Connections		*****			*******	******		
5	Labels					*			
6	Additional Graphics					*****			
11	Basic Graphics						******	****	
12	Idents	*******					********		Object Definition
13	Nodes	*****		****	****		****		
14	Connections	******		****			********	****	Diagram Definition
15	Labels						****		
21	Basic Graphics					*****			
22	Command Code Nodes				****	*****			
23	Connections				****	*****			
24	Labels				****	*****			Command Interface Definition
25	Interpreter					*****			
26	Additional Graphics					*****			
27	Additional Graphics					*****		****	
28	Additional Graphics					*****		****	
29	Additional Graphics					*****		****	

Figure 13 Level code definitions of the TSIM block diagram database.

tion, and the command interface definition; there are three display contexts and four interpreter contexts as shown in Figure 13. The three display contexts, namely, the **object level, system level,** and **run time** contexts, determine which of the elements of an object are shown to the user at different levels of block diagram definition. For example, when the user is formulating a block diagram from a library of predefined system block objects at **system level,** the underlying TSIM definitions of an individual block object are not of primary concern to the user and are, therefore, excluded from the screen display, that is, made invisible to the user. In the same vein, the four interpreter contexts, namely, the **syntax check, lexical analysis, semantic analysis,** and **run code** contexts, define the levels of information to be accessed by the graphical interpreter which together with the four interpreter contexts will be described in greater detail in the next section.

22.5 A GRAPHICAL INTERPRETER

The primary function of a graphical interpreter is to extract the functional and topological information embedded in a diagram and translate this information into a form that is understood by a target application. In the context of control system analysis and design, the target application is a simulation package or a set of control algorithms. A graphical notation has three levels of abstraction and so an interpreter which is designed to support this structure should also be able to extract the three levels of information. This interpretation function can be implemented as a three-pass operation similar to that of a compiler for a programming language. The first pass is the lexical analysis which identifies all the occurrences of different types of functional elements. Examples of these functional elements in a block diagram are the input, output, dynamic, and nonlinearity blocks. The second pass is the syntactic check through which the structure and topology of the functional elements of a diagram are determined. In a block diagram, these will be represented by the interconnections of system blocks. The final pass is concerned with the extraction of the semantics of each of the functional elements of a graphical notation. In other words, the function behind a graphical notation, not the pictorial representation itself, is extracted. In the case of a block diagram, the system dynamic equation embedded in a system dynamic block or the nonlinear relationship that describes a nonlinearity block are extracted. The interpretive process is illustrated in the example in Figure 12, where lexical analysis identifies all the occurrences of system block symbols defined in the block diagram notation lexicon. The semantics of a dynamical block in a block diagram are translated into a group of TSIM statements which define a certain dynamical or computational function such as an integrator or a summer. The signal flow in the diagram is translated into the program flow of the TSIM program. As an example, the overall interpretive process for the TSIM block diagram notation is summarized in Figure 14, where the **lexical analysis** is performed on a previously defined block diagram to extract the functional description of each of its constituent components, the output of which is stored in a raw code file. This is then followed by a **syntactic analysis** which determines the structures and topology of the block diagram and this topology organized as an interconnection matrix, is placed in the nodal database. To allow the appropriate levels of information to be accessed by the various analysis processes, the level code information stored in a previously defined level table is used to select the required level of information to access. Finally, a **semantic analysis** is performed to combine the topological information in the nodal database and the raw code to generate the TSIM simulation file and the node coordinate file which stores the position of each node in the form of X-Y coordinates. Having generated the TSIM simulation file, the next phase will be to build and perform a simulation to obtain analysis results and present them to the user. During the interpretive process, different levels of information are accessed depending on which one of the four interpreter contexts, as shown in Figure 13, the process is in, namely, the **syntax check, lexical analysis, semantic analysis,** and **run code** contexts. The implementation of the contextual interpretation mechanism results in an efficient interpretive process because only the relevant levels of in-

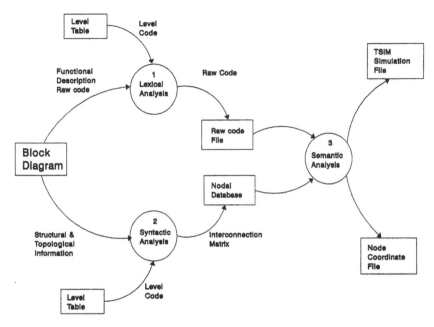

Figure 14 The block diagram interpretive process—a summary.

formation are presented to the interpreter. For example, when the interpreter is performing **semantic analysis,** only the appropriate information concerning the input-output nodes of the individual blocks and the connections of the block diagram is accessible to the interpreter, and, consequently, the interpreter does not have to scan through irrelevant information.

22.6 EXAMPLES

The work as carried out in collaboration with the Rolls Royce Plc. [14] and, therefore, the definition of the block diagram elements is linked to that used within the working practice of the collaborator in the development of complex nonlinear aero-engine controllers. The first example, as shown in Figure 15, illustrates the type of formal description used to describe an aero-engine controller. The model in input as a block diagram to the graphical environment as shown in Figure 16, and the time response characteristics of the system can be obtained by pointing to the node on the diagram where the analysis results are desired.

In the second example, the model of a DC motor with feedback, shown in Figure 17 is considered. An icon interface, as shown in Figure 18, is defined

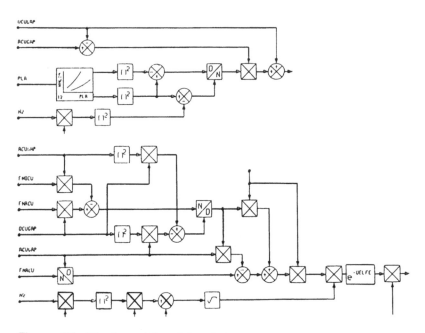

Figure 15 The form of formal description used to describe a controller.

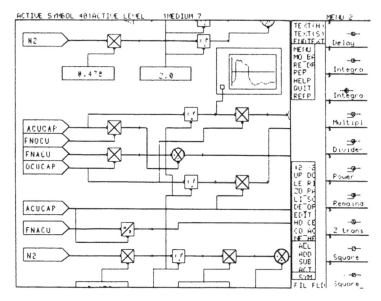

Figure 16 An aero-engine controller model.

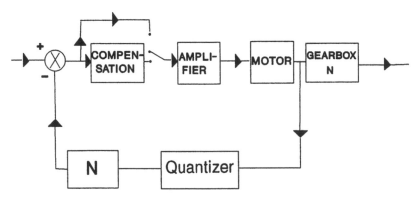

Figure 17 A schematic diagram of the DC motor example.

which allows the user to drive the TSIM simulation package from the graphical interface. As shown in Figure 18, to obtain simulation results from the model shown in Figure 17, the user has to click on the oscilloscope icon followed by selecting the node on the block diagram for simulation results output illustrated in Figure 19. The TSIM simulation package will then be executed as a background process and the simulation results generated fed back to the graphical environment. After appropriate scaling and windowing, the output is displayed at the specified position on the screen as shown in Figure 20. The simulation results can be saved as a file for archive purposes, an example of which is shown in Figure 21.

The generic editor concepts mentioned previously enhances the CAD environment in terms of supporting different notations. To demonstrate this concept, an entirely different library of block diagram modules is defined as shown in Figure 22. A dynamical model formulated from this library is shown in Figure 23, where analysis results are displayed in proximity to the nodes where the results are requested.

22.7 EVALUATION

The implementation of the graphical interface has facilitated the use of the TSIM simulation package as the designer can now describe his mental conceptual models in terms of block diagrams without the need to mentally translate the block diagram into a form compatible with TSIM language format. The efficiency of the designer is greatly increased as the error and inconsistency due to the mental translation process are now eliminated and the control system building, editing, analysis, reanalysis, and documentation cycle can be performed in a homogeneous environment. The capability of the designer to handle complexity is enhanced by the implementation of the inheritance, aggregation, and in

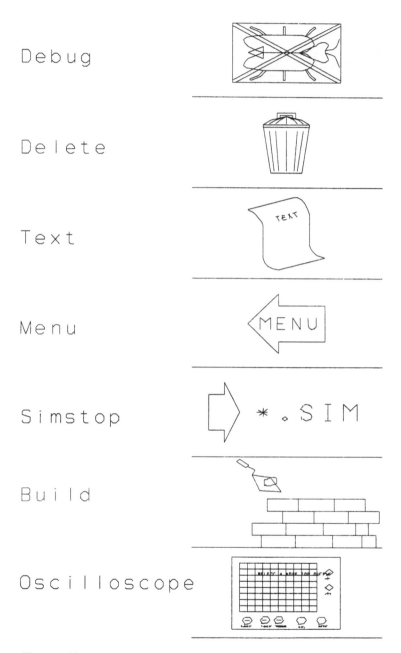

Debug

Delete

Text

Menu

Simstop

Build

Oscilloscope

Figure 18   The icon menu.

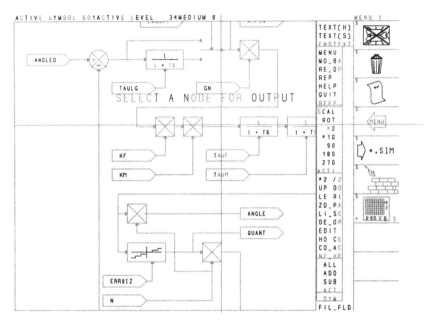

Figure 19 The user is prompted to select a node for simulation output.

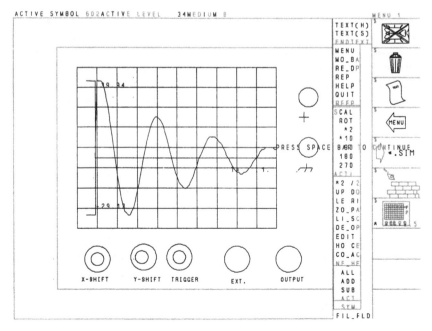

Figure 20 Time response result shown as a trace on an oscilloscope.

556   *Li and Gray*

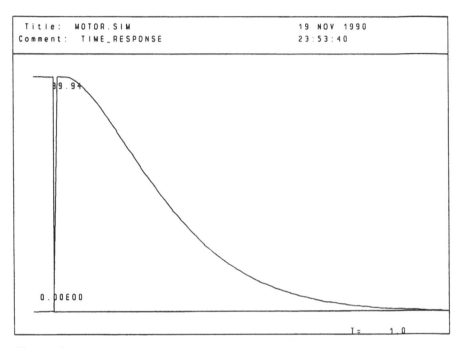

```
Title:   MOTOR.SIM                          19 NOV 1990
Comment:   TIME_RESPONSE                     23:53:40
```

Figure 21   The time response saved in a graphics file.

formation latency mechanisms. On an organization level, the implementation of
the graphical interface has facilitated the integration of various design functions
as they now share a common language in the form of block diagrams, through
which engineering information can be communicated in a standardized and precise manner. The results are clearly generic in the sense that they are applicable
to different simulation and computer-aided control system design packages.

22.8   CONCLUSION AND FUTURE WORK

In this chapter the input requirements for a CACSD system were outlined which
are graphical editing, graphical database management, and graphical interpretation. The use of a generic editor is the preferred approach to the implementation of the graphical function for reasons of its flexibility and extensibility. A
graphical database management system should support the three levels of abstraction of a graphical notation, namely, the lexical, the syntactic, and semantic
levels. The composite and hierarchical nature of engineering design requires that
the underlying mechanism of the database management system will support the
functions of inheritance, information latency, and aggregation. Finally, graphical interpretation is a three-pass process which extracts correspondingly the se-

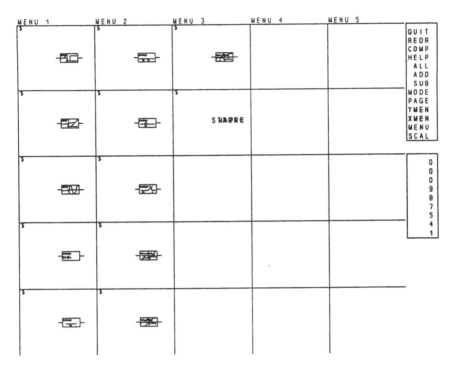

Figure 22 A different library of block diagram elements.

mantic, lexical, and syntactic levels of information from a diagram constructed from a graphical notation.

So far, the discussion has been focusing on single-user environments; however, on an organizational level, the issues of database sharing and concurrent access have to be considered. This requires the integration of the object-oriented and the database technologies and the incorporation of the resultant object-oriented database model within the present graphical database framework [15]. The paradigm of object orientation offers tools for the representation of complex engineering objects in terms of their topology, subcomponent relationships, and properties, whereas the database management system provides facilities for concurrent access, constraint management, and transaction management.

The growing interest in the use of expert system methodology [16–19] to enhance a CACSD system in terms of offering design advice and providing guidance through the design process, has prompted the drive for tighter coupling between the design expert systems and the conventional engineering design packages. However, the data structures employed in these engineering design systems, such as simulation, system analysis, and design management packages, are so diverse that building conversion software for each of the individual

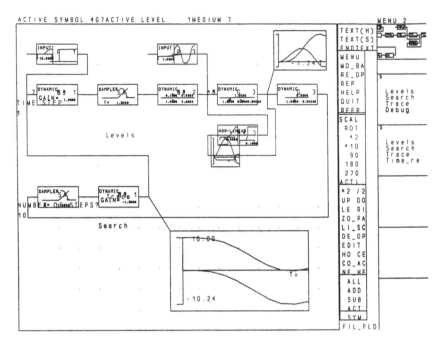

Figure 23   A dynamical model constructed from the library in Figure 22.

packages seems impractical, and, furthermore, the semantic contents of the information are very often lost during the process of transfer. When engineers interact with each other, information instead of plain data is very often being referred to, that is, it is the interpretation of the data rather than the data itself that is important. Therefore, what is needed as well as the data conversion requirements is a mechanism which will allow the semantic content of the information to be preserved. The object-oriented database framework with its rich semantics is believed to be able to provide a unifying mechanism for information transfer and its capability of representing complex structure will surely facilitate the implementation of model-based reasoning in addition to the rule-based reasoning employed by current expert systems [20].

REFERENCES

1. Pew, R. W., Human skills and their utilization (Lecture Note), *MAN/MACHINE Interface Symposium, C. E. I. EUROPE*, June 1983.
2. Card, S. K., Moran, T. P., and Newell, A. A., *The Psychology of Human Computer Interaction*, Lawrence Erlbaum Associates, Hillsdale, NJ, 1983.
3. *TSIM User Guide*, Cambridge Control Ltd., Cambridge, 1987.

4. Winter, J. S., Corbin, M. J., and Murphy, L. M., Description of TSIM2: A software package for CAD of Flight Control Systems, RAE Technical Report 83007, 1983.
5. Shneiderman, B., *Designing the User Interface*, Addison-Wesley, Reading, MA, 1987.
6. McLeod, H. A. D., and Parker, D. K. A., "An extensible object-oriented approach to database for VLSI/CAD, *Readings in Object-Oriented Database*, Morgan Kauffman, 1990.
7. Munro, N., ECSTASY—An environment for control system theory, analysis and synthesis, *IFAC Proc. Series*, No. 7, 225–230 (1989).
8. Shneiderman, B., *Designing the User Interface*, Addison-Wesley, Reading, MA, 1987.
9. Hekmatpour, S., and Woodman, M., Formal specification of graphical notation and graphical software tools, *Proc. ESEC'87*, Strasbourg, France, 1987, pp. 317–325.
10. *Simulab*, The Mathworks Inc., South Natwick, MA, 19 .
11. *VISIM*, Visual Solutions Inc.,
12. Breuer, P. T., Maciejowski, J. M., and Phaal, P., Definition and implementation of a data model for computer-aided control engineering, preprint, *Tenth IFAC World Congress, Munchen*, 1987, Vol. 7, pp. 261–266.
13. *ACSL, Advanced Continuous Simulation Language*.
14. Li, P. H. M., A graphical input system for computer aided control system design, Ph.D. thesis, Electrical and Electronic Department, University of Salford, Salford, United Kingdom, 1992.
15. Stein, J., Object oriented programming and database, *Dr. Dobb's Journal*, (March 1988).
16. Taylor, J. H., Frederick, D. K., and James, J. R., An expert system scenario for computer-aided control engineering, *Proceedings of the American Control Conference*, San Diego, California, 1984.
17. James, J. R., Considerations Concerning the Construction of an Expert System for Control System Design, PhD. Thesis, Rensselaer Polytechnic Institute, Troy, NY, 1986.
18. Boyle, J. M., Pang, G. K. H., and MacFarlane, A. G. J., The development and implementation of MAID: a knowledge based support system for use in control system design, *Trans. Inst. MC, 11* (No. 1) (1989).
19. Trankel, T. L., Markosian, L. Z., and Sheu, P., Expert system architecture for control system design, *Proceedings of the American Control Conference*, Seattle, Washington, 1986.
20. Knuz, J. C., Stelzner, M. J., and Willianms, M. D., From classic expert systems to models: Introduction to a methodology for building model-based systems, *Topics in Expert System Design*, North-Holland, Amsterdam, 1989.

23

Graphical User Interfaces in Computer-Aided Control System Design

H. A. Barker, M. Chen, P. W. Grant, C. P. Jobling, and P. Townsend

University of Wales at Swansea
Swansea, Wales

23.1 INTRODUCTION

The SKETCHPAD system [1] developed at the Massachusetts Institute of Technology (MIT) in 1963 is regarded as the point of origin for computer-aided design (CAD) [2] as well as computer graphics [3]. Although digital computers had been used for making analytical calculations in engineering since they emerged in the 1940s, the novelty of SKETCHPAD was that the user could for the first time interact with the computer graphically, via the medium of a display screen and light pen. Nevertheless, only relatively recently the development of graphical workstations has reached the point at which a genuine graphical user interface can be provided for a wide range of CAD applications, including computer-aided control system design (CACSD). It is now possible to match the considerable amount of effort invested in implementing modern system analysis and simulation tools with work of equal importance in developing graphical user interfaces to offer support for integration. It is for this reason that it is timely to consider the effects which recent developments are making, and will continue to make, on graphical user interfaces for CACSD.

23.2 HISTORICAL DEVELOPMENT

23.2.1 Programming Languages and Subroutine Libraries

The history of using computers in control system design can be traced as far back as the late 1940s when simple electronic analog computers were used to simulate servomechanisms for gunnery control and radar tracking [4]. During the subsequent 30 years, much effort was employed to write programs to solve particular problems. Nevertheless, almost all programs developed between the late 1940s and early 1970s were batch-oriented implementations and very difficult to use. The interface between control engineers and the computer was via programming, mostly in an assembly language or Fortran. Many of these programs had to be written by professional programmers acting as intermediaries.

The development of general-purpose software for control system design and analysis began in the early 1970s. The software was initially in the form of subroutines and the emphasis of the development was on implementing control algorithms and collecting them into libraries. Early products included the Scandinavian Control Library [5] and AUTLIB [6]. Experience gained in the development of these control libraries, and particularly the achievements in the development of mathematical libraries such as EISPACK [7] and LINPACK [8], laid the foundations for those control libraries which emerged in the 1980s, such as SLICE [9, 10], RASP [11], and SLICOT [12]. These libraries provided control engineers with access to robust numerical algorithms and reliable subroutines to be used with Fortran.

23.3.2 Specialized Modeling Languages and Command-Driven Packages

The introduction of the general simulation language CSSL [13] in 1967 was a prelude to the development of CACSD packages. However, there were few successful products until a decade later. Most CACSD packages began to appear in the 1970s and early 1980s. Examples of these early CACSD packages include the UMIST Control Suite at Manchester [14], CLADP developed at Cambridge [15], IDPAC, SIMNON, and several related packages developed at Lund [16–18], ACSL at Concord [19], DELIGHT at Berkeley [20, 21], and MATRIX$_x$ at Palo Alto [22].

A factor that had a great influence on the field of CACSD was the introduction of MATLAB [23] in 1980. One important feature of MATLAB is its extensibility allowing users to code their own algorithms. Shortly after its introduction, a number of derivatives were aggressively marketed throughout the world and soon formed a MATLAB family. Examples include CTRL-C [24], MATRIX$_x$ (the commercial version) [25], IMPACT [26], and PC-MATLAB/PRO-MATLAB [27]. These so-called second-generation CACSD packages [28] inherited the matrix modeling environment and the command-driven interface from their ancestor MATLAB and set a de facto standard for CACSD user interaction [29].

Interactive mechanisms were introduced into most of these packages, mainly in the form of question-and-answer dialogues, graphical plotting, command languages, and specialized programming languages. Advanced graphical hardware and software was not available to control engineers at that time, and techniques such as menu-driven systems, form filling, and direct manipulation were rarely seen in these packages before 1987. Most used only numerical data models such as state-space matrices and transfer functions, which are traditionally represented by arrays. Therefore, to model a large or complicated system, especially in its natural pictorial form, remained a difficult and time-consuming task.

23.3.3 Graphical User Interfaces

The necessity for graphical system modeling tools was addressed by Barker and Linn in 1979 [30]. Though similar software had been available for some time in electrical and electronic engineering, civil engineering, and mechanical engineering, numeric modeling methods dominated almost all software in control engineering before 1984. The United Kingdom took the initiative in this area and exploratively developed ARGOS [31], INTEGRA [32], and CES [33]. In 1987, commercial products started to appear along with popular CACSD packages, such as System Build for Matrix$_x$ and Model-C for CTRL-C, and more recently, SIMULINK was developed as a simulator communicating with MATLAB. However, sophisticated general-purpose user interfaces for CACSD are still not in general use by many control engineers, who still communicate with computers using numerical, and only numerical, data models. The topology and hierarchy of a control system representation is not accepted by most packages unless it is first converted into either numerical form or programming code. Improvement of user–computer interaction is becoming an essential requirement which all CACSD software users will desire.

The development of modern computer graphics workstations holds promise for a more user-oriented modeling environment which allows graphical, symbolic, and numeric data representations to be constructed for general system models. Traditional graphics-based analysis methods will also be enhanced by introducing new features such as three-dimensional plotting, dynamic display, animation, and color hardcopy.

23.3 THE CONTROL SYSTEM DESIGN PROCESS

In creating a control system, the control engineer must construct models for the dynamic system that is to be controlled, analyze their behavior, design appropriate control strategies, and evaluate the overall performance. Eventually, the design will be implemented using appropriate devices, computer hardware, and software. Although there is not a unique design strategy that is universally applicable to every control system or acceptable to every control system designer, most designers adopt a common systematic design procedure illustrated in block diagram form in Figure 1. This procedure contains a number of design activities and several feedback paths necessary because the results and the products of a design activity may cause one or more previous actions to be revised. In practice, there are feedback paths around all the activities and all the detailed steps inside each activity. Without these feedback paths, the design procedure would be incomplete and unworkable. The efficiency with which the design proceeds relies heavily on the availability of the tools that bring to each design activity speed and ease of use.

The development of computer-aided control system design has already provided a range of software tools for numerical analysis and simulation of linear and nonlinear, continuous and/or discrete control systems. Modern technology holds promise for more advanced tools to be used in the design procedure, including tools for:

- The input of graphical and symbolic system representations, which enable more general control system models to be constructed and modified graphically
- The manipulation of hierarchical system models, which allow large systems to be easily structured into subsystems
- The transformation between pictorial models, numerical models, and computer models, which reduce unnecessary human involvement in tedious and error-prone activities such as model translation and programming
- Symbolic computation, which present symbolic properties of a control system enabling more general and possibly new analytical methods to be developed
- Realistic presentation of simulation results, which could be displayed using advanced three-dimensional graphics and computer animation

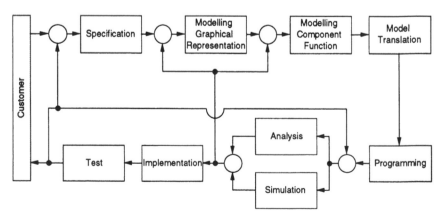

Figure 1 A systematic design procedure.

- Automatic implementation of microprocessor-based controllers, which hopefully are then able to release substantial resources from time-consuming activities such as programming controllers with low-level languages
- Knowledge acquisition, or expert systems, which provide novice designers with design experience and information on existing methods, and which also help experienced designers attack more complicated and multidisciplinary control problems

It is, therefore, necessary to create an environment, such as shown in Figure 2, within which all tools for CACSD can be integrated. One of the key components of such an environment is a sophisticated general-purpose graphical user interface, which serves all parts of the environment, and through which the user is able to conduct most, if not all, design activities by accessing a variety of CACSD tools.

23.4   PICTORIAL SYSTEM DESCRIPTIONS

The traditional medium for the "man-to-man" communication of a control engineer's ideas is some form of conceptual representation—perhaps symbolic in the form of mathematical equations or graphical in the form of a block diagram or a signal flow graph. Though it may not be so difficult to represent pictorial system descriptions in the computer in terms of linked datastructures or matrices, it is certainly very difficult for the control engineer to use computer representations directly. For most control engineers, the question is not the suitability of the pictorial representations in system modeling, but the existence of an efficient way for entering them into the computer.

In the past, users of the many CACSD packages have found that the human–computer interaction of these packages are mainly via textual media. For example, a pictorial representation of a control system such as a signal flow graph has to be converted to either a matrix, a syntax-oriented description, or a program fragment. The application of computer graphics to CACSD has been restricted to the presentation of output data, which has normally taken the form of time and frequency responses, root-locus diagrams, and the like. The disadvantages of using textual media for pictorial information input is rather obvious. To convert a pictorial representation to a textual representation and vice versa, the user has to cope with:

- Inconvenience of reading and constructing unconventional data representations
- Effort of learning specialized programming languages
- Need to cope with a diversity of mathematical models and programming languages
- Difficulty in handling large and complicated representations

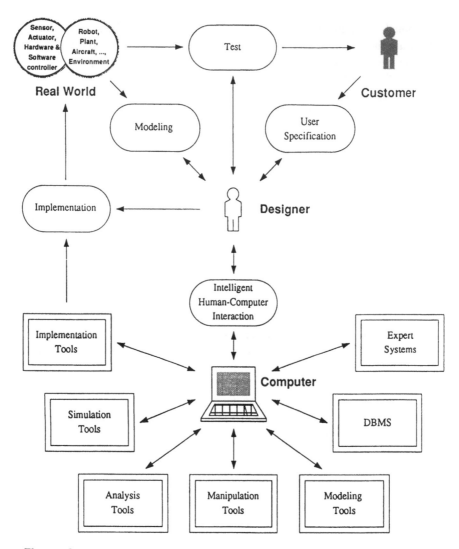

Figure 2   A computer-aided control system design environment.

In the past, one was faced with the unavailability of modern graphical devices and techniques, but with high-speed and high-resolution display devices and easy-to-use input devices, such as the mouse, new opportunities have arisen for improving the user interface for CACSD.

   An advanced graphical user interface for a CACSD environment allows control engineers to communicate with computers pictorially as well as textually; it improves on conventional interactive methods for inputting symbolic mathematical expressions and numerical data; it also enhances the presentation of output data by the use of sophisticated methods such as animation; and last but not least, it enables the organization of all the human–computer interaction in a user-friendly, efficient, and consistent manner.

23.5 DESIGN PRINCIPLES OF GRAPHICAL USER INTERFACES

Perhaps the most important function of a graphical user interface for CACSD mentioned in the previous section is to enable a control engineer to construct and modify pictorial system descriptions as well as textual ones.

23.5.1 Constructing Pictorial Representations

Pictorial representations such as block diagrams and signal flow graphs can be entered into a computer in a number of ways:

1. Converting into intermediate representations such as lists or matrices, and entering into the computer as data or expressions.
2. Converting into programs in specialized programming languages. The programs are then translated by the computer into appropriate internal representations.
3. Drawing on paper and reading into the computer by scanning. The images are then translated into structured pictorial representations by the computer using image processing techniques.
4. Drawing directly on the computer screen using general-purpose drawing packages, and saving as unstructured pictorial representations. The drawings are translated into structured pictorial representations by the computer using image processing techniques.
5. Drawing directly on the computer screen using specialized drawing packages. The computer maintains structured pictorial representations according to the user's drawing operations.

Methods (1) and (2) put an extra burden on the control engineer and the task of converting a sizable pictorial representation is always time-consuming and error prone. Although methods (3) and (4) may seem quite acceptable, the flexibility of handdrawing or building pictures using general-purpose drawing packages leads to difficulties in the task of identifying pictorial elements and translating from unstructured pictorial representations into structured ones. Method (5) has been adopted by almost all graphical editors currently used in CACSD. With this method, visual objects such as blocks and connection paths are manipulated using a mouse, and the computer recognizes the drawing operation and maps it into appropriate operations on the corresponding pictorial representation. For example, when a user draws a connection between two blocks in a block diagram, the computer determines which terminal of which block has been specified according to the geometrical information of the end points of the connection. The correctness of the input is then checked to prevent errors, such as an input-to-input connection. Finally, new data items are created for the signal and connection path and linked to the data records for related blocks and ports.

23.5.2 Constructing Textual Representations

Textual representations, which include data, expressions, and programs, can be entered into a computer in a number of ways:

a. Typing into the computer using ordinary text editors and stored in flat text files.
b. Entering into the computer using a form-filling method. Users are provided with forms with predefined fields, each of which may accept numbers or character strings in certain format.
c. Typing into the computer using a specialized editor. The text is displayed in conventional two-dimensional format.

Method (a) is widely available on most computer systems and suitable for entering programs and large volumes of data. The user is usually responsible for making sure that specific formats have been followed. A syntax-directed text editor [34] is useful to reduce format errors during the input but not widely available. For entering a simple mathematical expression or a set of parameters, this method is generally sufficient but user-unfriendly. Method (b) has been adopted by a number of graphical user interfaces in CACSD for entering parameterized functions and mathematical expressions. For example, when a user wants to define a "limiter," a form which contains three fields for upper limit, lower limit, and slope is displayed, and the user need only enter a number in each field. This method has also been found quite suitable for editing matrices, provided that the number and dimensions of the matrices can be predetermined. The aim of method (c) is to provide users with a more natural way of presenting information. If a user wishes to enter the transfer function

$$G(s) = \frac{1}{s(s + 1)(s + 2)^2}$$

it is perhaps easier for the user to view and edit the expression in the same way it would be written on paper, rather than typing the expression as a string, $G(s) = 1/(s*(s + 1)*(s + 2)^2)$, or as two arrays of coefficients, $[0,0,0,0,1]$ and $[1,3,6,4,0]$. Another example is the input of some parameterized functions that may be represented graphically, such as "limiters" and "backlashes." When the parameters of such functions are entered or modified, the corresponding curves change accordingly. It is clear that for entering mathematical information in this way, a sophisticated text editor supported by graphical display and direct manipulation is required.

25.5.3 Defining and Managing Icons

In pictorial representations used in control engineering, each graphical element, such as a block or a connection, is associated with a set of properties, one of which is the operational type of the element. In a block diagram, for example, a block may be of type "transfer function," "state space," "limiter," and so on. The operational type of an element usually determines the form of the other properties of the element including its shape, its connectability, and the form of the corresponding subsystem if any. There are two main approaches which are commonly provided in graphical user interfaces for defining the operational type of an element. With one of these, the user simply works with the generic elements or elements with uniform default properties to construct a pictorial representation. For each element, the user has to define a type explicitly and the form of other properties is changed accordingly. With the second approach, the type of an element is determined before it is added into a pictorial representation. A selection device using a menu or browser is usually provided, with which a type is assigned to an element implicitly. An explicit approach is suitable for elements with a very small set of variations in types, when the majority of elements used are of just one particular type, defined as the default type. The implicit approach, on the other hand, is more flexible and efficient to use though it requires an additional mechanism. The selection device can be implemented in a number of ways:

a. As a menu that lists a set of commands as text or icons.
b. As a browser that displays a set of available types as text or icons. A scrolling mechanism is usually incorporated into the browser to display more choices.
c. As a hierarchical browser that classifies available types into a number of categories and subcategories.

The choice of which implementation is used depends mainly on the number of types available and the style of a graphical user interface. For example, in a simple signal flow graph editor with two types of vertices and one type of edge, it is quite justifiable to use a menu of two commands Add Node and Add Branch. For a block diagram editor with dozens of block types, a hierarchical browser is obviously preferable.

Probably in many cases, the available types of elements can be predefined and are unlikely to need any change. For example, the types of nodes required for signal flow graphs need only include Input Nodes, Output Nodes, Continuous Variable Nodes, and Discrete Variable Nodes. The icons for these four types of nodes may be precoded into the corresponding graphical user interfaces. However, in some cases, new types often need to be added into the existing sets of available elements, by either users or the software developer. It is very useful to provide an icon editor to define new types of elements for specific applications. For each new type of element, a user can specify the shape of an icon, the constraints on the manipulation of these elements, and a template of properties with default settings. Extra facilities are also required for allowing newly defined icons to be added into the selection device, either a menu or a browser. To achieve this, an icon browser should be upgraded to an icon library manager with facilities for icon addition, deletion, and classification. Even for a software developer, the combination of an icon editor and an icon library manager is a very convenient way for introducing new types of elements without adding new code into a graphical user interface.

23.5.4 Other Facilities

In addition to the above essential facilities for constructing and modifying pictorial representations, it is desirable for a graphical user interface to provide high-level functions for manipulating system representations, which include:

- **Hierarchical structuring**—for combining, decomposing, and reorganizing system representations. Most pictorial system representations can be decomposed into subsystems in a recursive manner. For example, a hierarchical block diagram is a block diagram with some of its blocks defined as subsystems, that is, other block diagrams. Therefore, facilities are needed for allowing control engineers to organize a pictorial system representation by dividing it into several subsystems or embedding subsystems into it.
- **System transformation**—for automatically transforming one type of system representations into another. In control engineering, system descriptions in different forms are often interrelated and can be transformed from one to another. For example, a block diagram representing a linear system can be transformed into a signal flow graph. When a graphical user interface supports more than one type of system description, the facilities for the transformation between different system descriptions is required to reduce the cost and inconsistency of producing and storing different models of the same system [35]. Transformation facilities are needed not only for the transformation between pictorial representations but also for that between pictorial and textual representations. For instance, a linear system represented by a block diagram or a signal flow graph may be converted into an overall transmittance represented as a transfer function [36], and a state-space representation may be converted into a signal flow graph.
- **Symbolic manipulation**—for dynamically modifying pictorial features of a system representation while preserving its functionality. For block diagram system representations, for example, facilities may be provided to perform

cascade, loop, and parallel combinations of blocks and the relocation of summers and take-off points. To maintain the functionality of a block diagram, these facilities not only change the graphical appearance of the block diagram, but also symbolically modify the operations associated with its blocks following a well-defined set of rules for block diagram manipulation [37].

• **Automated layout**—for automatically generating pictorial representations in a form that is aesthetically acceptable and conforms to the conventions used in drawing such representations by hand. Many high-level functions for manipulating system representations will more or less alter the topology of a pictorial representation so that the existing graphical information such as the position of a block and the geometric path of a connection is required to be modified accordingly. It is, therefore, very useful for a graphical user interface to be equipped with facilities for automatically generating or amending the layout of a pictorial representation based on its topological and functional information [38].

23.6 MODERN GRAPHICAL USER INTERFACES

23.6.1 Graphical User Interface Standards

Since 1976 when the first call was made for a graphics standard, several standards have been developed including GKS [39] and PHIGS [40], which contain no basic window management concepts. Although some of their implementations can now run under window management systems, they mostly embody single-window approaches, providing no facilities to directly handle window-based objects such as icons and pop-up/pull-down menus.

At present, there are several window management systems in existence, each having some differences from one another. Therefore, portability of user-interface applications implemented under proprietary window management systems is not feasible without a major software rewrite.

In an attempt to overcome the portability problems, the X window system was developed at MIT [41]. It has a unique device-independent architecture based on a client-server network model and it has now been adopted as a de facto standard by nearly all the major workstation manufacturers. With the X window system, there is a standard programming interface Xlib [42], which provides complete access and control over the display, windows, and input devices. There are also a wide range of toolkits based on Xlib [43–45], each of which provides a rich set of high-level functions for creating and managing various types of windows, menus, and window elements. This inevitably leads to the decision that the X window system, rather than a graphics package or a proprietary window system, should be chosen as part of the core programming environment of a modern graphical user interface.

The usability of a graphical user interface depends on a number of factors including:

- **"Look and Feel"**—every application conforms to a standard style, so that all applications following the guidelines have similar features.
- **Customization**—which allows the application's interface to be tailored to the needs and requirements of the user.
- **On-line help**—help at different levels is always available from any point in the program's execution.

These considerations are seldom reflected in the traditional programming environments for graphical user interfaces. It the past, most graphical packages provided neither guidelines nor functions to assist the design of user interfaces, leaving the burden totally on the programmer. A proprietary window management system usually imposes a specific "look and feel" on applications by providing a very restricted programming interface. Similar to the portability problem of its programming interface, the "look and feel" is confined to specific hardware and often conflicts conceptually with the "look and feel" imposed by another window system.

In a separate but linked development, considerable attention has been focused on attempts to standardize the "look and feel" of graphical user interfaces for workstations. Both Unix International and the Open Software Foundation (OSF) have laid down standards for "look and feel," namely, OPEN LOOK [46] and Motif [47], respectively. Both specify the visual appearance of objects such as windows, menus, and scrollbars and the way in which these objects react to inputs. Both are based on the X window system and inherit from it all the good features, including resource management for customization, interclient communication, and event-driven programming. Both provide application programmers with toolkits and high-level development tools, together with comprehensive guidelines on the style of application software. Both are supported by a set of powerful tools for file management, program debugging, and a mechanism for on-line help generation.

23.6.2 eXCeS—Extended X-based Control System workStation

In the late eighties, the Control and Computer Aided Engineering Research Group at Swansea developed a software system called CES (Control System workStation) to provide a general-purpose graphical user interface for modeling and manipulating control system representations both graphically and symbolically, together with links to software for numerical simulation, analysis, and controller implementation [48–51].

In the early stages of the CES project when there were no accepted standards for window management or user-interface design, the graphics standard GKS was adopted for system implementation. A typical CES screen is shown in Figure 3. CES does not use any windowing facilities, and only one software tool chosen from a set of such tools can be displayed at a time. To switch from one tool to another in CES, a user has to press a number of command buttons to exit from the current tool and enter a new one. Moreover, due to the rigid process control in GKS, no more than one process is allowed to access a GKS screen. All CES tools thereby have to be physically combined into one program, which results in a great deal of difficulty in maintaining the software modularity and reliability.

In the light of these shortcomings, a rigorous specification and design exercise was undertaken to produce a full functional and visual definition of the successor to CES, eXCeS (Extended X-based Control System workStation) [52], which is now being implemented. eXCeS has been designed to provide a similarly coherent and comprehensive working environment for control engineers, which covers all stages of the control system design process including modeling, manipulation, analysis, simulation, and implementation. The modularized infrastructure of eXCeS provides the environment with flexibility, extensibility, and configurability. Each tool can be used independently as well as in cooperation with others. Foreign packages and tools can be integrated into the

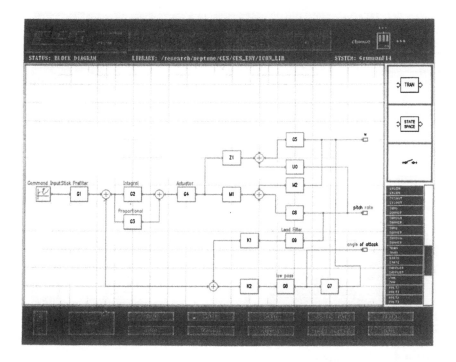

Figure 3 CES screen.

environment. A data description language is used throughout for interacting with a common database and for data communication between tools. From a user's perspective, all tools in the environment conform to the same "look and feel." Once a control engineer has used one of the tools, other tools as well as other applications can be used without the need for an extensive learning effort.

Figure 4 shows a typical eXCeS screen. In this figure, the user is running simultaneously three eXCeS tools—LibraryManager, BlockEdit, and Operation-Edit, and an OPEN LOOK "look and feel" is maintained consistently across all three tools. BlockEdit is a block diagram editing tool in eXCeS and contains a scrolling canvas on which block diagrams may be drawn. LibraryManager is provided for the quick selection of icons used in the construction of block diagrams and for the management of icon libraries. OperationEdit is invoked by selecting and opening a block in the block diagram and is used to define operations in the form of mathematical expressions, parameterized functions, and subsystems.

As an integrated computer-aided design environment for control engineering, eXCeS contains a CACSD toolbox, a common database, an intertool communication mechanism as well as a standard "look and feel" [53]. The toolbox

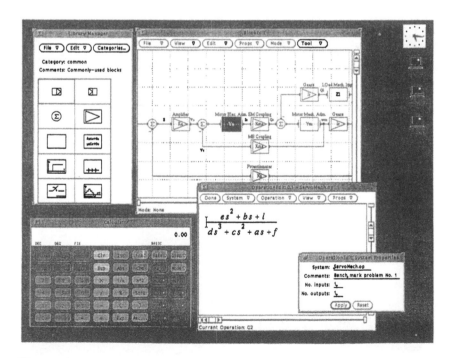

Figure 4 eXCeS screen.

contains a basic set of tools for system modeling, manipulation, analysis, simulation, and implementation. The basic data management system used in eXCeS has been designed and developed in-house to achieve flexibility and consistency in the data environment. It provides a data description language based on Prolog syntax, a programming interface for creating, retrieving, and modifying data objects in a database, and a simple user interface for application programmers and experienced users to access the database directly. The intertool communication mechanism has been built on the standard X communication protocols. It provides an efficient and consistent way for the data exchange among a variety of CACSD tools in the eXCeS environment and, therefore, enables eXCeS to incorporate any new developments in the field. The AT&T OPEN LOOK widget set [44] and the Motif widget set [47] have been chosen for the development of the graphical user interfaces in eXCeS, as it is not yet clear which of these two "look and feel" standards will eventually dominate.

23.7 CONCLUSIONS

In this chapter, the historical developments in user-interface design for CACSD have been described and the design principles of graphical user interfaces for CACSD have been discussed. The requirements and various techniques for the construction and manipulation of pictorial and textual system representations have also been studied. With the recent emergence of standards for window management systems and for the "look and feel" of user interfaces, the development of eXCeS has shown that it is necessary and feasible to incorporate this latest technology into a modern CACSD environments.

REFERENCES

1. Sutherland, I. E., SKETCHPAD: A man-machine graphical communication system, *Proc. IE & AFIPS Joint Computer Conference*, Spartan Books, Baltimore, 1963, pp. 329–346.
2. Bowman, D. J., and Bowman, A. C., *Understanding CAD/CAM*, Howard W. Sams & Co., Indianapolis, 1987.
3. Townsend, P., A Thousand Words: The Development of Computer Graphics, Inangural Lecture 0-86076-086-1, University of Wales, Swansea, United Kingdom, 1988.
4. Williams, F. C., and Ritson, F. J. U., Electronic servo simulators, *J. I. E. E.*, *94*:IIA (1947).
5. Elmqvist, H., Tysso, A., and Wieslander, J., Scandinavian Control Library Programming, Report of Dept. of Automatic Control, Lund Institute of Technology, Lund, Sweden, 1976.
6. Cellier, F. E., Grepper, P. O., Rufer, D. F., and Toedtli, J., Educational aspects of development and application of a subprogram-package for control, *Proc of 10th IFAC Symp. on Trends in Automatic Control Education*, Barcelona, Spain, 1977, pp. 151–159.

7. Garbow, B. S., Boyle, J. M., Dongarra, J. J., and Moler, C. B., *Matrix Eigensystem Routines—EISPACK Guide Extension*, Springer-Verlag, New York, 1977.
8. Dongarra, J. J., Bunch, J. R., Moler, C. B., and Stewart, G. W., *LINPACK User's Guide*, Society for Industrial and Applied Mathematics, 1979.
9. Denham, M. J., and Benson, C. J., Implementation and documentation standards for the software library in control engineering (SLICE), Internal Report 81-3, Control Systems Research Group, Kingston Polytechnic, 1981.
10. Denham, M. J., A software library and interactive design environment for computer-aided control system design, in *Computer-Aided Control Engineering*, (M. Jamshidi and C. J. Herget, eds.), North-Holland, Amsterdam, 1985.
11. Grubel, G., Die Regelugstechnische Programmbibliothek RASP, *Regelungstechnik*, *31*, 75–81 (1983).
12. van den Boom, A., Brown, A., Dumortier, F., Geurts, A., Hamarling, S., Kool, R., Vanbegin, M., and van Huffel, S., SLICOT—a subroutine library in control and systems theory, *CADCS's 91—Proc. 5th IFAC Symposium on Computer Aided Design in Control Systems*, Swansea, United Kingdom, 1991, pp. 89–95.
13. Augustin, D. C., Strauss, J. C., Fineberg, M. S., Johnson, B. B., Linebarger, R. N., and Sansom, F. J., The SCi continuous system simulation language (CSSL), *Simulation*, *9*, 281–303 (1967).
14. Munro, N., The UMIST control systems design and synthesis suites, *Proc. of IFAC Symposium on Computer Aided Design of Control Systems*, Zurich, 1979, pp. 343–348.
15. Edmunds, J. M., Cambridge linear analysis and design programs, *Proc. of IFAC Symposium on Computer Aided Design of Control Systems*, Zurich, 1979, pp. 253–258.
16. Wieslander, J., *IDPAC User's Guide, Revision 1*, Report 7605, Dept. of Automatic Control, Lund University, Lund, Sweden, 1979.
17. Elmqvist, H., Astrom, K. J., and Schonthal, T., *SIMNON—User's Guide for MS-DOS Computers*, Dept. of Automatic Control, Lund Institute of Technology, Lund, Sweden, 1986.
18. Astrom, K. J., Computer aided modeling, analysis and design of control systems—a perspective, *IEEE Control Syst. Mag. 3:2*, 4–16 (1983).
19. *Advanced Continuous Simulation Language (ACSL)*, Mitchell and Gauthier Associates, Inc., Concord, MA, 1981.
20. Nye, W., Polak, E., Sangiovanni-Vincentelli, A., and Tits, A., DELIGHT: an optimization-based computer-aided design system, *CAD86—Proc. IEEE International Symposium on Circuits and Systems*, Chicago, Illinois, 1981.
21. Polak, E., Siegel, P., Wuu, T., Nye, W. T., and Mayne, D. Q., DELIGHT.MIMO: An interactive, optimization-based multivariable control system design package, *IEEE Control Syst. Mag.*, *2:3*, 9–14 (1982).
22. Walker, R., Gregory, C., Jr., and Shah, S., MATRIXx: A data analysis, system identification, control design and simulation package, *IEEE Control Sys. Mag.*, *2:3*, 30–37 (1982).
23. Moler, C., *MATLAB: User's Guide*, Department of Computer Science, University of New Mexico, Albuquerque, NM, 1980.

24. Little, J. N., Emami-Naeini, A., and Bangert, S. N., CTRL-C and matrix environments for the computer-aided design of control systems, *Proc. 6th International Conference on Analysis and Optimization of Systems*, Berlin, pp. 191–205.

25. Shah, S. C., Floyd, M. A., and Lehman, L. L., MATRIXx: control design and model building CAE capabilities, in *Computer Aided Control Systems Engineering*, (M. Jamshidi and C. J. Herget, eds.), North-Holland, Elsevier Science Publishers, Amsterdam, 1985, pp. 181–207.

26. Rimvall, M., Man-machine Interfaces and Implementational Issues in Computer-Aided Control System Design, Number ETH 8200, Swiss Federal Institute of Technology (ETH), Zurich, 1986.

27. Moler, C., Little, J., and Bangert, S., *PRO-MATLAB: User's Guide*, The MathWorks, Inc., Sherborn, MA 1987.

28. Schmid, Chr., Techniques and tools of CADCS, *CADCS'88—Proc. 4th IFAC Symposium on Computer Aided Design in Control Systems*, Beijing, China, 1988, pp. 67–75.

29. Rimvall, M., Interactive environments for CACSD software, *CADCS'88—Proc. 4th IFAC Symposium on Computer Aided Design in Controg Systems*, Beijing, China, 1988, pp. 17–26.

30. Barker, H. A., and Linn, C. Y., A process control system based on a graphical language, *Trends in On-line Control Systems*, IEE, London, 1979, pp. 45–48.

31. King, R. A., and Gray, J. O., A methodology for the design and implementation of graphical man machine interfaces, *CAD 86—Proc. 7th International Conference on the Computer as a Design Tool*, Butterworths, London, 1986, pp. 419–420.

32. Munro, N., and Griffiths, M., Graphical input of system descriptions, *Control 85*, IEE London, 1985, pp. 192–197.

33. Barker, H. A., Chen, M., Grant, P. W., Jobling, C. P., and Townsend, P., The development of a graphical man-machine interface for computer-aided control system design, *CAD86—Proc. 7th International Conference on the Computer as a Design Tool*, Butterworths, London, 1986, pp. 411–418.

34. Reps, T. W., and Teitebaum, T., *The Synthesizer Generator*, Springer-Verlag, New York, 1989.

35. Barker, H. A., Chen, M., and Townsend, P., Algorithms for transformations between block diagrams and signal flow graphs, *CADCS'88—Proc. 4th IFAC Symposium on Computer Aided Design in Control Systems*, Beijing, China, 1988, pp. 231–236.

36. Jobling, C. P., and Grant, P. W., A rule-based program for signal flow graph reduction, *Eng. Appl. Artificial Intelligence*, *1*, 22–33 (1988).

37. Barker, H. A., Chen, M., Grant, P. W., Jobling, C. P., Simon, D. A., and Townsend, P., The manipulation of graphical and symbolic models of dynamic systems, *Proc. 3rd IFAC Symposium of Man-Machine Systems*, Oulu, Finland, 1988.

38. Chen, M., Graphical man-machine interface for computer aided control system design, Ph.D. Thesis, University of Wales, 1991.

39. *Graphical Kernel System (GKS)—Function Description*, ISO DIS 7942, International Standard Organization (ISO), 1982.

40. *Programmer's Hierarchial Interactive Graphics System (PHIGS) Functional Specification*, ISO/TC97/SC21 N 819, International Standard Organization (ISO), 1985.
41. Scheifler, R. W., and Gettys, J., The X window system, *ACM Trans. Graphics*, *5:2*, 79–109 (1986).
42. Nye, A., *Xlib Programming Manual*, Vol. 1 of *The Definitive Guides to the X Window System*, O'Reily & Associates, Inc., Sebastopol, CA, 1988.
43. Heller, D., *XView Programming Manual*, Vol. 7 of *The Definitive Guides to the X Window System*, O'Reily & Associates, Inc., Sebastopol, CA, 1989.
44. Miller, J. D., *An OPEN LOOK at UNIX—A Developer's Guide to X*, M&T Publishing, Inc., Redwood City, CA, 1990.
45. Young, D. A., *X Window Systems Programming and Applications with Xt*, Prentice-Hall, Englewood Cliffs, NJ, 1989.
46. *OPEN LOOK Graphical User Interface Application Style Guidelines*, Sun Microsystems, Inc., and Addison-Wesley, Reading, MA, 1990.
47. *OSF/Motif Style Guide, Revision 1.0*, Open Software Foundation, Cambridge, MA, 1990.
48. Barker, H. A., Chen, M., Jobling, C. P., and Townsend, P., Interactive graphics for the computer-aided design of dynamic systems, *IEEE Control Sys. Mag. 7:3*, 19–25 (1987).
49. Barker, H. A., Townsend, P., Chen, M., and Harvey, I., CES—a workstation environment for computer-aided design in control systems, *CADCS'88—Proc. 4th IFAC Symposium on Computer Aided Design in Control Systems*, Beijing, China, 1988, pp. 248–251.
50. Barker, H. A., Chen, M., Grant, P. W., Jobling, C. P., and Townsend, P., Development of an intelligent man–machine interface for computer-aided control system design and simulation, *Automatic, 25*, 311–316 (1989).
51. Grant, P. W., Jobling, C. P., and Ko, Y. W., A comparison of rule-based and algebraic approaches to computer-aided manipulation of linear dynamic system models, *Proc. IEE Conference Control'88*, IEE CP 285, IEE, Oxford, 1988.
52. *Specification of Control System workStation (CES) Version II*, CCAE Research Group, CCAE-TR89-2, University of Wales, Swansea, United Kingdom, 1989.
53. Barker, H. A., Chen, M., Grant, P. W., Harvey, I. T., Jobling, C. P., Parkman, A. P., and Townsend, P., The making of eXCeS—a software engineering perspective, *CADCS'91—Proc. 5th IFAC Symposium on Computer Aided Design in Control Systems*, Swansea, United Kingdom, 1991, pp. 27–32.

Index